Sustainable Agriculture towards Food Security

Arulbalachandran Dhanarajan
Editor

Sustainable Agriculture towards Food Security

Springer

Editor
Arulbalachandran Dhanarajan
Department of Botany, School of Life
Sciences
Periyar University
Salem, Tamil Nadu, India

ISBN 978-981-10-6646-7 ISBN 978-981-10-6647-4 (eBook)
https://doi.org/10.1007/978-981-10-6647-4

Library of Congress Control Number: 2017958740

Printed on acid-free paper

This Springer imprint is published by Springer Nature
The registered company is Springer Nature Singapore Pte Ltd.
The registered company address is: 152 Beach Road, #21-01/04 Gateway East, Singapore 189721, Singapore

This book is dedicated to the memory of my father who instilled in me and my beloved brother....

I feel immense pleasure and privilege to write about the matchless qualities of such a distinguished person, my brother Shri D. Ravichandran, who is paving my life path since my childhood. He is my first mentor and I learned many things from him such as working hard and following ethics in life, and he has taken care of all the moments in my academic activities till now. I am honored to dedicate this book to my brother Shri D. Ravichandran. He is my inspiration and motivation and thoughtful love, who enlightened me to continue to improve my knowledge and move my career forward.

Editor

Foreword

The priority process of sustainable agriculture is driven by the need to produce sufficient food, feed and fibre to meet the growing demands of the world's burgeoning population. Within this is the need to appreciate, recognise and reshape agricultural systems to meet future demands. It is clear to me through my work on dissecting the genetic basis for cereal crop response and adaptation to production environments that this is a complex and multifaceted challenge. When considering the wider context of farming systems, this becomes all the more complicated, not least because the range of farming systems we work within differ widely. An additional element in the sustainable intensification of agriculture is the role of orphan or alternative food sources and how traditional plant and animal products may play a role in ensuring future food and nutritional security.

I am therefore very pleased to introduce this book edited by Dr. D. Arulbalachandran of Periyar University, Salem, Tamil Nadu, India. As an agricultural plant scientist pursuing molecular breeding strategies to understand and improve the performance of crop and medicinal plants, Dr. D. Arulbalachandran has assembled an exciting range of contributions providing a new dimension to the discussion of issues, challenges and strategies for addressing agricultural sustainability.

The book introduces the context and urgency of sustainable agriculture and looks at improvements underway in both conventional and organic farming contexts. The latter is particularly important as societal preferences change and it presents both a challenge and an opportunity to the maintenance of food production. Of particular interest is the use of seaweed as a superior and renewable source of food and crop amendment, which could drive yield increases not linked to petrochemical-based inputs.

Alternative food sources are also considered in depth in the book, and the importance of food or dietary diversification is relevant and timely. Whilst this can be achieved through the manipulation of nutrients or their absorptive qualities in existing plant parts, it is also relevant to consider primary sources of food, whether they be plant, algae or mushroom based. The role of indigenous food and medicinal plants is also important to consider. Here, clear gains are possible through breeding and agronomy as many have been poorly characterised and subjected to only limited empirical selection.

The advent of new technology is increasingly heralded for its potential to revolutionise the agricultural sector. In this book, the role of tissue culture, manipulation of biologically active compounds from plants and somatic embryogenesis are all considered in detail. At the other end of the scale, the soil that supports and underpins agricultural production is discussed and the importance of soil security is clearly made. The more we understand about how plants interact with and respond to their local environment, the more important we recognise the soil to be. This is particularly so in regions where soil toxicities limit productive land area and here bioremediation is crucial to land area that can support productivity.

In conclusion, the challenges of sustainable agriculture are great and exceedingly diverse. This diversity demands multidisciplinary thinking and application, and this book brings together a wide range of contributions to address the biological challenges and potential change underpinning this. In the future, it is my hope that all of these areas can integrate their research objectives and can work together with social, physical, economic and political sciences to address our pressing food security challenges.

National Institute of Agricultural Botany (NIAB) Alison Bentley
Cambridge, UK

Preface

The world's population is projected to grow from around 7.2 billion today to 9.3 billion in 2050. Agricultural production on land and aquatic systems already dominates much of the global terrestrial surface and has major negative impacts on the earth's ecosystems. At the same time, rural areas are still home to the majority of the world's poor and vulnerable populations who rely heavily on natural capital for their livelihoods and lack secure access to these resources. Agricultural production systems and the policies and institutions that underpin global food security are increasingly inadequate. The world's food systems are heading towards an unprecedented confluence of pressure for the next 40 years (Foresight UK 2011). The major cause of unsustainable agriculture is due to the unprecedented burgeoning environmental crisis which is one of the seeds for mounting pressure on food security. The combined effects of climate change, land degradation, cropland losses, water scarcity and species infestations may cause projected yields to be 5–25% short of demand by 2050. Besides, agriculture is practised in so many climates and in different cultural contexts, so sustainable agriculture cannot possibly imply a special way of thinking or of using farming practices. The agricultural challenges of the future will, as today, differ according to their geopolitical and socio-economic contexts, and moreover they are not linked to any particular technological practices. Indeed, sustainable agriculture is defined as the ability of farmland to produce food and other agricultural products that satisfy human needs indefinitely as well as have sustainable impacts on the broader environment. This requirement in agriculture to avoid severe or irreversible damage to endogenous or external ecosystem services upon which it depends, notably soil fertility, irrigation of water, genetic variability and pollinators, which have acceptable impacts on the broader environment. Significant change in sustainable agriculture and food security is needed, without this, the challenges will be exacerbated in the near future. The world needs a paradigm shift in agriculture development for sustainable food production and security through green revolution and eco-friendly approaches. Sustainable agriculture technologies and practices must be locally adapted. Sustainable agriculture should have adaptability and flexibility over time to respond to demands for biomass production. New strategies are needed that respond to the daunting challenges posed by climate change mitigation and adaptation, water scarcity, the decline of petroleum-based energy, biodiversity loss and persistent food insecurity in growing populations. The book *Sustainable Agriculture Towards Food Security* will generate awareness to the larger part of issues as it deals with food security and addresses perspectives and insights of sustainability of food production and security through sustainable agriculture towards the future in the way of classical and recent advancements of technologies and strategies by sustainable production through plant and animal origin; productivity growth by pest management and transgenic techniques; and mitigation of toxicity in soil and environment by bioremediation, environmental stress resistance, plant growth-promoting microbes, regulators, breeding strategies, tissue cultures, bio-fertilizers and integrated approaches of food nutrition. It is assured that the chapters of this book provide a new dimension to discuss the issues, challenges and strategies of agricultural sustainability in a comprehensive manner and glimpses to students and advanced and budding researchers to make a novel approach for sustainability with environmentally sound practices.

Salem, TN, India Arulbalachandran Dhanarajan

Acknowledgments

The completion of a task is never a one-man effort which is often a result of direct or indirect contribution of many of the individuals. This sort of work of editing books comes out to be a great source of learning process and experience. It will be appropriate to acknowledge each and every member of the persons who rendered me all possible support and assistance directly and indirectly while compiling the book.

I would like to express my special thanks of gratitude to Dr. Souframanien, Scientist F of the Nuclear Agriculture and Biotechnology Division, BARC, Trombay, Mumbai, India, for his valuable suggestions toward the preparation of this book.

I have to express my appreciation to Fernando P. Carvalho of the Radiological Protection and Safety Unit (UPSR), Centre for Nuclear Science and Technology, Technological and Nuclear Institute (ITN), Sacavém, Portugal, for reviewing in spite of his very busy schedule. He made several important suggestions which nourished the information of this book.

I am very grateful to Dr. Febina Mathew, assistant professor at the Department of Plant Pathology, South Dakota State University, USA, for reviewing this book proposal with fruitful comments.

I am pleased to express my gratitude to my spouse Mrs. D. Sukanya for her profuse support throughout the work while preparing the manuscript.

I would like to express my gratitude to Dr. K. Murugesan, professor of environmental science at Periyar University, Salem, Tamil Nadu, India, for his critical and valuable suggestions and developing ideas.

I extend my thanks to Periyar University, Salem, Tamil Nadu, India, for giving the platform to achieve such kind of academic activities.

I am immensely grateful to the publishing company, Springer (India) Private Limited, New Delhi, for enabling me to publish this edited book. The completion of this work would have been impossible without the publisher's support and guidance.

D. Arulbalachandran

Contents

Contributors

Wan Azlina Ahmad Biotechnology Laboratory, Department of Chemistry, Faculty of Science, Universiti Teknologi Malaysia, Johor Bahru, Malaysia

S. Alagendran Department of Biochemistry, Crop Improvement Division, Adhiyamaan College of Agriculture and Research, Krishnagiri, Tamil Nadu, India

G. Archunan Department of Animal Science, Bharathidasan University, Trichy, Tamil Nadu, India

D. Arulbalachandran Department of Botany, School of Life Sciences, Periyar University, Salem, Tamil Nadu, India

M. Ayyappan Department of Botany, Government Arts College, Thiruvalluvar University, Thiruvannamalai, Tamil Nadu, India

Rishi K. Behl CCS Haryana Agricultural University, Hisar, Haryana, India

V. Bhuvaneshwari Department of Zoology, Chikkaiah Naicker College (Affiliated to Bharathiar University), Erode, Tamil Nadu, India

Felipe Caballero-Briones Materials and Technologies for Energy, Health and Environment (GESMAT), Instituto Politécnico Nacional-CICATA, Altamira, Mexico

Amal Thomas Cheeran Molecular Biology Laboratory, Department of Botany, Bharathiar University, Coimbatore, Tamil Nadu, India

S. Dhanam Department of Botany, Arignar Anna Government Arts College (Affiliated to Thiruvalluvar University), Villupuram, Tamil Nadu, India

Gerhard Flachowsky Institute of Animal Nutrition, Friedrich-Loeffler-Institute (FLI), Federal Research Institute of Animal Health, Braunschweig, Germany

S. Gangadharan Department of Agronomy, Crop Improvement Division, Adhiyamaan College of Agriculture and Research, Krishnagiri, Tamil Nadu, India

Dhandapani Gurusamy Department of Entomology, University of Kentucky, Lexington, KY, USA

K. Jothimani Division of Crop Stress and Molecular Biology, Department of Botany, School of Life Sciences, Periyar University, Salem, Tamil Nadu, India

P. Lakshmanakumar Department of Agronomy, Crop Improvement Division, Adhiyamaan College of Agriculture and Research, Krishnagiri, Tamil Nadu, India

S. Lalitha Department of Botany, School of Life Science, Periyar University, Salem, Tamil Nadu, India

R. Latha Department of Zoology, Chikkaiah Naicker College (Affiliated to Bharathiar University), Erode, Tamil Nadu, India

S. Latha Department of Botany, School of Life Sciences, Periyar University, Salem, Tamil Nadu, India

R. Lawrence Xavier Department of Botany, Raja Doraisingam Government Arts College, Sivagangai, Tamil Nadu, India

Sivakumar Loganathan Division of Toxicology and Molecular Oncology, Department of Environmental Science, School of Life Sciences, Periyar University, Salem, Tamil Nadu, India

Manoharan Manosathiyadevan Department of Zoology, Chikkaiah Naicker College (Affiliated to Bharathiar University), Erode, Tamil Nadu, India

Ulrich Meyer Institute of Animal Nutrition, Friedrich-Loeffler-Institute (FLI), Federal Research Institute of Animal Health, Braunschweig, Germany

L. Mullainathan Department of Botany, Annamalai University, Chidambaram, Tamil Nadu, India

Tamilselvi Murugan Department of Zoology, Govt. Arts College (Affiliated to Bharathiar University), Coimbatore, Tamil Nadu, India

S. Murugesan Department of Botany, School of Life Science, Periyar University, Salem, Tamil Nadu, India

M. Nagarajan Department of Botany, A.V.C. College (Autonomous), Mannampandal, Tamil Nadu, India

Devarajan Natarajan Natural Drug Research Laboratory, Department of Biotechnology, School of Biosciences, Periyar University, Salem, Tamil Nadu, India

Thillaigovindhan Nedumaran Plant Biology and Plant Biotechnology, Sir Theyagaraya College (Affiliated to University of Madras), Chennai, Tamil Nadu, India

N. Puspha PG & Research Department of Microbiology, Cauvery College for Women, Trichy, Tamil Nadu, India

Chinnasamy Ragavendran Natural Drug Research Laboratory, Department of Biotechnology, School of Biosciences, Periyar University, Salem, Tamil Nadu, India

D. Rajarajan PG & Research Department of Botany, Periyar University, Salem, Tamil Nadu, India

Department of Plant Breeding & Genetics, Crop Improvement Division, Adhiyamaan College of Agriculture and Research, Krishnagiri, Tamil Nadu, India

M. Ramachandran Department of Botany, School of Life Sciences, Periyar University, Salem, Tamil Nadu, India

K. Sankar Ganesh Department of Botany, Government Arts College (Autonomous), Kumbakonam, Tamil Nadu, India

Kamaraj Sathish-Kumar Tecnológico Nacional de México / Unidad Especializada en Energías Renovables, Blvd. San Pedro, 500, Parque Industrial, Ferropuerto Mieleras, Torreón, Coahuila, Mexico

V. Soundarya Department of Botany, School of Life Sciences, Periyar University, Salem, Tamil Nadu, India

Ganapathi Sridevi Department of Biotechnology, School of Biotechnology, Madurai Kamaraj University, Madurai, Tamil Nadu, India

P. Sundaramoorthy Department of Botany, Annamalai University, Annamalainagar, Tamil Nadu, India

Palaniswamy Thangavel Department of Environmental Science, School of Life Sciences, Periyar University, Salem, Tamil Nadu, India

S. Vanmathi Department of Botany, School of Life Sciences, Periyar University, Salem, Tamil Nadu, India

Krishnan Vasanth Molecular Biology Laboratory, Department of Botany, Bharathiar University, Coimbatore, Tamil Nadu, India

Institute for Wine Biotechnology, Department of Viticulture and Oenology, Stellenbosch University, Stellenbosch, South Africa

Chidambaram Kulandaisamy Venil Biotechnology Laboratory, Department of Chemistry, Faculty of Science, Universiti Teknologi Malaysia, Johor Bahru, Malaysia

Venkatasamy Vignesh Government of India, Animal Quarantine and Certification Service, Department of Animal Husbandry, Dairying and Fisheries, Ministry of Agriculture & Farmers Welfare, Mumbai, India

Melané A. Vivier Institute for Wine Biotechnology, Department of Viticulture and Oenology, Stellenbosch University, Stellenbosch, South Africa

Dirk von Soosten Institute of Animal Nutrition, Friedrich-Loeffler-Institute (FLI), Federal Research Institute of Animal Health, Braunschweig, Germany

K. Yasmin Division of Crop Stress and Molecular Biology, Department of Botany, School of Life Sciences, Periyar University, Salem, Tamil Nadu, India

Nur Zulaikha Binti Yusof Biotechnology Laboratory, Department of Chemistry, Faculty of Science, Universiti Teknologi Malaysia, Johor Bahru, Malaysia

About the Editor

D. Arulbalachandran is an assistant professor in the Department of Botany, School of Life Sciences, Periyar University, Periyar Palkalai Nagar, Salem, Tamil Nadu, India. His area of specialization in research is plant molecular breeding and stress physiology. He completed his doctoral degree in botany at Annamalai University, Tamil Nadu, and he has more than 10 years of teaching experience. He is focusing on research in the field of cytogenetics, mutation breeding, and molecular analysis in crops. He completed a research project under UGC and is involved in the administrative work pertaining to the academic side as a board member for curricula in various institutions and member of various scientific societies. He is a reviewer of various national and international journals, and he has published research articles of national and international repute. He has also published two books, viz., *Ecology: Pollution and Biodiversity* and *Mutation Breeding in Crop Plants*, and also contributed book chapters in Springer. He delivered lectures in refresher courses and presented his research findings in national-level conferences/symposia.

Part I
Agriculture and Food Security

Chapter 1
Food Security and Sustainable Agriculture

D. Arulbalachandran, L. Mullainathan, and S. Latha

Abstract Food security is a tedious process, subject to risks of various natures. The risks can impact directly the various dimensions such as production of agriculture, food access utilization, and stability. According to the United Nations, still more than 836 million people in the world are living in extreme poverty. Progress is being continued fight against hunger, yet an unacceptably large number of people still, lack the food, and they need an active, healthy, and wealthy life. The latest available estimates indicate that about 795 million people in the world just over one in nine were undernourished in 2014–2016. The share of undernourished people in the population or the prevalence of undernourishment has decreased from 18.6% in 1990–1992 to 10.9% in 2014–2016, reflecting fewer undernourished people in a growing global population. Changes in large populous countries, notably China and India, play a large part in explaining the overall hunger reduction trends in the developing regions. To meet the global population demand, sustainable agriculture is the immediate remedy to produce increasing productivity. In this review, perspectives of various agricultural practices and focusing the insights of sustainability in agricultural toward food security were discussed.

1.1 Introduction

Food security issituation that exists when all people, at all times, have physical, social, and economic access to sufficient, safe, and nutritious food that meets their dietary needs and food preferences for an active and healthy life (FAO 2001). This

D. Arulbalachandran (✉)
Division of Crop Stress and Molecular Biology, Department of Botany, School of Life Sciences, Periyar University, Salem 636011, Tamil Nadu, India
e-mail: arul78bot@gmail.com

L. Mullainathan
Department of Botany, Annamalai University, Annamalainagar, Chidambaram 608002, Tamil Nadu, India

S. Latha
Department of Botany, School of Life Sciences, Periyar University, Periyar Palkalai Nagar, Salem, Tamil Nadu, India

A. Dhanarajan (ed.), *Sustainable Agriculture towards Food Security*,
https://doi.org/10.1007/978-981-10-6647-4_1

definition comprises four key dimensions of food supplies: availability, stability, access, and utilization. The first dimension relates to the availability of sufficient food, i.e., to the overall ability of the agricultural system to meet food demand. Its subdimensions include the agroclimatic fundamentals of crop and pasture production (Tubiello et al. 2007).

Food security is a broad concept, whose meaning and scope has evolved over the years, and traditional concepts of food security included simple measures such as national food production, food grain storage, national food self-sufficiency, and food aid. These were mainly macro indicators reflecting food supply, and which were used as a basis for developing conventional early warning systems against famine. In designing these systems, it was believed that such indicators could predict acute food insecurity and cause the relevant authorities to respond adequately through centralized distribution of national food reserves or food aid (Davies et al. 1991).

Increased food consumption by a growing human population presents large challenges for agriculture, which are amplified by the effects of climate change, land degradation, and declining resources such as freshwater, phosphates, fossil fuels, and fertile topsoil. Over the next decades, annual crop yield is projected to grow with less than 1%, and there is very limited space for expansion of arable land (Alexandratos and Bruinsma 2012).

Overall demand for food is affected by population growth, while economic development and rising incomes tend to shift diets toward meat and animal products that are more expensive and resource intensive to produce (FAO 2010).

1.2 Food Demand

Global food demand for agricultural crops is increasing and may continue to do so for decades, propelled by a 2.3 billion person increase in global population and greater per capita incomes anticipated through midcentury (Godfray et al. 2010). Agriculture already has major global environmental impacts: land clearing and habitat fragmentation threaten biodiversity (Dirzo and Raven 2003). Land clearing and more intensive use of existing croplands could contribute to the increased crop production needed to meet such demand, but the environmental impacts and trade-offs of these alternative paths of agricultural expansion are unclear (Godfray et al. 2010; GOS report 2011). About one-quarter of global greenhouse gas (GHG) emissions result from land clearing, crop production, and fertilization (Burney et al. 2010), and fertilizer can harm marine, freshwater, and terrestrial ecosystems (Vitousek et al. 1997).

A threefold challenge now faces the world (von Braun 2007): match the rapidly changing demand for food from a larger and more affluent population to its supply; do so in ways that are environmentally and socially sustainable; and ensure that the world's poorest people are no longer hungry. This challenge requires changes in the way food is produced, stored, processed, distributed, and accessed that are as radical as those that occurred. During the eighteenth- and nineteenth-century Industrial and Agricultural Revolutions and the twentieth-century Green Revolution, increases in production will have an important part to play, but they will be constrained as never before by the finite resources provided by Earth's lands, oceans, and atmosphere (Conway 1997).

1.3 Causes for Food Production

1.3.1 Population

Majority of recent reports on the food crisis focus principally on population growth and an increasing demand for food. However, population growth is one of several demographic factors likely contributing to the current food crisis. Urbanization is one of the key factors for fragmentation of agricultural lands due to population density. Most of the countries with the highest numbers of people facing food insecurity, however, have high fertility rates and rapid population growth which is cause of increases in the challenge of adequately meeting nutritional needs (UNPD report 2009).

The largest population increase in Asia, particularly in China, India, and Southeast Asia, accounts for about 60% and will increase the world's population by 2050 (UNPD 2007). However, the rate of population growth is still relatively high in Central America and highest in Central and part of Western Africa. Africa will experience the most rapid growth, over 70% faster than in Asia: annual growth of 2.4% versus 1.4% in Asia, compared to the global average of 1.3% and only 0.3% in many industrialized countries (UNPD 2007).

Sub-Saharan Africa has the highest population growth rate in the world. By 2050, even if fertility rates decline, the population of the region is projected to more than double. This area also holds the largest proportion of food-insecure people, with one in four people undernourished (United Nations Population Division report 2009).

Sub-Saharan Africa has the lowest agricultural productivity in the world and the highest percentage of people living in poverty (World Bank report 2009). Food production depends on croplands and water supply, which are under strain as human populations increase. Pressure due to limited land resources, driven in part by population growth, can mean expansion of cropland. This often involves destruction of vital forest resources and overexploitation of arable land. Globally, the world is becoming more urban and although urban residents have access to a wider array of foods, without land to farm, their food security is dependent on their income and ability to purchase food products (FAO 2010). More than 200,000 people are added to the world food demand every day. The human population in the world has increased almost fourfold in the past 100 years (UNPD 2007).

1.3.2 Land Acquisition and Agriculture

Land is the basic for every form of physical development and constitutes the primary medium for food production for the provision of shelter and manufacturing utilities and the establishment of institutions to support the basic needs of modern communities (Lasun and Olufemi 2006). Land is the farmer's most important asset and plays essential role in increasing as well as sustain the agricultural production (Ukaejiofo 2009). Besides, the total land area available for agricultural production will be increasingly constrained by land requirements for other purposes, like infrastructure development, urbanization, bioenergy production, and biodiversity protection (Sands and Leimbach, 2003), but also by soil degradation (McNeill and Winiwarter 2004; Oldeman et al. 1990).

The demand for agricultural land can be used for various purposes such as, urban, residential, industrial, commercial, recreational, educational, and other uses. Demand for agricultural land is mainly dependent on the price of agricultural land. The demand will be more when the price of agricultural land is low, and demand will be less when the price of agricultural land is high. Thus, there is an inverse relationship between demand for and price of agricultural land (Walter and Barnhart 2013).

1.3.3 Climate Change

Agriculture is inherently sensitive to climate variability and change, due to either natural causes or human activities. Climate change is caused by various anthropogenic activities such as emissions of greenhouse gases, industrial pollution, and deforestation which is expected to directly influence crop production systems for food, feed, or fodder, to affect livestock health, and to alter the pattern and balance of trade of food and food products. These impacts change differential pattern of warming and associated changes in rainfall. Climate change could have a range of direct and indirect effects on all four dimensions of food security. Hence, global climate change is an additional constraint to agricultural production in the second half of the twenty-first century. Increasing of atmospheric carbon dioxide levels and a corresponding rise in global temperatures will not only affect plant growth and yields but also alter the regional patterns of precipitation and water availability, as well as land erosion and fertility. Regional impacts of climate change vary quite significantly, with tropical regions potentially suffering from drought impact with combined effects of various changes are still highly uncertain.

Impact of climate change on agricultural productivity will reduce the food crops and thus the expected land scarcity in 2050 (Nelson et al. 2010). Many other simulations such as the effects of climate change with and without adaptation (induced technological progress, domestic policy change, international trade liberalization, etc.) and mitigation, for example, CO_2 stabilization, variants for temperature, rainfall change, and distribution (Darwin et al. 1995).

1.4 Effect of Climate Change on Food Production and Availability

Climate change impacts on agriculture and food production in complex. It affects food production directly through changes in agroecological conditions and indirectly by affecting growth and distribution of incomes and thus demand for agricultural produce. Impacts are quantifiable in numerous studies and under various sets of assumptions. Changes in temperature and precipitation associated with continued emissions of greenhouse gases will bring changes in land suitability and crop yields.

1.5 Effect of Climate Change on Food Utilization

Climate change affects the ability of individuals to use food effectively by altering the conditions for food safety. The main concern about climate change and food security is that changing climatic conditions can initiate a vicious circle where infectious disease causes or compounds hunger, which in turn, makes the affected populations more susceptible to infectious disease and the result may be substantial decline in labor productivity and an increase in poverty and even mortality. Essentially, all manifestations of climate change, they may be drought, higher temperatures, or heavy rainfalls, have an impact on the disease pressure, and there is growing evidence that these changes affect food safety and food security (IPPC Report 2007a, b).

1.6 Increasing Water Demand

Population growth and economic development are driving significant increases in agricultural demand for water. Agriculture accounts for more than two-thirds of global water use including as much as 90% in developing countries. Freshwater consumption worldwide has been more than doubled since World War II and is expected to rise another 25% by 2030 (Daniel Wild et al. 2007).

1.7 Crucial Role Water in Agricultural Production

Freshwater and food production are indirectly connected; producing one ton of grain requires 1000 tons of water. Food production is so wholly dependent on water that agriculture can use 75–90% of freshwater in a region (WWO report 2010). Water scarcity creates food shortages, raises food prices, and increases a countries' dependence on food imports.

1.8 Management of Land for Sustainable Agriculture

The challenges of global needs in facing agriculture are how to provide food for the increasing world population. The human population is projected to reach nine billion people by the year 2050, and at the same time, the need to conserve the environment is another task (Spore 2012). Sustainable increase of agriculture involves the use of agricultural practices that are economically and environmentally sustainable which offers a useful approach to tackling food in security facing the world as the result of increase in population and environmental degradation which have long-term effect on agriculture globally (Simon et al. 2013). In recent decades, agricultural land that was formerly productive has been lost to urbanization and other human uses, as well as to desertification, salinization, soil erosion, and other consequences of unsustainable land management (Nellemann et al. 2009). Research into sustainable land management seeks to improve our understanding of the complex interdependencies between economic, environmental, and social conditions which affect the organization of the use of land and natural resources (Eppink et al. 2012).

1.8.1 Cropping Systems in Arid and Semiarid Regions

The main objectives for ecologically and economically sustainable agriculture are maintaining soil fertility and improving crop productivity and stability. Management options are site- and time-specific nutrient and water management, crop protection measures, and the choice of adapted, high-yielding cultivars. The effects of the various measures that are of importance for the maintenance and use of the resource base cannot easily be assessed within one growing cycle but should be evaluated over a sequence of crops. Crop rotation is an important component of an integrated approach of sustainable agriculture and resource conservation. Short- and long-term effects of a cropping sequence and related management practices can be expressed in physical soil properties such as water-holding capacity and bulk density; chemical soil properties such as pH, carbon content, and nutrient contents and biological soil properties such as microbial activity (Lal 2008; Shibu et al. 2006). Growing special crops in a rotation can improve the sustainability of the cropping system (Struik and Bonciarelli 1997).

1.9 Mixed Farming Systems in Agriculture

Integration of crop and animal production on the farm and regional scales may be an opportunity to increase eco-efficiency apart from crop rotation (Wilkins 2008). Nitrogen is mobile in the soil-plant-animal system and with the required N inputs for high crop yields and intensive livestock production the risk of N losses increases

(Van Keulen et al. 2000). Traditionally, nutrient management has been concerned with optimizing the economic return from nutrients used for crop production. The main emphasis was on the expected crop response from adding nutrients to the soil.

1.10 Nutrient-Balanced Farm for Agriculture

Nutrient imports are approximately equal to exports. Because these farms are often at the upper limit of being able to safely handle all the nutrients in the production system, nutrient management planning may offer potential environmental benefit. Agronomically, farmers should aim at the minimum input of each production resource required to allow maximum utilization of all other resources (De Wit 1992).

1.11 Farming of Sustainable Agriculture

The sustainable agriculture generally classifies into three main objectives: economic profitability, environmental health, and ethical soundness. It is often presented as a conceptual three Ps framework: People-Planet-Profit (Fig. 1.1). The changes in agriculture from a purely profit-oriented activity into a three P-based production sector, trying to meet productivity, efficiency, and efficacy aims have been of considerable importance for sustainability (Matson et al. 1997).

The demand of resources such as water, land, and biomass and the rate of environmental degradation in a cumulative sense are increasing manifested by land salinization, groundwater pollution due to excessive leaching of fertilizer residues, and the pollution of surface water due to the poor or lack of treatment of trade effluents disposed into natural water resources (Government of India 2001).

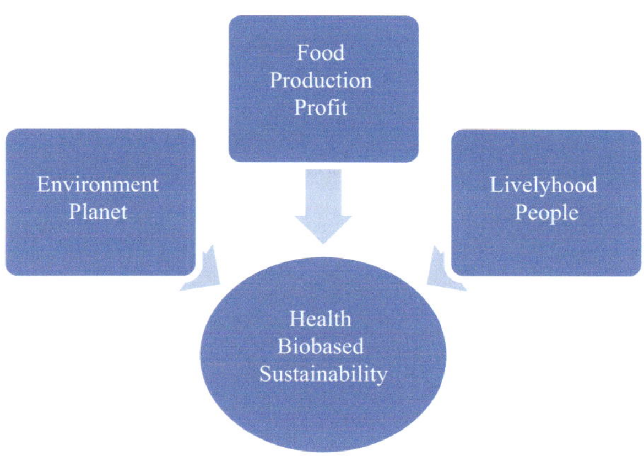

Fig. 1.1 Efficient and efficacy of sustainable agriculture

1.12 Water Scarcity and Its Implications for Agriculture

Increasing water shortage would be a major challenge to achieving global food security (Leisinger 1996). The future demand for water in India is from all the four competitive use sectors, viz., agricultural, industrial, domestic, and livestock drinking for the year 2025 (Seckler et al. 1998; GOI 1999; Ballabh et al. 1999; Kumar 2001).

1.13 Technologies for Increasing Sustainable Agricultural

Sustainability in agriculture relates to the capacity of an agroecosystem to predictably maintain production through time. A key concept of sustainability is stability under a given set of environmental and economic circumstances that can only be managed on a site-specific basis. To a large extent, the rate of technology development and the degree of innovation in future technologies will greatly influence the stability and certainly the productivity of agriculture (Hutchins and Gehring 1993). Technology, in the classical sense, includes the development and use of nutrients, pest control products, crop cultivars, and farm equipment, but it also includes the vision of genetically modified crops providing greater nutritional efficiency (more calories per yield or more yield), manipulation of natural pest control agents, and use of farm management techniques that focus on whole farm productivity over time, not just annual production per hectare (Stone and Pedigo 1972).

Several on-farm management practices in the Indian farmers can adopt in agriculture. These agricultural practices are particularly important for the semiarid regions which have already taken to intensive farming with irrigation water, both from canals and aquifers in some states of India (Kumar 2002). Such practices, if carried out consistently, it can progressively reduce the water requirement of the existing crops and improve primary productivity of the cultivated land. For alteration and gradual reduction of chemical fertilizers, Indian farmers are increasingly using the organic manure, vermin-culture technologies, and agronomic practices such as mulching, crop rotation, and the use of bio-pest control measures. Organic manure can help regain structure and texture of soils and enhance their moisture retention capacity along with improving soil nutrients. The use of farm management practices such as mulching can reduce the evaporation from soil surface, thereby increasing the efficiency of irrigation water utilization practiced by Indian farmers (Kumar 2002).

1.14 Conclusion

Global action toward food security is needed for the present scenario to ensure and sustain the food production to access adequate food for every human in the world. Although many conventional techniques have existed since antique to make food production in agriculture, other new development strategies are also needed for sustainable agricultural practices. Hence, depending on the demand of the food toward increasing world population in the future, the active and sustainable agricultural research is needed to fight against hunger and sustain the food security.

References

Alexandratos N, Bruinsma J (2012) World agriculture towards 2030/2050: 2012 revision, ESA working paper No. 12–03. Food and Agriculture Organization of the United Nations, Rome

Ballabh V, Kumar MD, Talati J (1999) Food security and water management in India challenges for the first quarter of 21st century. Paper presented at the national symposium on building and managing organisations for rural development, Institute of Rural Management, 13–14 December 1999

Burney JA, Davis SJ, Lobell DB (2010) Greenhouse gas mitigation by agricultural intensification. Proc Natl Acad Sci U S A 107:12052–12057

Conway G (1997) The doubly green revolution. Penguin Books, London

Darwin R, Tsigas M, Lewandrowski J, Raneses A (1995) World agriculture and climate change, economic adaptations, Agricultural Economic Report 703. Department of Agriculture, Washington, DC

Davies S, Leach M, David R (1991) Food security and the environment: conflict or complementarity? University of Sussex, Brighton

Dirzo R, Raven PH (2003) Global state of biodiversity and loss. Annu Rev Environ Resour 28:137–167

Eppink FV, Werntze A, Mas S et al (2012) Management and ecosystem services. How collaborative research programmes can support better policies. Gaia Ecol Perspect Sci Soc 1:55–63

Food and Agriculture Organization (2001) State of food security in the world. Rome

Food and Agriculture Organization (2002) The state of food insecurity in the world 2001. Food and Agriculture Organization, Rome

Food and Agriculture Organization of the United Nations (2010) The impact of global change and urbanization on household food security, nutrition and food safety. Accessed 8 Mar 2011

Godfray HCJ et al (2010) Food security: the challenge of feeding 9 billion people. Science 327:812–818

Government of India (GOI) (1999) Integrated water resource development: a plan for action. Report of the national commission on integrated water resources development, vol I. Ministry of Water Resources

Government of India (GOI) (2001) National human development report. New Delhi

Hutchins SH, Gehring PJ (1993) Perspective on the value, regulation, and objective utilization of Pest control technology. Am Entomol 39:1215

Intergovernmental Panel on Climate Change (2007a) Climate change: the scientific basis. Contribution of working group I to the fourth assessment report of the Intergovernmental Panel on Climate Change. Cambridge University Press, Cambridge/New York

Intergovernmental Panel on Climate Change (2007b) Climate change: impacts, adaptation and vulnerability, contribution of working group II to the fourth assessment report of the Intergovernmental Panel on Climate Change. Cambridge University Press, Cambridge. in press

Kumar MD (2001) Demand management in the face of growing scarcity and conflicts over water use in India. Future option. IRMA working paper 153

Kumar MD (2002) Integrated water management in the face of growing demand and threatened resource base in North Gujarat: constraints and opportunities: a pilot project to protect North Gujarat's groundwater ecology and agriculture economy. Paper presented at the annual partners' meet of the IWMI-Tata water policy research programme. 19–20 February 2002

Lal R (2008) Soils and sustainable agriculture. A Rev Agron Sustain Dev 28:57–64
Lasun Mykail Olayiwola, Olufemi Adeleye (2006) Land reform experience from Nigeria: promoting land administration and good governance: 5th FIG regional conference, Accra, Ghana
Leisinger KM (1996) Food security for a growing world population. Paper presented at the Saguf Symposium, "How will the Future World Feed Itself?". October 9–10
Matson PA, Parton WJ, Power AG et al (1997) Agricultural intensification and ecosystems properties. Science 277:504–509
McNeill J, Winiwarter V (2004) Breaking the sod: humankind, history, and soil. Science 304:1627–1629
Nellemann et al (eds) (2009) The environmental food crisis. United Nations Environment Programme (UNEP), Nairobi
Nelson GC et al (2010) Food security, farming, and climate change to 2050: scenarios, results, policy options. International Food Policy Research Institute, Washington, DC
Oldeman LR, Hakkeling RTA, Sombroek WG (1990) World map of the status of human induced soil degradation. An explanatory note. International Soil Reference and Information Centre, Wageningen
Reference is made here to the developing regions as defined by the M49 country classification of the United Nations (see http://unstats.un.org/unsd/methods/m49/m49regin.htm). Countries included in these regions are also reported in Annex 1, Table A1
Reserve Bank of India database. https://www.rbi.org.in/Scripts/PublicationsView.aspx?id16443
Sands RD, Leimbach M (2003) Modeling agriculture and land use in an integrated assessment framework. Clim Chan 56:185–210
Seckler D, Amarasinghe U, Molden D et al (1998) World water demand and supply 1990–2025: scenarios and issues, Research report 19. International Water Management Institute, Colombo
Shibu ME, Leffelaar PA, Van Keulen H et al (2006) Quantitative description of soil organic matter dynamics – a review of approaches with reference to rice-based cropping. Geoderma 137:1–2
Simon BP, Garba A, Bunu GM (2013) Determinants of sustainable agricultural land management practices among arable crop farmers in northern part of Taraba State, Nigeria ARPN J Sci Tech
Spore (2012) Sustainable Intensification, A crucial revolution, Spore 1(158). April–May 2012 edition:13
Stone JD, Pedigo LP (1972) Development of economic injury level of the green clover worm on soybean in Iowa. J Econ Entomol 65:197–201
Struik PC, Bonciarelli F (1997) Resource use at the cropping system level. Eur J Agron 7:133–143
The Government Office for Science (2011) Foresight. The future of food and farming. Final project report. The Government Office for Science, London
Tubiello FN, Soussana JF, Howden SM (2007) Crop and pasture response to climate change. Proc Natl Acad Sci U S A 104:19686–19690
Ukaejiofo (2009) Identifying appropriate tools for land governance in Nigeria. Spatial data serving people. Land governance and environment building the capacity, Hanoi
UNPD (2007) http://www.un.org/en/development/desa/population/
United Nations Population Division (2009) World population prospects: the 2008 revision. UN population division, New York. FAO. 2010. "Food security statistics: prevalence of undernourishment in total Population." http://www.fao.org/economic/ess/food-security-statistics/en/. Accessed 6 Jan 2011
Van Keulen H, Aarts HFM, Habekotte B et al (2000) Soil-plant- animal relations in nutrient cycling: the case of dairy farming system 'De Marke. Eur J Agron 13:245–261
Vitousek PM, Aber JD, Howarth RW et al (1997) Human alteration of the global nitrogen cycle. Ecol Appl 7:737–750
Von Braun J (2007) The world food situation: new driving forces and required actions. International Food Policy Research Institute, Washington, DC
Walter C, Barnhart J (2013) Agricultural land prices, supply, demand and current trends. University of Kentucky, Lexington

Wild D, Francke C-J, Menzli P, Schon U (2007) Water: a market of the future – global trends open up new investment opportunities. Sustainability Asset Management (SAM) Study, Zurich
Wilkins RJ (2008) Eco-efficient approaches to land management: a case for increased integration of crop and animal production systems. Phil Tran Roy Soc B 363:517–525
World Bank (2009) The sunken billions: the economic justification for fisheries reform. World Bank, Washington, DC
WWO (2010) http://www.unwater.org/

Chapter 2
Nutritive Value, Sustainable Agriculture and Rural Development: An Integrated Approach

S. Alagendran, G. Archunan, N. Puspha, D. Rajarajan, P. Lakshmanakumar, and S. Gangadharan

Abstract Sustainable agriculture is an approach of agriculture which focuses on the production food products to mediate the food supply to the human and animal without any harm of environment. Sustainable agriculture along with balanced nutrition in rural people is very challenging due various factors influencing to fulfill in rural areas throughout the world. The challenge of sustainable agriculture for rural development is urbanization in countryside, low economy in rural people, lagging of service provision, disaster, etc. Besides, in rural region, nutrition of food they consume is not well balanced due to poor income. Hence, nutritive value with sustainable agriculture integrated with rural people is very much needed to improve them in socioeconomic status. In this review, the integrated management of sustainable food productivity, utilization with nutrition to improve the rural development in sustainable manner, has been discussed.

S. Alagendran (✉)
Department of Biochemistry, Crop Improvement Division, Adhiyamaan College of Agriculture and Research, Athimugam, Shoolagiri TK, Krishnagiri 635 105, Tamil Nadu, India
e-mail: proteinblg@gmail.com

G. Archunan
Department of Animal Science, Bharathidasan University, Trichy 620024, Tamil Nadu, India

N. Puspha
PG & Research Department of Microbiology, Cauvery College for Women, Trichy 620017, Tamil Nadu, India

D. Rajarajan
PG & Research Department of Botany, Periyar University, Salem, Tamil Nadu, India

Department of Plant Breeding & Genetics, Crop Improvement Division, Adhiyamaan College of Agriculture and Research, Athimugam, Shoolagiri TK, Krishnagiri 635 105, Tamil Nadu, India

P. Lakshmanakumar • S. Gangadharan
Department of Agronomy, Crop Improvement Division, Adhiyamaan College of Agriculture and Research, Athimugam, Shoolagiri TK, Krishnagiri 635 105, Tamil Nadu, India

A. Dhanarajan (ed.), *Sustainable Agriculture towards Food Security*,
https://doi.org/10.1007/978-981-10-6647-4_2

2.1 Introduction

Agriculture is an important instrument in reducing hunger in rural areas, by increasing farmers' income and assets (Kennedy and Bouis 1993). However, there is plenty of evidence of agricultural projects increasing farmers' income and assets, which did not succeed in improving child nutrition (Braun and Kennedy 1994). Only those programs which included additional components such as nutrition education produced improved nutrition outcomes (Ruel 2001).

2.1.1 Policy in Improving Nutrition Diet

Socioeconomic factors such as people's education, women's empowerment, traditional beliefs, infant and young child feeding practices, intra-household food distribution, and social norms are considered determinants of people's access to passable and nutritious food, in addition to their dietary behaviors, and can themselves be affected by changes in diets along with nutritional impact (Newton and Freyfogle 2004).

Food-based solutions to nurture the food security have surely subsidized to our facts on the efficiency of integrated approaches as these case studies validate and it is essential to emphasize these types of approaches when designing involvements (Demment et al. 2003). These food systems that ecological concepts can be of clear-cut use to study and improve the sustainability and nutritional value of the underlying food systems. Sustainable agriculture integrates three principal criteria such as environmental health, economic profitability, and social and economic equity (Pretty 2008).

2.1.2 Systematic Approach for Food Security

A systems perception is necessary to perceptive sustainability and is proposed in its comprehensive sense, from the individual farm, adapted to local ecosystem and to populations exaggerated by this farming system. So, systemic approach for sustainable food security provides an idea and tools to explore the network among the rural and other aspects of the environment (Pretty 2008).

The custom and universal approaches besides indicate the interdisciplinary determinations in systemic research and education which impose not only the input of researchers from various disciplines but also farmers, farm workers, consumers, and policymakers (Swanson 2008). For agriculturalists, the progression to sustainable agriculture generally requires a sequence of trifling, realistic steps (Smith et al. 1996).

The achievement on the way to the goal of sustainable agriculture is the obligation of all contestants in the system, including planters, work hands, legislators, researchers, vendors, and clients (Spiertz and Oenema 2005).

2.2 Farming and Natural Resources

2.2.1 Natural Resources of Water

Earlier studies encrypt the deterioration of ancient evolution in Mesopotamia, the Mediterranean area, and Pre-Columbian Southwest USA, and Central America is believed to have been strongly influenced by natural resource degradation from non-sustainable farming and forestry practices. Water is the principal resource that has helped agriculture and society to prosper, and it has been a major limiting factor when mismanaged (Howell 2001).

2.2.2 Water Supply and Use

Water is life solution and plays pivotal role in all the organisms living in the earth till their life span. India has extensive water storage and traditional irrigation systems for agriculture to produce crop productivity. However, due to various unpreventable natural issues like drought, the water is limited for supply necessary need for life especially for agriculture. Periodic droughts, some lasting up to 50 years, have occurred in South part of India. Several steps should be taken to develop drought-resistant farming systems, including both policy and management actions such as water conservation and storage measures (Sriskandarajah et al. 1991).

2.2.3 Diversity Approach for Sustainable Agriculture

Diversified farms are usually more economically and ecologically resilient. While monoculture farming has advantages in terms of efficiency and ease of management, the loss of the crop in any 1 year could put a farm out of business and/or seriously disrupt the stability of a community dependent on that crop. By growing of variety of crops, farmers spread economic risk and are less susceptible to the price fluctuations associated with changes in supply and demand (Engel 1995).

2.2.4 Soil Management Approach for Agriculture

Indeed, healthy soil is a key component of sustainable agriculture, which produces healthy crop plants that have optimum vigor and is less susceptible to pests. In sustainable agricultural systems, the soil is viewed as a fragile and living medium that must be protected and nurtured to ensure its long-term productivity and stability (Owens 1994). Methods to protect and enhance the productivity of the soil include using cover crops, compost and/or manures, reducing tillage, avoiding traffic on wet soils, and maintaining soil cover with plants and/or mulches (Palm et al. 1997).

2.3 Sustain the Efficient Approach for Farmer Goals

Many inputs and practices used by conventional farmers are used in sustainable agriculture for crop productivity, and sustainable farmers, however, maximize reliance on natural, renewable, and on-farm inputs. Equally important are the environmental, social, and economic impacts of a particular strategy. Converting to sustainable practices does not mean simple input substitution. Frequently, it substitutes enhanced management and scientific knowledge for conventional inputs, especially chemical inputs that harm the environment on farms and in rural communities. The goal is to develop efficient, biological systems which do not need high levels of material inputs (FAO 1996).

2.3.1 Need of Economic, Social, and Political Context in Agricultural Farming

Although addition of strategies for preserving natural resources and changing production practices, sustainable agriculture requires a commitment to changing public policies, economic institutions, and social values. Strategies for change must take into account the complex, reciprocal and ever-changing relationship between agricultural production and the broader society (Tozer and Isbister 2007). A wide diversity of strategies and approaches are needed to create a more sustainable food system. These will range from specific and concentrated efforts to alter specific policies or practices to the longer-term tasks of reforming key institutions, rethinking economic priorities, and challenging widely held social values (Strochlic and Hamerschlag 2006).

2.3.2 Perspective of Food and Agricultural Policy

As mentioned above, the agriculture farming is needed to be intensified to meet future demands for commodities and goods and to avoid further expansion onto marginal lands and encroachment on fragile ecosystems. Increased use of external inputs and development of specialized production and farming systems tend to increase vulnerability to environmental stresses and market fluctuations (Kantor 2001). Where intensification of farming systems is not possible, other on-farm and off-farm employment opportunities should be identified and developed, such as cottage industries, wildlife utilization, aquaculture and fisheries, and non-farm activities (Feenstra 2002).

2.4 Objectives of Nutritive Value: Physiological Changes of Sensorial Variability for Sustainability

Following principles are the necessary actions to be planned for sustainable agriculture farming:

(a) To improve farm productivity in a sustainable manner, as well as to increase diversification, efficiency, food security and rural incomes, while ensuring that risks to the ecosystem are minimized
(b) To enhance the self-reliance of farmers in developing and improving rural infrastructure and to facilitate the transfer of environmentally sound technologies for integrated production and farming systems, including indigenous technologies and the sustainable use of biological and ecological processes, including agroforestry, sustainable wildlife conservation and management, aquaculture, inland fisheries, and animal husbandry

Simple appropriate technology for the preservation of micronutrient-rich foods would need further development and promotion for their year-round availability. Linking community development policies to national programs for the alleviation of hunger and malnutrition, with an emphasis on increasing the variety of foods consumed is probably the best strategy for improving micronutrient malnutrition sustainably (Ramachandran 2007).

2.5 Usage of Lands in Farming Management System

Existing farmland conversion patterns often discourage farmers from adopting sustainable practices and a long-term perspective on the value of land. By helping farmers to adopt practices that reduce chemical use and conserve scarce resources, sustainable agriculture research and education can play a key role in building public support for agricultural land preservation (Fresco et al. 1994).

2.5.1 Consumers and the Food System

Consumers can play a critical role in creating a sustainable food system. The challenge now is to find strategies that broaden consumer perspectives, so that environmental quality, resource use, and social equity issues are also considered in shopping decisions. At the same time, new policies and institutions must be created to enable producers using sustainable practices to market their goods to a wider public. Coalitions or other public forums can be important vehicles for clarifying issues, suggesting new policies, increasing mutual trust, and encouraging a long-term view of food production, distribution, and consumption (Garnett 2013).

2.5.2 *Sustainable Farming Focuses on Raising Food for Consumers*

Industrially raised food is grown with many pesticides and chemicals and is processed with additives and preservatives. These toxins have been linked to a range of diseases and disorders including infertility and birth defects and can potentially create damage to the nervous system and cause cancer. Industrial food is also refrigerated and shipped from long distances, decreasing its nutritional value (Drewnowski and Popkin 1997).

2.6 Protect Domestic Plantations

It is not always easy to predict how a species will perform based on the ecological conditions in its natural habitat. Some species are adaptable and have the plasticity to grow under a wide range of conditions in new and different environments. Species selection is complicated by the great variability in some species, so process careful testing is an essential part of the domestication in relation of plants species (Turnpull).

2.6.1 *Support and Protect Rural Communities*

The rural areas' less-favored regions, still largely dependent on agriculture, are frequently source of negative growth, soaring unemployment and mounting rural poverty (Lerman 2000). This is a decrease in agricultural production as it prevented development of individual farms by depriving them of significant investment. The negative social and economic consequences upon the rural population are detailed in Vranken et al. 2004.

2.6.2 *Food Distribution*

India is the world's second largest producer of food next to China and has the potential of being the biggest in the World. Food and food products are the biggest consumption category in India, with spending on food accounting for nearly 21% of India's GDP and with a market size of $181 billion. The Indian domestic food market is expected to grow by nearly 40% of the current market size to $258 billion by 2015 and $344 billion by 2025 (Merchant 2008).

2.6.3 Political-Agricultural Practices

Various political-agricultural practices contribute to food insecurity worldwide. These include substituting commodity crops for food crops (e.g., growing corn instead of vegetables) and heavy exportation of food crops at the expense of food security of the exporting country. In addition, the recent demand for biofuels, currently produced primarily from corn and soy, has further decreased the amount of viable arable land being used for food production (Wood et al. 2000).

2.7 Food Insecurity and Sustainable Agriculture

National food security stated that we need to grow sufficient food within the country. At the same time, for domestic food security, we need to sustain economic growth to raise the income levels and purchasing power of the poor people. These apart, agricultural regulation through fixation of food grain procurement prices, regulation of consumer prices, and public distribution have an important role in ensuring food security at the domestic level, even if self-sufficiency is achieved in food grain at the national level.

2.8 Improving Agricultural Biodiversity

More diverse ecosystems, with more species or more genetic diversity within species, often have higher overall productivity than simpler systems. At larger scales, there is a widespread recognition of the importance of maintaining crop variety diversity in production systems (Hector and Hooper 2002). Biotic or abiotic stress on a genetically uniform monoculture Agricultural production of regulating and supporting ecosystem services that include nutrient cycling, regulation of water flow and storage, regulation of soil movement and properties, and regulation of biological populations (Swift et al. 2004).

2.8.1 Nutrition and Sustainable Agricultural Food Products

Sustainable agriculture is a way of growing or raising food, including animals, in an ecologically and ethically responsible manner using practices that protect the environment, safeguard human health, are humane to farm animals, and provide fair treatment to workers (Mason 2003).

2.8.2 Environmental Preservation Toward Sustainable Farming

Sustainable farms produce crops and raise animals without relying on toxic chemical pesticides, synthetic fertilizers, genetically modified seeds or practices that degrade soil, water, or other natural resources (Roling 1994). By growing a variety of plants and using techniques such as crop rotation, conservation tillage, and pasture-based livestock husbandry, sustainable farms protect biodiversity and foster the development and maintenance of healthy ecosystems (Edwards 2005).

2.8.3 Protection of Public Health

Since sustainable crop farms avoid hazardous pesticides, they are able to grow fruits and vegetables that are safer for consumers, workers, and surrounding communities. Through careful, responsible management of livestock waste, sustainable farmers also protect humans from exposure to pathogens, toxins, and other hazardous pollutants (Herrero and Thornton 2013).

2.8.4 Sustaining Vibrant Communities

An important component of sustainable agriculture is its ability to remain economically viable, providing farmers, farm workers, food processors, and others employed in the food system with a livable wage and safe, fair working conditions. Sustainable farms also bolster local and regional economies, creating good jobs and building strong communities (Hume et al. 2011).

2.8.5 Organic Agricultural Products for Sustainable Agriculture

Organic agriculture, one of the most viable sustainable agricultural practices that is cost-effective. Besides, organic production improves soil health which helps to absorb vital nutrients of plants. For example, a recent study demonstrated that organically grown tomatoes have higher levels of flavonoids and potent antioxidants found in plants.

2.9 Global Perspectives of Food Production

Over 800 million people have inadequate access to safe, nutritious food. The demand for food continues to grow with the global population and is predicted to increase by 70% by 2050. The challenge is to provide a sustainable and secure supply of good quality food. Currently, there is much focus on volume and nutritional quality, but there is a key role for sensory science in ensuring that solutions to maintain the world's food supply are sensorially acceptable (Reis 2012). Hence, food security in sustainable manner is essential to meet the global demand.

2.10 Improving Foods for the Elderly

The field of neuroscience is helping to explain flavor perception as functional magnetic resonance imaging can identify areas of the brain that respond to a particular stimulus (Bi 2006). As techniques in this area improve, we are learning more about how our brain processes information concerning the sensory properties of food (Meilgaard et al. 2007). Deficiencies in dietary intake, digestion and absorption, metabolism, excretion, as well as alterations in the metabolic requirements of dietary energy, protein, and other macronutrients related to specific conditions can determine undernourishment (Ng et al. 2013). However, the loss of olfaction can be particularly insidious and escape detection because, unlike the loss of sight or hearing, it is not readily apparent to others. A good example of this difficulty of detection is that patients with congenital anosmia in our population did not discover their olfactory loss until after age 10 years (Bockreis and Steinberg 2005).

2.11 Conclusion

The food insecurity directly correlates with human health. Consequently, the present review is aimed with food production in sustainable manner. India like developing countries should have much awareness to for food production and food security with balanced nutrition. Developing countries have many barriers to sustain the food production in various perspectives toward global marker geographically and economically. Nutritive depletion in food production is exerted various diseases leading to cause cognitive impairment such as depression, fatigue, and dementia and by loss of minerals possess selenium, magnesium, and zinc low distribution in their mass index intended for food safety or under nourished nutrition. Along with food production, sustainable nutritive improvement is essential as integral approach for food security toward human health. Hence, sustainable agriculture should focus not only the increase the food production but also should concentrate in the nutritive value of food produced through various conventional and conventional agricultural practices.

References

Bi J (2006) Sensory discrimination tests and measurements: statistical principles, procedures and tables. Blackwell Publishing Ltd., Iowa

Braun J, Kennedy E (1994) Agricultural commercialization, economic development, and nutrition. Johns Hopkins University Press, Baltimore

Bockreis A, Steinberg I (2005) Measurement of odor with focus on sampling techniques. Waste Manage 25:859–863

Demment MW, Young MM, Sensenig RL et al (2003) Providing micronutrients through food-based solutions: a key to human and national development. J Nutr 133:3879S–3885S

Drewnowski A, Popkin BM (1997) The nutrition transition: new trends in the global diet. Nutr Rev 55:31–43

Edwards SR (2005) Sustaining uses of biological diversity. CBD

Engel P (1995) Facilitating innovation: an action-oriented approach and participatory methodology to improve innovative social practice in agriculture. Published doctoral dissertation. Agricultural University, Wageningen

FAO (1996) Global plan of action for the conservation and sustainable utilization of plant genetic resources for food and agriculture. Adopted by the International Technical Conference on Plant Genetic Resources, Leipzig, Germany, 17–23 June 1996

Feenstra G (2002) Creating space for sustainable food systems: lessons from the field. Agric Human Values 19(2):99–106

Fresco LO, Stroosnijder L, Bouma J, van Keulen H et al (eds) (1994) The future of the land. Wiley, Chichester

Garnett T (2013) Food sustainability: problems, perspectives, and solutions. Proc Nutr Soc 72:29–39

Hector A, Hooper R (2002) Darwin and the first ecological experiment. Science 295:639–640

Herrero M, Thornton PK (2013) Livestock, and global change: Emerging issues for sustainable food systems. Proc Natl Acad Sci U S A 110(52):20878–20881

Howell TA (2001) Enhancing water use efficiency in irrigated agriculture. Agron J 93:281–289

Hume DA, Whitelaw CBA, Archibald AL et al (2011) The future of animal production: improving productivity and sustainability. J Agr Sci 149(S1):9–16

Kantor LS (2001) Community food security programs improve food access. Food Rev 24(1):20–26

Kennedy E, Bouis HE (1993) Linkages between Agriculture and Nutrition. Implications for Policy and Research, IFPRI, Washington, DC, p 28

Lerman Z (2000) Agriculture in transition economics; from common heritage to divergence. Agric Econ 1481:1–20

Meilgaard M, Civille GV, Carr BT et al (2007) Sensory Evaluation Techniques, 4th edn. CRC Press, New York

Mason J (2003) Sustainable Agriculture [2nd Edition], Landlinks Press

Merchant A (2008) India- Food Processing Industry OSEC Business Network land

Newton LJ, Freyfogle ET (2004) Sustainability: a dissent. Conserv Biol 19:23–32

Ng M, Chaya C, Hort J et al (2013) Beyond liking: comparing the measurement of emotional response using essense profile and consumer defined check-all-that-apply methodologies. Food Qual Prefer 26(2):121–134

Owens LB (1994) Impacts of soil N management on the quality of surface and sub-surface water. In: Lal R, Stewart BA (eds) Soil Processes and Water Quality. Lewis Publishers, Boca Raton, pp 137–162

Palm CA, Myers RJK, Nandwa SM et al (1997) Combined use of organic and inorganic nutrient sources for soil fertility maintenance and replenishment. In: Buresh RJ, Sanchez PA, Calhoun F (eds) Replenishing Soil Fertility in Africa, SSSA special publication no. 51. Soil Science Society of America, Madison, pp 193–217

Pretty J (2008) Agricultural sustainability: concepts, principles and evidence. Philos Trans R Soc B 363:447–465

Ramachandran P (2007) Poverty nutrition linkages. Indian J Med Res 126:249–261
Reis M (2012) Food insecurity and the relationship between household income and children's health and nutrition in Brazil. Health Econ 21(4):405–427
Roling N (1994) Platforms for decision making about ecosystems. In: Fresco L (ed) The future of the land. Wiley, Chichester
Ruel MT (2001) Can food-based strategies help reduce vitamin a and iron deficiencies? A review of recent evidence. Washington, DC, International Food Policy Research Institute
Smith SV, Chambers RM, Hollibaugh JT et al (1996) Dissolved and particulate nutrient transport through a coastal watershed-estuary system. J Hydrol 176(1/4):181–203
Spiertz H, Oenema O (2005) Resource use efficiency and management of nutrients in agricultural systems. ERSEC Ecological Book Series (Wageningen) 1: 171–183. http://en.scientificcommons.org/1164284
Sriskandarajah N, Bawden RJ, Packham RG et al (1991) Systems agriculture: a paradigm for sustainability. Assoc Farming Syst Res Ext Newslett 2(2):1–5
Strochlic R, Hamerschlag K (2006) Best farm labor practices on twelve farms: toward a more sustainable food system. California Institute for Rural Studies, Davis
Swanson BE (2008) Global review of good agricultural extension and advisory service practices. Food and Agriculture Organization of the United Nations, Rome, 82 p
Swift MJ, AMN I, Van Noordwijk M (2004) Biodiversity and ecosystem services in agricultural landscapes: Are we asking the right questions? Agric Ecosyst Environ 104:113–134
The role of food, agriculture, forestry and fishers in human nutrition vol -II tree domestication and the history of plantations
Tozer PR, Isbister BJ (2007) Is it economically feasible to harvest by management zone? Precis Agric 8(3):151–159
Vranken L, Noev N, Swinnen J (2004) Fragmentation abandonment and co-ownership: transition problems of the Bulgarian land market. Q J Inter Agric 43:391–408
Wood S, Sebastian K, Scherr SJ et al (2000) Pilot analysis of global ecosystems (Agroecosystems). International Food Policy Research Institute and World Resources Institute, Washington, DC, 110 pp

Part II
Conventional Farming

Chapter 3
Insights of Novel Breeding Strategies in Sustainable Crop Production

K. Yasmin, D. Arulbalachandran, K. Jothimani, V. Soundarya, and S. Vanmathi

Abstract Differential environmental conditions in various geographical regions are not ample for growing crop plants to meet the required production. Such an environmental hindrance towards agriculture are varied, viz. biotic and abiotic stresses which limit productivity of crops. The scarcity of food production in 2050 is estimated 70% more production for growing world population. The efforts have been taken for crop production for requirement; however, sustainability in agriculture is always a challenge. One of the crucial ways to sustainable agriculture since long ago is breeding strategies which proved still for increasing crop production. In this context, various breeding techniques have been discussed for production crops in sustainable manner through novel and advanced strategies.

3.1 Introduction

Agricultural sustainability which is defined as production of food crops is at least proportional to the rate of population growth. The United Nations has estimated and projected the world population by the year 2050 will be 9.2 billion and the bulk of which will be from developing and least developed countries. Estimation of increasing population will be 4.3%, 61.2% and 156.4% for developed, developing and least developing nations, respectively (United Nations 2007). The productivity of crops is mainly affected by various abiotic and biotic stresses. Approximately 70% of potential yield is lost due these stress; the major abiotic stresses that affect food production are drought, salinity and acidity (Gale 2002). According to the Food and Agriculture Organization (FAO), about 80% of future increases in crop production

K. Yasmin (✉) • D. Arulbalachandran • K. Jothimani
Division of Crop Stress and Molecular Biology, Department of Botany, School of Life Sciences, Periyar University, Salem 636 011, Tamil Nadu, India
e-mail: yasbot14@gmail.com

V. Soundarya • S. Vanmathi
Department of Botany, School of Life Sciences, Periyar University, Salem 636 011, Tamil Nadu, India

A. Dhanarajan (ed.), *Sustainable Agriculture towards Food Security*,
https://doi.org/10.1007/978-981-10-6647-4_3

in developing countries are predicted to come from agricultural intensification (FAO 2002). Developing sustainable agriculture in environmentally sensitive systems is the great challenge of the coming decades. More food, animal feed, fibre, fuel and forest products must be produced with less available land, water and nutrients to meet basic human needs and improve the sustainability of production (Edgerton 2009).

In these perspectives and goals, crop breeders focus towards achieving improved cultivars that produce higher yields and at the same time tolerate to the suboptimal soil and climatic conditions. By utilizing various breeding techniques, a number of improved cultivars from different species have reached farming communities and contributed to increases in global food production (Maluszynski et al. 2000). However, plant breeders have a look for new innovative tools, such as genetic engineering together with traditional breeding for sustaining food production to feed the world (Mohan Jain 2010) to sustain the crop production. Plant breeding plays a pivotal role in this coordinated effort for increased food production. Given the context of current yield trends, predicted population growth and pressure on the environment, traits relating to yield stability and sustainability should be a major focus of plant breeding efforts to meet the forthcoming population. These traits include durable disease resistance, abiotic stress tolerance and nutrient and water-use efficiency (Mackill et al. 1999; Slafer et al. 2005; Trethowan et al. 2005).

3.1.1 Sustainable Crop Production

With an increasing population, the production of food needs to increase with it. It is estimated that a 70% increase in food production is needed by 2050 in order to meet the Declaration of the World Summit on Food Security. But with the natural degradation of agricultural land, simply planting more crops is no longer a viable option. This triggers, one of the ways to sustainability, developing new varieties of plants/crops through breeding strategies that generate an increase of yield without relying on an increase in land area (Haddad et al. 2010). Moreover, new challenges such as climate change, human population growth, etc. are posing a big threat and challenge to sustain food production worldwide. The erratic rainfall pattern may either lead to shortage of water or increase in flooding which will ultimately have an adverse impact on shortage of food production and would increase the food price (Mohan Jain 2010).

Plant breeders and agronomists are under pressure to sustain food production under the climatic changes. The food prices have already gone up worldwide, and both developed and developing countries are facing economic crunch due to food and fuel price rise; however, the developing countries are hardest hit with current crisis. Conventional breeding in combination with other techniques such as mutagenesis, biotechnology, genetic engineering or molecular breeding utilize local genetic resources for developing new cultivars that could handle frequent climatic changes, and targeted breeding varieties are very much helpful (Mohan Jain 2010).

3.2 Plant Breeding

It is the art and science of changing the traits of plants in order to produce desired characteristics (Sleper and Poehlman 1995). Plant breeding is playing a key role in this coordinated effort for increased food production. Traits relating to yield stability and sustainability should be a major focus of plant breeding efforts. These traits include durable disease resistance, abiotic stress tolerance and nutrient and water-use efficiency (Mackill et al. 1999; Slafer et al. 2005; Trethowan et al. 2005).

3.2.1 Types and Evolutionary Stages in Breeding

Evolutionary breeding involves four stages: In the first stage, which in principle is not different from conventional breeding, genetic diversity is created, e.g. by hand crossing parent plants or by mixing multiple cultivars. The second is a cycle of multiplication of seeds from each cross or from varieties separately; seeds of each cross are then equally mixed to produce the first generation. After the initial crosses, the entire offspring is sown to grow which in the case of generating via crosses. In the third stage, as the number of plants in the population increases, a proportion of the harvested seed is saved for sowing, again without active selection of individual plants. The fourth stage concerns the output of the evolutionary breeding process: while the grain can be used as food or feed, it can also be used to provide input into plant breeding, via selecting single plants that can be used as new genetic material in further breeding programmes or as parental material for new composite crosses or other evolving populations (Fig. 3.1). The evolution of crop plants can be described as having three phases: (1) gathering from the wild, (2) domestication and agronomy and (3) plant breeding (Forster and Shu 2012).

3.2.2 Objectives of Crop Breeding Towards Sustainable Improvement

The common objectives of all breeding programme by breeders are focusing on the following characteristics for sustainable crop production:

- Appreciable amount of yield qunatity
- Stability of yield traits
- Endurance of quality and nutritive value
- Sustainability against environmental impact
- Better adaptation at specific climatic condition (Caligari 2001)

Fig. 3.1 Evolution of plant breeding and its application

3.2.3 Methods of Selection for Sustainable Breeding

The key and fundamental basis of selection of the specific plants/crops in plant breeding is with desirable traits. The selection is typically involved by evaluating a breeding population for one or more traits in field or glasshouse trials (e.g. agronomic traits, disease resistance or stress tolerance) or with chemical tests (e.g. grain quality). The goal of plant breeding is to assemble more desirable combinations of genes in new varieties of crops (Deppe 2000).

3.2.4 Phenotypic and Genotypic Selection of Crops

Phenotypic selection has been very important in developing current germplasm resources and probably has to be considered one of plant breeding's greatest accomplishments. Phenotypes are the visual trait(s) observed and are the first features that are obvious whether expressed, e.g. beautiful ornamental flower, maturity, plant stature, resistance to pests or any other observable trait.The effectiveness of phenotypic selection depends on the relative heritabilities of the traits; greater progress naturally was made for traits with the greater heritability (Hallauer 2011). Physical or mechanical selection can be used efficiently to determine the shape, size, weight, density of seeds, etc. using appropriate sieving machinery. Visual screening is the most effective and efficient method for identifying mutant phenotypes (Roychowdhury et al. 2012). While, genotypes observation is indirect method of selection which are basic units of selection, whether by either natural or human selection. During the millennia that included domestication, phenotypic selection, evaluation of breeding values and presently when molecular genetics are employed in the choice of parental genotypes and development of genetically modified organisms (GMO's), the genotypes of individuals remain the unit of selection. The effectiveness of genotypic selection depends on the heritability of the trait, which is determined by how the environment affects genotype expression (Vassel 2001).

3.3 Techniques and Strategies for Improving Crops

There are varieties of techniques such as micropropagation, haploid production, protoplasts, embryo culture, apical culture, somatic embryogenesis and polyploidization that have been developed under that can assist in improving selection response and some of them described here for increasing crop yield.

3.3.1 Polyploidization

Polyploidy is the presence of more than two complete sets of chromosomes per cell nucleus, which has been considered a ubiquitous phenomenon in plant evolution and diversification (Soltis et al. 2009). Polyploidization is considered a major evolutionary force in plants. Polyploidy is by either duplication of a single genome (autopolyploidy) or from the combination of two or more differentiated genomes (allopolyploidy) (Grant 1981). Different mechanisms have been proposed to explain how polyploids arise in nature. Two major pathways are known to lead to polyploidy in plants: somatic doubling and formation of unreduced reproductive cells. Somatic doubling is associated with mitotic events such as endomitosis or endoreduplication, which may occur either in a zygote cell or in apical meristematic tissues, giving rise to mixoploids or even completely polyploid organisms. Despite being constantly used to attain artificial polyploids, somatic doubling is supposed to have a minor role in the origin of natural polyploid organisms (Ramsey and Schemske 1998).

3.3.2 Production of Haploids

Haploid plants are of great interest to geneticists and plant breeders. Haploids offer geneticists the opportunity to examine genes in the hemizygous condition and facilitate identification of new mutations. Plant breeders value haploids as a source of homozygosity following chromosome doubling from which efficient selection of both quantitative and qualitative traits is accomplished (Griffing 1975). Application of haploidy in plant breeding is dependent on the ability to produce a haploid population of sufficient size to accommodate selection of desired gene combinations (Keller et al. 1987; Morrison and Evans 1988). The most common method for chromosome doubling is treatment of haploid meristems with colchicine. Devaux (1989) reported efficiency in producing double haploids was derived from a number of barley hybrids via colchicine treatment of haploids plants.

3.3.2.1 Significance of Haploids in Crop Improvement

The ability to produce homozygous lines after a single-round recombination saves a lot of time for the plant breeders. Studies conclude that random double haploids are comparable to the selected lines in pedigree inbreeding (Winzeler et al. 1987). The use of doubled haploids in breeding programmes can thus greatly reduce the time required for development of improved cultivars. Heritability studies are simplified due to haploid plant having only one set of chromosome; hence recessive mutation is easily identified (Bhojwani and Razdan 1996).

3.3.3 Hybridization Techniques

Although it is a classical technique in genetics which become a fundamental science of plant breeding after Gregor Johann Mendel discovered the laws of heredity in the nineteenth century, further advancements in plant breeding have to be taken place in hybridization methodology. Its aim is to combine desirable genes found in two or more different varieties in order to produce pure-line progeny superior to the parental types in many respects (Novak and Brunner 1992).

3.3.4 Somatic Hybridization

It is the cell fusion which believed to revolutionize plant improvement research. In somatic hybridization (i.e. protoplast fusion), the nuclear DNA as well as the extra chromosomal DNA of the cell organelles is recombined. During regeneration and replication of the somatic hybrids, chromosomes and cell organelles of both parents will be rearranged, and multiple new combinations may be formed. In addition the fusion of two somatic cells enables the recombination of genomes from plants which cannot reproduce sexually.

3.3.4.1 Somatic Hybridization Method

In plants, this fusion is divided into two types: they are protoplast and cytoplast fusion. Nuclear and cytoplasmatic genes are recombined by the fusion of two somatic cells. When cell fusion is followed by fusion of the two nuclei, the somatic hybrids will have the combined chromosome number of both parents (i.e. allopolyploidy). Depending on the species, either the chromosome number of the parents or of the hybrid combination might have to be reduced (e.g. using colchicine). In cytoplast fusion, the nucleus of one of the cells is destroyed prior to fusion with another cell; only the extrachromosomal DNA of the cell organelles is transferred without changing the nuclear DNA.

3.3.4.2 Application of Somatic Hybridization in Breeding

In plant breeding intraspecific cell fusion is not widespread, but for instance applied in vegetable breeding, somatic hybrids have been used to introduce disease resistance genes from sexually incompatible wild species to rice and potato varieties (Helgeson et al. 1998). Protoplast fusion is for instance widely used in vegetable breeding to introduce CMS from radish into cabbage species (e.g. white cabbage, cauliflower, broccoli, cabbage turnip) (Thommen 2008).

3.3.4.3 Embryo Rescue Method

Embryo culture, sometimes called embryo rescue, is an in vitro technique that has been used for more than half a century to save the hybrid products of fertilization when they might otherwise degenerate (Norstog 1979). Embryo culture is one of the earliest forms of in vitro culture applied to practical problems and is probably the tissue culture technique that has proven of greatest value to breeders (Dunwell 1986). Embryo culture can be used to localize sites of germination promoters and inhibitors, for studies of embryogenesis and for cryopreservation (Grout 1986). Embryo culture involves isolating and growing an immature or mature zygotic embryo under sterile conditions on an aseptic nutrient medium with the goal of obtaining a viable plant. The technique depends on isolating the embryo without injury, formulating a suitable nutrient medium and inducing continued embryogenic growth and seedling formation. The culture of immature embryos is used to rescue embryos. Success with this type of culture depends strongly on the developmental stage of the embryo when it is isolated (Monnier 1978; Raghavan 1980).

3.4 Mutation Breeding

Mutation breeding is one of the conventional breeding methods in plant breeding. It is relevant with various fields like morphology, cytogenetics, biotechnology, molecular biology, etc. Mutation breeding has become increasingly popular in recent times as an effective tool for crop improvement (Acharya et al. 2007). Mutagens may cause genetic changes in an organism, break the linkages and produce many new promising traits for the improvement of crop plants (Mlcochova et al. 2004). In Plant breeding, mutation induction has become an effective way of supplementing existing germplasm and improving cultivars (Micke et al. 1987).

3.4.1 Mutagenic Agents

Agents that induce mutation are called mutagens; they are generally grouped into two broad categories, namely, chemical mutagens and physical mutagens (Mba et al. 2010). Mutagenesis is the process whereby sudden heritable changes occur in the genetic information of an organism not caused by genetic segregation or genetic recombination but induced by chemical, physical or biological agents (Roychowdhury and Tah 2013). Chemical and physical mutagens are used to induce mutations, among them gamma rays and ethyl-methane sulphonate (EMS) are widely used for mutation induction (Jain 2005; Ahloowalia et al. 2004). Induced mutagenesis is a significant tool to break through the limitations of variability and to create variability in a short period of time (Yaqoob and Rashid 2001).

3.4.1.1 Physical Mutagenesis

In the past 80 years, physical mutagens mostly ionizing radiations have been used widely for inducing hereditary aberrations, and more than 70% of mutant varieties were developed using physical mutagenesis (Mba 2013). Mutation induction with radiation has been the most frequently used method to develop direct mutant varieties, accounting for about 90% of obtained varieties (64% with gamma rays, 22% with X-rays) (Jain 2005).

3.4.1.2 Chemical Mutagenesis

Mutations may arise spontaneously or they may be induced by using chemical mutagen. Among the chemical mutagens, chemical EMS induces a vastly higher proportion of point mutations (Minocha and Arnason 1962; Hajra 1979). In plants, EMS usually causes point mutations, but loss of a chromosome segment or deletion can also occur (Thurling and Depittayanan 1992). Chemical mutagenic groups is the group of alkylating agents (these react with the DNA by alkylating the phosphate groups as well as the purines and pyrimidines) (Acquaah 2006). The chemical mutagens mostly used for mutation induction belong to the class of alkylating agents ethyl methanesulphonate (EMS), diethyl sulphate (DES), ethyleneimine (EI), ethyl nitroso urethane (ENU), ethyl nitrosourea (ENH), methyl nitrosourea (MNH) and sodium azides (Bhagwat and Duncan 1998).

3.4.2 Breeding Strategies Through Mutagenesis

Mutagenesis through induced mutation involves the development of new varieties by generating and utilizing genetic variability through chemical and physical mutagenesis; it is now a pillar of modern plant breeding, along with recombinant breeding and transgenic breeding (Shu et al. 2012). Genetic variation has led to an increase in the quantitative traits of crops, and variability on genome is induced by mutation, which enhances the crop productivity (Arulbalachandran et al. 2010).

3.4.2.1 Spontaneous Mutation

Mutation is a sudden heritable change in organism generally the structural change in gene. It's produced by change in the base sequence of genes, and it can be induced either spontaneously or artificially both in seed and vegetative propagated crops (Lee et al. 2002). The term spontaneous has been used to describe mutations that have arisen in the absence of any specified treatment (Glickman et al. 1994).

3.4.2.2 Induced Mutation

Induction of mutations has been used to improve agronomic traits of many crops, and the use of ionizing radiations, such as X-rays, gamma rays, beta rays, neutrons and chemical mutagens such as EMS DES, etc. for inducing genetic variation, is well established (Mlcochova et al. 2004). Mutation can be induced by physical and chemical mutagens; Gamma rays affect the plant growth by altering the genetical, physiological, biochemical and morphological features of the cells (Gunckel and Sparrow 1961). Induced mutation techniques serve as an important tool for creating usable genetic variability in crop plants and as a supplement to conventional breeding to alter one or two characters in good varieties, thus reducing time and money spent by classical methods for this change, especially in self-pollinated crops.

3.4.3 *Types of Mutations*

Mutations can be broadly divided into intragenic or point mutations (occurring within a gene in the DNA sequence); intergenic or structural mutations within chromosomes (inversions, translocations, duplications and deletions) and mutations leading to changes in the chromosome number (polyploidy, aneuploidy and haploidy). In addition, it is important to distinguish between nuclear and extranuclear or plasmon (mainly chloroplast and mitochondrial) mutations, which are of considerable interest to agriculture (Pathirana 2012).

3.4.4 *Development of Mutant Varieties*

The widespread use of mutation techniques in plant breeding programmes throughout the world has generated thousands of novel crop varieties in hundreds of crop species (FAO/IAEA Database). The recent database of Food and Agriculture Organization of the United Nations (IAEA 2015) indicates that 3222 mutant varieties with improved characters have been released officially as summarized in Table 3.1.

3.5 Future and Novel Techniques of Breeding of Crops

Molecular biology-based techniques hold promise for enhancing the efficiency levels of plant breeding activities and the use of these technologies and techniques in developing novel crop varieties. The recent developments in genomics have provided new tools for discovering and targeting novel alleles and genes which can enhance efficiency of plant breeding by using molecular marker-assisted selection (MAS) (Guimaraes et al. 2007). Marker-assisted selection can also be employed as a diagnostic tool to facilitate selection of progeny that possesses the desired traits, greatly speeding up the breeding process (Lammerts van Bueren et al. 2010). Biochemical markers (proteins and isozymes) were among the first marker used in genetic studies and plant breeding (Koebner et al. 1988). Biochemical markers were limited and have been eclipsed by DNA markers over the past decades; the most widely used markers in plant breeding have been simple sequence repeat (SSRs) also known as microsatellites (Gupta et al.1999).

Marker-assisted selection is particularly important for improving complex, quantitatively inherited traits that alter yield and for speeding up the breeding process (Dias 1989). Crop genomics have also been improving in the last decade, and today there are faster and cheaper systems being increasingly used in gene banks, genetic research and plant breeding, e.g. for studying interactions between loci and alleles such as heterosis, epistasis and pleiotropy or analysing genetic pathways, and

Table 3.1 Officially released mutant varieties in the FAO/IAEA mutant varieties database, July 2015

Country	Registration date	No. of released varieties	Country	Registration date	No. of released varieties
Albania	1996	1	Korea	1970–2008	35
Algeria	1979	2	Malaysia	1993–2002	7
Argentina	1962–1987	6	Mali	1998–2000	15
Australia	1967–2010	9	Mexico	0	5
Austria	1959–1995	17	Moldova	2004–2007	7
Bangladesh	1970–2010	44	Mongolia	1984–2004	4
Belgium	1967–1987	22	Myanmar	1975–2004	8
Brazil	1974–2005	13	Netherlands	1954–1988	176
Bulgaria	1972–2010	76	Nigeria	1980–1988	3
Burkina Faso	1978–1979	2	Norway	1978–1988	2
Canada	1964–2000	40	Pakistan	1970–2009	53
Chile	1981–1990	2	Peru	1995–2006	3
China	1957–2011	810	Philippines	1970–2009	15
Congo	1972	3	Poland	1977–1995	31
Costa Rica	1975–1996	4	Portugal	1983	1
Cote D'Ivoire	1976–1987	25	Romania	1992	1
Cuba	1990–2007	12	Russia	1965–2011	216
Czech Republic	1965–1996	18	Senegal	1968	2
Denmark	1977–1990	21	Serbia	1974	1
Egypt	1980–2011	9	Slovakia	1964–1995	19
Estonia	1981–1995	5	Spain	2010	1
Finland	1960–1981	11	Sh Lanka	1970–2010	4
France	1970–1988	38	Sudan	2007	1
Germany	1950–2005	171	Sweden	1950–1988	26
Ghana	1997	1	Switzerland	1985	1
Greece	1969–1970	2	Syrian	2000	1
Guyana	1980–1983	26	Taiwan	1967–1973	2
Hungary	1969–2001	10	Thailand	2006	20
India	1950–2010	330	Tunisia	1977–2007	1
Indonesia	1982–2011	29	Turkey	1994–2011	9
Iran	2004–2008	4	Ukraine	1997–2007	10
Iraq	1992–1995	23	Unrted Kingdom	1966–1990	34
Italy	1968–1995	35	United States	1956–2006	139
Japan	1961–2008	481	Uzbekistan	1966–1991	9
Kenya	1985–2001	3	Viet Nam	1975–2011	55
			Total		3222

Courtesy FAO/IAEA

advances in crop genomics are providing useful data and information for identifying DNA markers, which can be further used for both germplasm characterization and marker-assisted breeding; genomics-assisted breeding approaches along with bioinformatics capacity and metabolomics resources are becoming essential components of crop improvement programmes worldwide (Dias 2011).

3.6 Significance of Molecular-Assissted Markers in Plant Breeding

In plant breeding has witnessed a revolution due to emergence of molecular breeding. It is a subject which deals with all aspects of plant molecular biology that uses in crop improvement programmes, and molecular breeding consists of two major areas which are the transgenic crops and the molecular marker technology (Gupta et al. 2001). Molecular markers can be considered as constant landmarks in the genome, and they cannot be considered as a gene since they do not have any known biological function; they are only identifiable DNA sequences found at specific locations of the genome and transmitted by the standard laws of inheritance from one generation to the next (Semagn et al. 2006; Marica 2008).

Molecular markers are now widely used to track loci and genome regions in several crop breeding programmes, as molecular markers tightly linked with a large number of agronomic and disease resistance traits are available in major crop species (Phillips and Vasil 2001, Jain et al. 2002, Gupta and Varshney 2004). Molecular makers have proven to be powerful tools in the assessment of genetic variation and in elucidation of genetic relationships within and among species (Chakravarthi and Naravaneni 2006). Markers can also be used for dissecting polygenic traits into their Mendelian components or quantitative trait loci (QTL), and this increasing understanding of the inheritance and gene action for such traits allows the use of marker selection procedures (Anderson et al. 1993).

Molecular markers are ‘landmarks’ on chromosomes that serve as reference points to the location of other genes when a genetic map becomes available; if genetic maps are constructed, then the plant breeder establishes association between markers and desirable phenotypic traits (Podlich et al. 2004; Goodman 2004).Other uses of molecular markers include gene introgression through backcrossing, germplasm characterization, genetic diagnostics, characterization of transformants, study of genome organization and phylogenetic analysis (Jain et al. 2002).

3.7 Molecular Markers Types

In general there are two types of molecular markers (Fig. 3.2): classical and DNA markers. The first one includes morphological markers, cytological markers and biochemical markers; DNA markers have developed into many systems based on different polymorphism-detecting techniques or methods such as RFLP, AFLP, RAPD, SSR and SNP (Collard et al. 2005) (Fig. 3.2). Genetic markers are used for labelling and tracking genetic variation in DNA samples, they are biological compounds which can be determined by allelic variation and can be used as experimental probes or labels to track an individual, tissue, cell, nucleus, chromosomes or gene, and they can be used to facilitate the study of heredity and variation (Avise 2004).

3.7.1 DNA Molecular Markers

Markers are codominant and available in unlimited number because only a small (1000 nucleotide base pair) fragment is used for cloning from genomes that may contain a billion or more base pairs linearly arranged along the chromosomes; another breakthrough was the emergence of polymerase chain reaction (PCR) (Willams et al. 1990). DNA-based molecular markers have acted as versatile tools and have found their own position in various fields like taxonomy, plant breeding, genetic engineering, etc. These include characterization of genetic variability, genome fingerprinting, genome mapping and gene localization, analysis of genome evolution, population genetics, taxonomy, plant breeding and diagnostics (Joshi et al. 2011).

DNA markers have several sets of markers and are divided into two main categories: PCR-based molecular markers and hybridization-based molecular markers (Fig. 3.2). RFLP is hybridization-based marker. Moreover, random amplified poly-

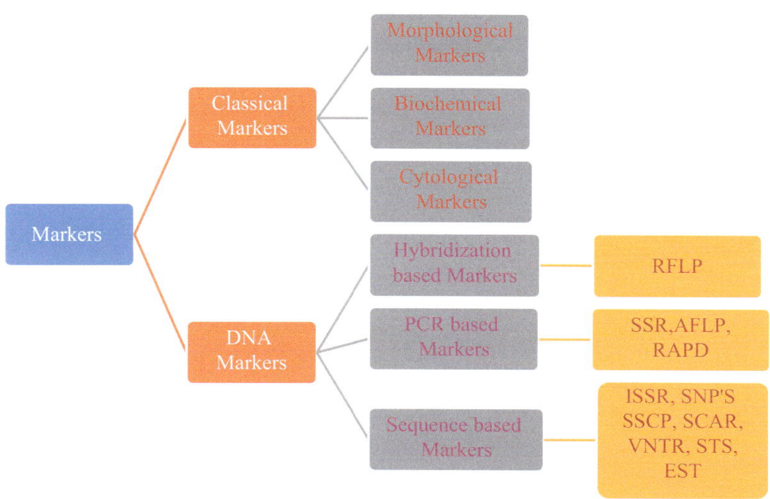

Fig. 3.2 Molecular Markers and its Techniques Utilised in Crops Improvement

morphic DNA, amplification length polymorphism (ALP), simple sequence repeat (SSR), amplified fragment length polymorphism (AFLP), sequence characterized amplified regions (SCA), sequence-tagged sites (STS), variable number of tandem repeats (VNTRs), DNA amplification fingerprinting (DAF), single-nucleotide polymorphism (SNP), microsatellites or short tandem repeats (STRs) and single strand conformation polymorphism (SSCP) are PCR-based molecular markers. DNA markers which are tightly linked to agronomically important genes can be used as molecular tool in marker-assisted selection in plant breeding (Rahimi et al. 2012; Rafalski and Tingey 1993).

3.7.2 Applications of Molecular Markers

RFLP is a technique which reveals the diversity in DNA and used to find phylogenetic studies ranging from individuals within populations or species to closely related species, and it is widely used in gene mapping studies (Miller and Tanksley 1990), as fingerprinting tool for diversity studies and for studies of hybridization and introgression as well as studies of gene flow between crops and weeds (Desplanque et al. 1999). AFLPs can be applied in studies involving genetic identity, parentage and identification of clones and cultivars and phylogenetic studies of closely related species because of the highly informative fingerprinting profiles and gene mapping studies (Vos et al. 1995). RAPD markers used gene mapping, population genetics, molecular evolutionary genetics and plant and animal breeding. This is mainly due to the speed, cost and efficiency of the RAPD technique to generate large numbers of markers in a short period (Kumari and Thakur 2014). STS markers have proved to be extremely valuable in the analysis of gene pool variation of crops during the process of cultivar development and classification of germplasm (Yang et al. 1994). SCARs are locus specific and have been applied in gene mapping studies and marker-assisted selection (Paran and Michelmore 1993). Molecular breeding approaches lie in advanced backcross QTL (AB-QTL) analysis, and marker-assisted selection (MAS) has been successfully employed in several crops, leading to improved cultivars; some other approaches such as marker-assisted recurrent selection (MARS) or genomic selection (GS) are being used in several crops (Varshney and Dubey 2009; Phillips 2010) (Table 3.2).

3.7.3 Quantitaive Trait Loci (QTL)

Genetic factors that are responsible for a part of the observed phenotypic variation for a quantitative trait can be called quantitative trait loci (QTLs), and QTL merely indicates a region on the genome and could be comprised of one or more functional genes (Falconer and Mackay 1996). QTL mapping is historically performed on structured panels, that is recombinant lines originating by the intercrosses of two or

Table. 3.2 Comparsion of the five widely used DNA markers in plants (Eagles et al. 2001)

	RFLP	RAPD	AFLP	SSR	SNP
Genomic coverage	Low copy coding region	Whole genome	Whole genome	Whole genome	Whole genome
Amount of DNA required	10 µg–50	100 ng–1	100 ng–1	120 ng–50	≥50 ng
Quality of DNA required	High	Low	High	Medium to high	High
Type of polymorphism	Single base changes, indels	Single base changes, indels	Single base changes, indels	Changes in length of repeats	Single base changes, indels
Level of polymorphism	Medium	High	High	High	High
Effective multiplex ratio	Low	Medium	High	High	Medium to high
Inheritance	Codominant	Dominant	Dominant/codominant	Codominant	Codominant
Type of probes/primers	Low copy DNA or cDNA clone	Usually 10 bp random nucleotides	Specific sequence	Specific sequence	Allele specific PCR primers
Technically demanding	High	Low	Medium	Low	High
Radioactive detection	Usually yes	No	Usually yes	Usually no	No
Reproducibility	High	Low to medium	High	High	High
Time demanding	High	Low	Medium	Low	Low
Radioactive detection	Usually yes	No	Usually yes	Usually no	No
Development/start-up cost	High	Low	Medium	High	High
Proprietary rights required	No	Yes and licenced	Yes and licenced	Yes and some licenced	Yes and some licenced
Suitable utility in diversity, genetics and breeding	Genetics	Diversity	Diversity and genetics	All purposes	All purposes

more inbred lines (Cui et al. 2013; Huang et al. 2012) also allowing the creation of a genetic reference map (Maccaferri et al. 2015). The frequency of genetic information obtained from genetic markers has a significant impact on the evolutionary biology and particularly in understanding the genetic basis of complex traits and quantitative trait loci (QTL) mapping followed by the development of statistical tools, have emerged in quantitative genetics to identify the genes involved in the genetic variability of complex traits, the complexity of these traits is influenced by the segregation of alleles at many loci, environmental factors, and their interactions (Rahimi et al. 2012; Yamamoto et al. 2009).

A benchmark article by Lander and Botstein (1989), describing a set of analytical methods for mapping QTLs with the help of molecular markers, led to increased interest in locating QTLs, and some authors recognize the dissection of quantitative genetic variation into genes at the molecular level as the greatest challenge geneticists are facing in the twenty-first century (Luo et al. 2002).The approximate numbers and locations of the QTLs underlying polygenic phenotypes can be estimated by experimental mapping approaches. Numerous molecular markers, scattered throughout the genome, are required for such QTL mapping approaches. In order to map QTLs, two homozygous inbred lines, differing in many phenotypic characteristics, are crossed to produce an F_1 progeny. The uniformly heterozygous F_1 is backcrossed to one or the other parental line resulting in a segregating mapping population, in which it can be monitored whether certain markers tend to co-segregate with specific phenotypes of interest that distinguish the parental lines (Avise 2004). If co-segregation between a marker of known chromosomal location and a phenotype is observed, it is assumed that the genes contributing to the phenotype and the marker must be closely linked (Avise 2004). In this way it is possible to construct a genetic map, showing the position of QTLs for a certain trait on the different chromosomes. After this first step, QTL analysis can be applied to plant breeding, and knowledge of QTL map locations is utilized for selection of improved varieties.

3.7.3.1 Composite Interval Mapping (CIM)

Composite interval mapping (CIM) has become popular for mapping QTLs. This method combines interval mapping with linear regression and includes additional genetic markers in the statistical model in addition to an adjacent pair of linked markers for interval mapping (Jansen 1993; Jansen and Stam 1994; Zeng 1993, 1994).

3.7.4 Genetic Mapping for Breeding

Molecular markers have revolutionized genome mapping over the last two decades, and the high density of markers that can now be generated from second-generation sequence data offers the potential for generating very-high-density genetic maps.

These markers can be used to develop haplotypes for genes or regions of interest, and complete genome mapping is now becoming a reality. Genetic mapping places molecular genetic markers in linkage groups based on their co-segregation in a population. The genetic map predicts the linear arrangement of markers on a chromosome, and maps are prepared by analysing populations derived from crosses of genetically diverse parents and estimating the recombination frequency between genetic loci. Many types of markers can be used for map construction, with population size and marker density being important for map resolution. SNP identified within whole genome sequence or large genomic fragments maintained within BAC can be applied for the genetic mapping of complex traits. This enables the genetic mapping of specific genes of interest and assists in the identification of linked or perfect markers for traits, as well as increasing the density of markers on genetic maps (Rafalski 2002). The development of these markers also allows the integration of genetic and physical maps. The use of common molecular genetic markers across related species permits the comparison of linkage maps. This allows the translation of information between model species with sequenced genomes and non-model species (Moore et al. 1995). Furthermore, the integration of molecular marker data with genomics, proteomics and phenomics data allows researchers to link sequenced genome data with observed traits, bridging the genome to phenome divide. These markers can then be used routinely in crop breeding programmes (Edwards and Batley 2009).

3.7.5 TILLING

TILLING (Targeting Induced Local Lesions IN Genomes) is a 'reverse genetics' process, and it relies on the ability of a special enzyme to detect mismatches in normal and mutant (or polymorphic) DNA strands when they are annealed, by selectively pooling the DNA and amplifying with fluorescently labelled primers; mismatched heteroduplexes were generated between wild-type and mutant DNA. Heteroduplexes were incubated with the plant endonuclease *CEL I*, (which cleaves heteroduplex mismatched sites), and the resultant products are visualized on a capillary sequencer, and the fluorescently labelled traces are analysed, and differential end labelling of the amplification products permits the two cleavage fragments to be observed and identifies the position of the mismatch or polymorphism.

3.7.5.1 TILLING Method

The TILLING strategy utilizes traditional mutagenesis followed by high-throughput mutation discovery (Mccallum et al. 2000; Colbert et al. 2001). The main steps in TILLING are mutagenesis, the development of a non-chimeric population, preparation of a germplasm stock, DNA extraction and sample pooling, screening the population for induced mutations and the validation and evaluation of mutants, and the methods required for each step can be applied to many species, making the TILLING process broadly applicable. Mutants discovered by TILLING can be used for gene-function studies and can be introduced into breeding programmes.

3.7.5.2 Mutant and Germplasm Collections in the Genomics Era: TILLING and EcoTILLING for Breeding Assistance

Plant breeding requires genetic variability to be selected in order to increase the frequencies of favourable alleles and genetic combinations. Sources of natural genetic variability can be found within the crop, mostly in the form of landraces, and also in the wild relatives. Although many landraces have been substituted by modern and uniform cultivars and genetic erosion has taken place in wild materials, gene banks preserve many of these materials, which constitute an important reservoir of genetic variation useful for breeding (Gepts 2006).

In order to facilitate the identification of the accessions of interest in these collections, a genetic reverse approach has been used (Till et al. 2003). TILLING is able to identify all allelic variants of a DNA region present in an artificial mutant collection. A similar procedure called ecotype TILLING (EcoTILLING) (Comai et al. 2004) can be used to identify allelic variants for targeting genes in natural collections. These two methods are based on the use of endonucleases, such as CEL I or Endo I, which recognize and cut mismatches in the double helix of DNA (Till et al. 2004; Triques et al. 2008). TILLING and EcoTILLING techniques identify all allelic variants for a certain genomic region; the phenotypic characterization effort can be concentrated in a reduced number of accessions with different variant, and the success of the identification of variation useful for breeding programmes will depend on the right selection of target genes. The availability of sequences coming from NGS sequencing projects and the information provided by gene expression studies is significantly increasing the number and quality of candidates for TILLING and EcoTILLING studies (Perez-de-Castro et al. 2012).

3.7.6 Transcriptomic Analysis in Crop Breeding

3.7.6.1 Transcriptome

The sequencing of cDNA rather than genomic DNA focuses analysis on the transcribed portion of the genome. Transcriptome sequencing has been used for applications ranging from gene expression profiling, genome annotation and rearrangement detection to noncoding RNA discovery and quantification. A unique feature of high-throughput transcriptome sequencing studies is the versatility of the data, which can simultaneously be analysed to provide insight into the level of gene expression, the structure of genomic loci and sequence variation present at loci (e.g. SNPs). To date, the 454 technology has dominated next-generation applications in transcriptomics; but at least one recent paper describes the use of the Illumina sequencer for profiling microRNAs (Morin et al. 2008).

Two technologies developed in cDNA microarrays fluorescently labelled cDNA from the organism of interest are hybridized to a DNA probe on a chip, and relative fluorescence is detected (Schena et al. 1995), and serial analysis of gene expression (SAGE) a library of joined cDNA fragments is generated and sequenced with the Sanger method (Velculescu et al. 1995).

3.7.7 DNA Microarray Technique

This technology was emerged in the early 1990s by convergence of two advances. DNA sequencing efforts and focused on the expressed component of the genome provided DNA sequence information and physical clones for thousands of human genes. A DNA microarray is composed of pieces of DNA ranging from 20 to 5000 base pairs concentrated into specific areas on a solid support such as a glass chip (Schena 1998). Microarray analysis allows scientists to understand the molecular mechanisms underlying normal and dysfunctional biological process; it has provided scientists with a tool to investigate the structure and activity of genes on a wide scale. Microarray technology could speed up the screening of thousands of DNA and protein samples simultaneously (Mattick 2008).

In DNA microarray, collected DNA probes are arrayed on a solid support and are used for assay, through hybridization in the presence of a complementary DNA that is present in a sample (Marmur and Doty 1961). DNA microarray is a chip of size of fingernail having 96 or more tiny wells, and each well has thousands of DNA probes or oligonucleotides arranged in a grid pattern on the chip (Sundberg et al. 2001; Afshari 2002). Thousands of different genes are immobilized at fixed locations on chip, and it means that a single DNA chip can provide information about thousands of genes simultaneously by base pairing and hybridization, and there are two types of DNA microarray: cDNA microarrays and oligonucleotide arrays (Ponder 2001; Rowley 1973; Durker et al. 2001). The three major steps of a microarray technology are preparation of microarray, preparation of labelled probes and hybridization and finally, scanning, imaging and data analysis (Labana et al. 2005; Esteve-Nunez et al. 2001).

3.7.8 Types of DNA Microarray

DNA microarray is a simple and natural tool to assess the genome. DNA microarray is a type of CDNA and oligonucleotide array (Brown and Botstein 1999).

3.7.8.1 Oligonucleotide Arrays

Oligonucleotide arrays for robotics punctuation of (printed) on a solid support (glass, coated glass, silicon or plastic), it contain short pieces of DNA (25 bp). Gene chips have been developed by companies like Affymetrix. Tens to hundreds of thousands of different oligonucleotide probes have been produced, in each array. Affymetrix is a pioneer in making cDNA microarray (Gibson 2002). The benefit of oligonucleotide arrays in quality and the proliferation as well as arrays printed on a unit is a cheap. In addition, long oligonucleotide probes can be produced directly in the surface array. Oligonucleotide array is limited to gene expression and analysis (Khadijeh Dadkhah et al. 2015).

3.8 Genomic Selection Breeding

Genomics has provided a vast amount of information linking gene activity with disease, but it does not predict PTM that most proteins undergo. Therefore, DNA sequence analysis does not predict the active form of a protein, and RNA quantitation does not always reflect the corresponding protein levels. It is believed that through genomics and proteomics, new disease markers and drug targets can be identified, which will ultimately help design products to prevent, diagnose and treat diseases (Belachew Beyene et al. 2016).

3.9 Next-Generation Sequencing (NGS)

It is a significant implication for crop genetics and breeding, the development of large-scale genomic resources, including transcript and sequence data, molecular markers and genetic and physical maps which are significant, in addition to other potential applications. Transcriptome and genome sequencing (both resequencing and de novo) using NGS technology is increasing for crop plants. The use of NGS technologies has already led to a quantum leap in the amount of genomic data available for crops for which not many genomic resources were previously available, such as chickpea and pigeon pea (Varshney et al. 2009). Moreover, the availability of large numbers of genetic markers developed through NGS technologies is facilitating trait mapping and making marker-assisted breeding more feasible. For instance, large-scale development of molecular markers using NGS can facilitate linkage mapping and WGS-based association genetics that are of practical use for MAS in marker-deficient crops. Metagenomics approaches and the sequencing of pooled amplicons generated for a large number of candidate genes across large populations offer possibilities to better understand population biology and to study genome-wide association genetics. Another important application of NGS is in gene expression studies, for which NGS has the potential to replace microarray experiments in the near future; in contrast to other gene expression approaches such as microarray and real-time PCR, NGS technologies can provide insights into the spatial and temporal control of gene expression owing to their ability to identify all RNA transcripts produced at a specific time (Ronaghi et al. 1996).

3.9.1 Application of Next-Generation Sequencing

- This technology can be used for draft sequencing via other methods, including pools of bacterial artificial chromosomes (BACs) clones, that can facilitate quick genome assembly, as shown for barley (Wicker et al. 2006).
- It is useful for rapid and efficient development of genomic resources for minor or so-called orphan crop species (Varshney et al. 2009).
- It is also a fast becoming method of choice for gene expression analysis, particularly for species for which reference genome sequences are already available (Weber et al. 2007; Cheung et al. 2006).
- Application of SOLiD includes whole genome resequencing, targeted resequencing, transcriptome research (including gene expression profiling, small RNA analysis, and whole transcriptome analysis) and epigenome (like ChIPSeq and methylation) (http://www.tecan.com/platform/apps/product/index.asp).

3.10 Conclusion

Sustainable food production is complex, which has various strategies. This review insights the plant breeding and its techniques which improve crops with high yield and other economically important traits which leads to sustainability. No doubt, the plant breeding techniques from conventional breeding to next-generation sequencing techniques are used to meet the world food security and sustainable food production (Fig. 3.3). Breeding techniques such as hybridization, haploid, micropropagation, mutation breeding and next-generation sequencing techniques are used to improve the crop production, high yield and creation of new varieties with genetic variation along with desirable traits which increase the crop production which leads to sustainable production in future. Hence, plant breeding plays a pivotal role to sustain food security with specific strategy of plant breeding to overcome the food demand.

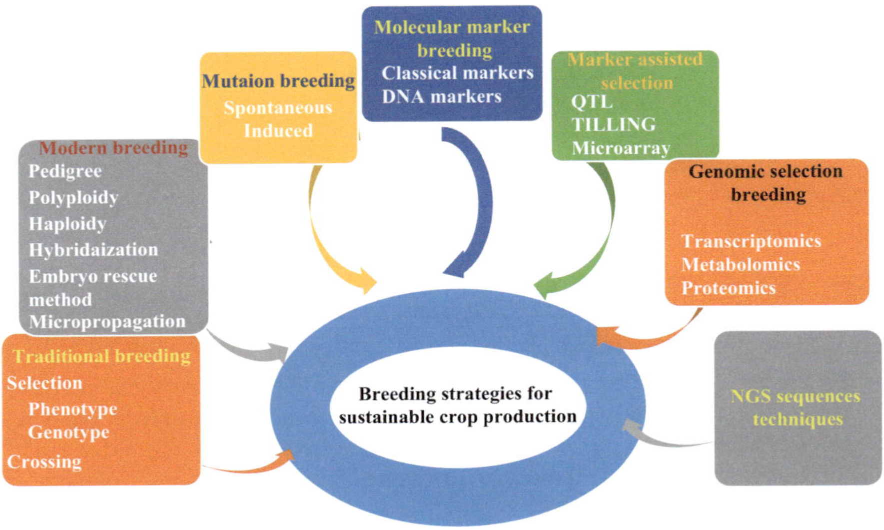

Fig. 3.3 Schematic representation of breeding stratagies for sustainable crop production (Yasmin et al.)

References

Acharya SN, Thomas JE, Basu SK (2007) Improvement in the medicinal and nutritional properties of fenugreek (*Trigonella foenumgraecum* L.) In: Acharya SN, Thomas JE (eds) Advances in medicinal plant research. India Research Signpost, Trivandrum

Acquaah G (2006) Principles of plant genetics and breeding.Wiley-Blackwell, Chichester; Wani MR, Kozgar MI, Tomlekova N et al (2002) Mutation breeding: a novel technique for genetic improvement of pulse crops particularly Chickpea (*Cicer arietinum* L). In: Parvaiz A, Wani MR, Afshari CA (eds) Perspective: microarray technology, seeing more than spots. Endocrinol 143(6):1983–1989

Afshari CA (2002) Perspective: Microarray Technology, Seeing More Than Spots. Endocrinology 143(6):1983–1989

Ahloowalia BS, Maluszynski M, Nichterlein K (2004) Global impact of mutation-derived varieties. Euphytica 135:187–204

Anderson JA, Sorrells ME, Tanksley SD (1993) RFLP analysis of genomic regions associated with resistance to pre-harvest sprouting in wheat (*Triticum aestivum*). Crop Sci 33:453–459

Arulbalachandran D, Mullainathan L, Velu S et al (2010) Genetic variability, heritability and genetic advance of quantitative traits in black gram by effects of mutation in field trail. Afr J Biotechnol 9(19):2731–2735

Avise JC (2004) Molecular markers, natural history, and evolution. Sinnauer Kluwer Aca Pub, Sunderland

Bhagwat B, Duncan EJ (1998) Mutation breeding of banana cv. Highgate (Musa acuminate AAA Group) for tolerance to *Fusarium oxysporum* f. sp. cubense using chemical mutagens. Sci Hort 73:11–22

Bhojwani SS, Razdan MK (1996) Plant tissue culture: theory and practice, a revised edition. Elsevier. Stud Plant Sci 5:167–214

Brown PO, Botstein D (1999) Exploring the new world of the genome with DNA microarrays. Nat Genet 1:33–37
Caligari PDS (2001) Plant breeding and crop improvement. Encyclopedia of life sciences. www.el
Chakravarthi BK, Naravaneni R (2006) SSR marker based DNA fingerprinting and diversity study in rice (*Oryza sativa* L). Afr J Biotechnol 5(9):684–688
Cheung F et al (2006) Sequencing *Medicago truncatula* expressed sequenced tags using life sciences technology. BMC Genomics 7:272–282
Colbert T, Till BJ, Tompa R, Reynolds S, Steine MN, Yeung AT, McCallum CM, Comai L, Henikoff S (2001) Highthroughput screening for induced point mutations. Plant Physiol 126:480–484
Collard BCY, Jahufer MZZ, Brouwer JB et al (2005) An introduction to markers, quantitative trait loci (QTL) mapping and marker-assisted selection for crop improvement the basic concepts. Euphytica 142:169–196
Comai L, Young K, Till BJ, Reynolds SH, Greene EA, Codomo CA, Enns LC, Johnson JE, Burtner C, Odden AR, Henikoff S (2004) Efficient discovery of DNA polymorphisms in natural populations by ecotilling. Plant J 37:778–786
Cui F, Zhao C, Ding A et al. (2013) Construction of an integrative linkage map and QTL mapping of grain yieldrelated traits using three related wheat RIL populations. Theor Appl Genet 127, 659–675. [PubMed]
Deppe C (2000) Breed Your Own Vegetable Varieties. Chelsea Green Publishing, pp 237–244
Desplanque B, Boudry P, Broomberg K, Saumitou-Laprade P, Cuguen J, van Dijk H (1999) Genetic diversity and gene flow between wild, cultivated and weedy forms of *Beta vulgaris* L. (Chenopodiaceae), assessed by RFLP and microsatellite markers. Theor Appl Genet 98:1194–120s1
Devaux P (1989) Variations in the proportions of fertile colchicine-treated haploid plants derived from winter barley plants. Plant Breed 103:247–250
Dias JS (1989) The use of molecular markers in selection of vegetables. SECH Actas de Horticult 3:175–181
Dias JS (2011) Biodiversity and vegetable breeding in the light of developments in intellectual property rights. In: Grillo O, Verona G (eds) Ecosystems biodiversity chapter 17. INTECH Publisher, Rijeka, pp 389–428
Druker BJ, Talpaz M, Resta DJ et al (2001) Efficacy and safety of a specific inhibitor of the BCR-ABL tyrosine kinase in chronic myeloid leukemia. N Engl J Med 344:1084–1086
Dunwell JM (1986) Pollen, ovule and embryo culture as tools in plant breeding. In: Withers LA, Alderson PG (eds) Plant tissue culture and its agricultural applications. Butterworths, London, pp 375–404
Eagles HA, Bariana HS, Ogbonnaya FC et al (2001) Implementation of markers in Australian wheat breeding. Crop Pasture Sci 52(12):1349–1356
Edgerton MD (2009) Increasing crop productivity to meet global needs for feed, food, and fuel. Plant Physiol 149:7–13
Edward D, Batley J (2009) Plant genome sequencing: application for crop improvement. Plant Biotech J 2–9
Esteve-Nunez A, Caballero A, Ramos JL (2001) Biological degradation of 2, 4,6trinitrotoluene. Microbiol Mol Biol Rev 65:335–352
Falconer DS, Mackay TF (1996): Introduction to Quantitative Genetics. 4th Ed. Essex, Longman, p 464
FAO (2004) Scientific facts on genetically modified cropsFAO Food and Agricultural Organization (2002) World agriculture: towards 2015/2030 FAO Rome FAO/IAEA database of mutant varieties. http://mvgs.iaea.org
Forster BP, Shu QY (2012) Plant mutagenesis in crop improvement: basic terms and applications. In: Shu QY, Forster BP, Nakagawa H (eds) Plant mutation breeding and biotechnology. CABI, Wallingford, pp 9–20

Gale M (2002) Applications of molecular biology and genomics to genetic enhancement of crop tolerance to abiotic stress – a discussion document. Food and Agriculture Organization of the United Nations, Rome

Gepts P (2006) Plant genetic resources conservation and utilization: the accomplishments and future of a societal insurance policy. Crop Sci 46:2278–2292

Gibson G (2002) Microarrays in ecology and evolution: a preview. Mol Ecol 11(1):17–24

Glickman BW, Saddi VA, Curry J (1994) International commission for protection against environment mutagens and carcinogens. Working paper no 2. Spontaneous mutations in mammalian cells. Mutat Res 304:19–32

Goodman MM (2004) Plant breeding requirements for applied molecular biology. Crop Sci 44:1913–1914

Grant V (1981) Plant speciation. Columbia Univ. Press, New York

Griffing B (1975) Efficiency changes due to use of doubled-haploids in recurrent selection methods. Theor Appl Genet 46:367–386

Grout BWW (1986) Embryo culture and cryopreservation for the conservation of genetic resources of species with recalcitrant seed, p. In: Withers LA, Alderson PG (eds) Plant tissue culture and its agricultural applications. Butterworths, London, pp 303–309

Guimaraes EC, Ruane J, Scherf BD, Sonnino A, Dargie JD (2007) Marker-assisted selection. Current status and future perspectives in crops, livestock, forestry and fish. FAO, Rome

Gunckel JE, Sparrow AH (1961) Ionizing radiation: biochemical, physiological and morphological aspects of their effects on plants. In: Ruhland W (ed) Encyclopedia of plant physiology. Springer, Berlin, pp 555–611

Gupta PK, Varshney RK (2004) Cereal genomics: an overview. In: Gupta PK, Varshney RK (eds) Cereal genomics. Kluwer Academic Press, Dordrecht, p 639

Gupta P et al (1999) Molecular markers and their applications in wheat breeding. Plant Breed 118:369–390

Gupta P, Roy J, Prasad M (2001) Single nucleotide polymorphisms: a new paradigm for molecular marker technology and DNA polymorphism detection with emphasis on their use in plants. Crop Sci 80:4–25

Haddad L, Godfray H, Charles J, Beddington JR, Crute Ian R, Lawrence D, Muir JF, Pretty J, Robinson S, Thomas SM, Toulmin C (2010) Food security: the challenge of feeding 9 billion people. Am Assoc Adv Sci 327(5967):812–818

Hajra NG (1979) Induction of mutations by chemical mutagens in tall indica rice. Indian Agric 23:67–72

Hallauer AR (2011) Evalution of plant breeding. Crop Breed Appl Biotechnol 11:197–206

Helgeson JP, Pohlman JD, Austin S et al (1998) Somatic hybrids between Solanum bulbocastanum and potato: a new source of resistance to late blight. Theor Appl Genet:738–742

Huang BE, George AW, Forrest KL et al (2012) A multi parent advanced generation inter-cross population for genetic analysis in wheat. Plant Biotechnol J 10:826–839

IAEA (2015) IAEA mutant database. International Atomic Energy Agency, Vienna (Accessed July 2015) Available from: http://mvd.iaea.org/

Jain SM (2005) Major mutation-assisted plant breeding programs supported by FAO/IAEA. Plant Cell Tiss Org Cult 82:113–123

Jain SM, Brar DS, Ahloowalia BS (2002) Molecular techniques in crop improvement. Kluwer Academic Publishers, Dordrecht

Jansen R (1993) Interval mapping of multiple quantitative trait loci. Genetics 135:205–211

Jansen R, Stam P (1994) High resolution of quantitative traits into multiple loci via interval mapping. Genetics 136:1447–1455

Joshi SP, Prabhakar K, Ranjekar PK et al (2011) Molecular markers in plant genome analysis. http/www.ias.ac.in/currsci/jul25/articles 15 htm pp 1–19

Keller WA, Amison PG, Cardy B (1987) Haploids from gametophytic cells-recent developments and future prospects. In: Green CE, Somers DA, Hackett WP, Biesboer DD (eds) Plant tissue and cell culture. Alan R. Liss, New York, pp 223–241

Khadijeh D, Seyed Hossein MA, Masoumeh K et al (2015) DNA microarray, types and its application in medicine. Sch Acad J Biosci 3(7):598–602

Koebner RMD, Miller TE, Snape JW et al (1988) Wheat endopeptidase: genetic control, polymorphism, intrachromosomal gene location and alien variation. Genome 30:186–192

Kumari N, Thakur SK (2014) Randomly amplified polymorphic DNA – a brief review. Am J Anim Vet Sci 9(1):6–13

Labana S, Pandey G, Paul D et al (2005) Plot and field studies on bioremediation of *p*-nitrophenol contaminated soil using *Arthrobacter protophormiae* RKJ100. Environ Sci Technol 39:3330–3337

Lammerts van Bueren ET, Backes G, de Vriend H et al (2010) The role of molecular markers and marker assisted selection in breeding for organic agriculture. Euphytica 175:51–64

Lander ES, Botstein D (1989) Mapping Mendelian factors underlying quantitative traits using RFLP linkage maps. Genetics 121:185–199

Lee YI, Lee IS, Lim YP (2002) Variation in potato regenerates from gamma-rays irradiated embryogenic callus. J Plant Biotech 4:163–170

Luo ZW, Wu CI, Kearsey MJ (2002) Precision and high-resolution mapping of quantitative trait loci by use of recurrent selection, backcross or intercross schemes. Genetics 161:915–929

Maccaferri M, Ricci A, Salvi S et al (2015) A high-density, SNP-based consensus map of tetraploid wheat as a bridge to integrate durum and bread wheat genomics and breeding. Plant Biotechnol J 13:648–663

Mackill DJ, Nguyen HT, Zhang J (1999) Use of molecular markers in plant improvement programs for rain fed lowland rice. Field Crop Res 64:177–185

Maluszynski M, Nichterlein K, van Zanten L, Ahloowalia BS (2000) Officially released mutant varieties-the FAO/IAEA database. Mut Breed Rev 12:1–84

Marica S (2008) The use of molecular markers for analyzing genes and genomes of livestock. Univerita deglistuidipadova, Padova

Marmur J, Doty P (1961) Thermal renaturation of deoxyribonucleic acids. J Mol Biol 3:585–594

Mattick J (2008) The human genome as an RNA machine. Febit Science LoungeWebinar June 12

Mba C (2013) Induced mutations unleash the potentials of plant genetic resources for food and agriculture. Agronomy 3(1):200–231

Mba C, Afza R, Bado S et al (2010) Induced mutagenesis in plants using physical and chemical agents. In: Davey MR, Anthony P (eds) Plant cell culture: essential methods. Wiley, Chichester, pp 111–130

McCallum CM, Comai L, Greene EA et al (2000) Targeting induced local lesions IN genomes (TILLING) for plant functional genomics. Plant Physiol 123:439–442

Micke A (1987) Induced mutations for crop improvement – a review. Trop Agri (Trinidad) 4(1987):259–278

Miller JC, Tanksley SD (1990) RFLP analysis of phylogenetic relationships and genetic variation in the genus Lycopersicon. Theor Appl Genet 80:437–448

Minocha JL, Arnason TJ (1962) Mutagenic effectiveness of ethyl methane sulfonate in barley. Nature 196:499

Mlcochova L, Chloupek O, Uptmoor R et al (2004) Molecular analysis of the barley cv. 'Valticky' and its Xray derived semidwarf mutant 'Diamant'. Plant Breed 123:421–427

Mohan Jain S (2010) Mutagenesis in crop improvement under the climate change. Rom Biotechnol Lett 15(2):88–105

Monnier M (1978) Culture of zygotic embryos. In: Thorpe TA (ed) Frontiers of plant tissue culture (1978). Univ of Calgary Press, Calgary, pp 277–286

Moore G, Devos KM, Wang Z, Gale MD (1995) Cereal genome evolution – grasses line up and form a circle. Curr Biol 5:737–739

Morin RD, Connor MD, Griffith M et al (2008) Application of massively parallel sequencing to microRNA profiling and discovery in human embryonic stem cells. Genome Res 18:610–621

Morrison RA, Evans DA (1988) Haploid plants from tissue culture: new plant varieties in a shortened time frame. Biotech 6:684–690

Norstog K (1979) Embryo culture as a tool in the study of comparative and developmental morphology. In: Sharp WR, Larsen PO, Paddock EF, Raghavan V (eds) Plant cell tissue culture. Ohio State Univ Press, Columbus, pp 179–202
Novak FJ, Brunner H (1992) Plant breeding: induced mutation technology for crop improvement. IAEA Bull 4:25–33
Paran I, Michelmore RW (1993) Development of reliable PCR based markers linked to downy mildew resistance genes in lettuce. Theor Appl Genet 85:985–993
Pathirana R (2012) Plant mutation breeding in agriculture. In: Hemming D (ed) Plant sciences reviews. CABI, Cambridge, pp 107–125
Perez-de-Castro AM, Vilanova S, Canizares J, Pascual L, Blanca JM, Diez MJ, Prohens J, Pico B (2012) Application of genomic tools in plant breeding. Curr Genomics 13:179–195
Phillips RL (2010) Mobilizing science to break yield barriers. Crop Sci 50:S99–S108
Phillips RL, Vasil IK (2001) DNA-based markers in plants. In: Phillips RL, Vasil IK (eds) DNA-based markers in plants. Kluwer Academic Press, Dordrecht, p 497
Podlich DW, Winkler CR, Cooper M (2004) Mapping as you go an effective approach for marker-assisted selection of complex traits. Crop Sci 44:1560–1571
Ponder BA (2001) Cancer genetics. Nature 411:336–341
Rafalski A (2002) Applications of single nucleotide polymorphisms in crop genetics. Curr Opin Plant Biol 5:94–100
Rafalski JA, Tingey SV (1993) Genetic diagnostics in plant breeding: RAPDs, microsatellites and machines. Tren Genet 9(8):275–280
Raghavan V (1980) Embryo culture In Vasil IK (ed.). Perspectives in plant cell and tissue culture. Intl. Rev. Cytol., Suppl. 11B. Academic, New York. p. 209–240.
Rahimi M, Dehghani H, Rabiei B et al (2012) Multi-trait mapping of QTLs for drought tolerance indices in rice. Cereal Res (in Farsi) 2(2):107–121
Ramsey J, Schemske DW (1998) Pathways, mechanisms, and rates of polyploid formation in flowering plants. Annu Rev Ecol Evol Syst 29:467–501
Ronaghi M, Karamohamed S, Pettersson B et al (1996) Real-time DNA sequencing using detection of pyrophosphate release. Anal Biochem 242:84–89
Rowley JD (1973) A new consistent chromosomal abnormality in chronic myelogenous leukemia identified by quina-crine fluorescence and Giemsa staining. Nature 243:290–293
Roychowdhury R, Tah J (2013) Mutagenesis a potential approach for crop improvement. In: Hakeem KR, Ahmad P, Ozturk M (eds) Crop improvement: new approaches and modern techniques. Springer, New York, pp 149–187
Roychowdhury R, Datta S, Gupta P et al (2012) Analysis of genetic parameters on mutant populations of mungbean (*Vigna radiata* L.) after ethyl methane sulphonate treatment. Not Sci Biol 4(1):137–143
Schena M (1998) Microarrays: biotechnology' discovery platform for functional genomics. Trends Biotechnol 16(7):301–306
Schena M, Shalon D, Davis RW et al (1995) Quantitative monitoring of gene expression patterns with a complementary DNA microarray. Science 270:467–470
Semagn K, Bjornstad A, Ndjiondjop N (2006) An overview of molecular marker methods for plants. Afr J Biotechnol 5(25):2540–2568
Shu QY, Forster BP, Nakagawa H (2012) Principles and applications of plant mutation breeding. In: Shu QY, Forster BP, Nakagawa H (eds) Plant mutation breeding and biotechnology. CABI, Wallingford, pp 301–325
Slafer GA, Araus JL, Royo C et al (2005) Promising eco-physiological traits for genetic improvement of cereal yields in Mediterranean environments. Ann Appl Biol 146:61–70
Sleper and Poehlman (1995) Breeding field crops. 3check
Soltis PS, Soltis DE (2009) The role of hybridization in plant speciation. Annu Rev Plant Biol 60:561–588
Sundberg SA, Chow A, Nikiforov T et al (2001) Microchip-based systems for biomedical and pharmaceutical analysis. Eur J Pharm Sci 14(1):1–12

Tadele Z, Chikelu MBA, Bradely JT (2010) TILLING for mutations in model plants and crops. In: Jain, Brar DS (eds) Molecular Techniques in crop improvement, pp 307–332
Thommen A (2008) Brauchen wir Bio-Kohl aus Protoplastenfusion? BNN-Nachrichten
Thurling N, Depittayanan V (1992) EMS induction of early flowering mutants in spring rape (*Brassica napus*). Plant Breed 108:177–184
Till BJ, Reynolds SH, Greene EA, Codomo CA, Enns LC, Johnson JE, Burtner C, Odden AR, Young K, Taylor NE, Henikoff JG, Comai L, Henikoff S (2003) Large-scale discovery of induced point mutations with high- throughput TILLING. Genome Res 13:524–530
Till BJ, Burtner C, Comai L, Henikoff S (2004) Mismatch cleavage by single-strand specific nucleases. Nucleic Acids Res 32:2632–2641
Trethowan RM, Reynolds M, Sayre K, Ortiz Monasterio I (2005) Adapting wheat cultivars to resource conserving farming practices and human nutritional needs. Ann Appl Biol 146:405–413
Triques K, Piednoir E, Dalmais M, Schmidt J, Le Signor C, Sharkey M, Caboche M, Sturbois B, Bendahmane A (2008) Mutation detection using ENDO1: application to disease diagnostics in humans and TILLING and EcoTILLING in plants. BMC Mol Biol 9:9
United Nations (2007) World population prospects, the 2006 revision. United Nations Depart Eco Social Affairs, New York
Varshney RK, Dubey A (2009) Novel genomic tools and modern genetic and breeding approaches for crop improvement. J Plant Biotechnol 18:127–138
Varshney RK et al (2009) Orphan legume crops enter the genomics era. Curr Opin Plant Biol 12:202–210
Vassel SK (2001) High quality protein corn. In: Hallauer AR (ed) Specialty corns. CRC Press, Boca Raton, pp 85–129
Velculescu VE, Zhang L, Vogelstein B, Kinzler KW (1995) Serial analysis of gene expression. Science 270:484–487
Vos P, Hogers R, Bleeker M et al (1995) AFLP, a new technique for DNA fingerprinting. Nucleic Acids Res 23:4407–4414
Weber APM et al (2007) Sampling the Arabidopsis transcriptomes with massively parallel pyrosequencing. Plant Physiol 144:32–42
Wicker T et al (2006) Sequencing put to the test using the complex genome of barley. BMC Genomics 7:275
Williams JGK, Kubelik AR, Livak KJ et al (1990) DNA polymorphism amplified by arbitrary primers are useful as genetic markers. Nucl Acids Res 18:6531–6535
Winzeler H, Schmid J, Fried PM (1987) Field performance of androgenetic doubled haploid spring wheat line in comparison with line selected by the pedigree system. Plant Breed 99:41–48
Yamamoto T, Yonemaru J, Yano M (2009) Towards the understanding of complex traits in rice: substantially or superficially. DNA Res 16(3):141–154
Yang GP, Saghai Maroof MA, Xu CG, Zhang Q, Biyashev RM (1994) Mol Gen Genet 245:187–194
Yaqoob M, Rashid A (2001) online. J Biol Sci 1:805–808
Zeng ZB (1993) Theoretical basis for separation of multiple linked gene effects in mapping quantitative trait loci. Proc Natl Acad Sci U S A 90:10972–10976

Chapter 4
Impact of Insects and Pests in loss of Crop Production: A Review

Manoharan Manosathiyadevan, V. Bhuvaneshwari, and R. Latha

Abstract Like abiotic stress, biotic stress plays a crucial role in loss of crop production worldwide. The damage caused by insect pest is one of the primary factors for reduced crop production. In this review, the production losses of major crops caused by insect pest have been discussed. There is an awareness and enthusiasm among farmers for taking plant protection measures. Introduction of new high-yielding varieties and hybrids with disease resistance towards insect pest is one of the major sustainable agricultural approaches. Pest and insect control is one of the unique ways to be departed from diseases. Uses of chemical control methods such as antifeedants, chemosterilants and irradiation techniques are customarily used. Such approaches lead to adopting and paving integrated pest management in crops. Moreover, the need for efficient pest surveillance service and quick forecasting of pests and biology of pests and diseases have become an essential tool that the pest population can be brought down immediately before any significant damage is done. Hence in the following context, types of crop pest, classification of insect pest, major crops and major pest, the pest status, scientific classification, biology of the pest and their prevention and control measures have been critically reviewed. Mechanical, chemical, and biological control measures have also been discussed in this review.

4.1 Introduction

Insect pests in agricultural systems are one of the major causes of damage to crop production and storage. In tropical countries, these pests are believed to cause losses approaching 60–70%, principally in stored products (Thomas 1999). Insects are the most diverse species of animals living on earth. Apart from the open ocean, insects can be found in all habitats, swamps, jungles and deserts, and even in highly harsh

M. Manosathiyadevan (✉) • V. Bhuvaneshwari • R. Latha
Department of Zoology, Chikkaiah Naicker College (Affiliated to Bharathiar University), Erode 638004, Tamil Nadu, India
e-mail: manosathiyadevan@yahoo.co.in

A. Dhanarajan (ed.), *Sustainable Agriculture towards Food Security*,
https://doi.org/10.1007/978-981-10-6647-4_4

environments (Imms 1964). Insects are undoubtedly the most adaptable form of life as their total numbers far exceed that of any other animal categories. The majority of insects are directly important to humans and the environment. Insect pests inflict damage to humans, farm animals and crops (Williams 1947).

A major focus of agricultural research is investigating the relationship between biodiversity and pest control. As natural systems are converted to agriculture, provisioning of some vital ecosystem services can decline (Kremen et al. 2004). Ecosystem services, such as pest control, are ecosystem processes that improve and sustain human life (Daily 1997). Many different forms of pest control are used, including cultural control, mechanical control, chemical control and biological control. Biological pest control involves attempts to use natural enemies (Huang and Yang 1987). Pests are the organisms that cause physical damage to man, animals and crops. In other words, pests can be described as any organism capable of causing damage to crop plant. Pest insects can have adverse and damaging impacts on agricultural production and market access, the natural environment and our lifestyle.

4.2 Important Crops in the World

There are several practical and conceptual reasons for being concerned with insect pest species interactions in agricultural systems. First, since the establishment of agriculture, planted and stored crops have always been infested by multiple pest species. This is still the rule today for many economically important crops all over the world such as rice (*Oryza sativa* L.) in Asia (Savary et al. 1994), corn (*Zea mays* L.) and sorghum (*Sorghum bicolor* L. Moench) in Africa (Le Ru et al. 2006), wheat (*Triticum aestivum* L.) in Europe (Daamen and Stol 1994), corn (*Z. mays*) in North America (Davidson et al. 2007) and potato (*Solanum tuberosum* L.) in South America (Dangles et al. 2008).

4.3 Insects Pests, Types and Their Classification

These can be classified into various groups based on their mode of feeding. These groups of insect pests include:

- Biting and chewing insects
- Piercing and sucking insects
- Boring insects

4.3.1 *Biting and Chewing Insects*

These insects' pests possess strong mandible and maxillae (mouthparts) which enable them to bite. These pests bite into and chew the leaves, stems, buds, flowers and even the roots of plants. Examples include chewing and biting pests which include Helicoverpa, caterpillars, beetles and slugs and snails. There are practical difficulties of achieving complete spray coverage in some at-risk crops (e.g., sweet corn, *Brassica* vegetables and lettuce).

The continuous presence of susceptible hosts in combination with the overuse of broad-spectrum synthetic chemicals (and resultant chemical resistance) has given rise to more reports of poor pest control, variable produce quality and the resultant loss of income (Masabni et al. 2009).

4.3.2 *Piercing and Sucking Insects*

These insect pests possess strong mouthparts called proboscis which enable them to pierce through plants and suck liquid materials from plants' tissues. Examples include aphids, cotton strainers, mealy bugs, scale insects, capsids, white flies, etc.

4.3.3 *Boring Insects*

Wood borers can be classified into several different groups: wood-boring beetles, ambrosia beetles, moths and horntail wasps. All of the species discussed here have four life stages and go through complete metamorphosis. The most damaging stage is usually the larval stage, and this is the stage that is often seen inside firewood or logs. Ambrosia beetles are the exception, where adults are the damaging stage (IDAHO Department of Lands 2014).

4.4 Major Crops and Major Pests

India is losing yearly 5000 crores worth of agricultural produce due to ravages of pests (weed diseases, insects, rodents and during storage). The average loss is about 18–25% for its total productivity. The pest control should be chalked out in such a way as to keep the insect population below economic tolerance levels by adopting the use of resistant varieties and biological control methods; use of antifeedants, chemosterilants and irradiation techniques; use of microorganisms; and minimal use of pesticides that too when it is considered very necessary. All such approaches will lead us towards adopting integrated pest management.

For achieving success in this approach, the need for efficient pest surveillance service and quick forecasting of pests and biology of pests and diseases have become essential so that the pest population can be brought down immediately before any significant damage is done (Thind 2015).

4.5 Pests of Rice

Rice production is largely concentrated in Asia, where it is considered to be the major food source. Oryza sativa is grown under different growth conditions with widely differing yield levels with irrigated and nonirrigated lowland rice and dryland rice being most important. In rice production, weeds, animal pests and pathogens, especially *Magnaporthe grisea*, *Thanatephorus cucumeris* and *Cochliobolus miyabeanus*, are regularly of economic importance. The estimates for the potential losses averaged 37, 25 and 13%, respectively, worldwide (Oerke 2006). The most destructive insects for rice are the lepidopteran stem borers (*Tryporyza incertulas* and *T. innotata*) and the rice leaf folder (*Cnaphalocrocis medinalis*) which cause annual losses in the order of 10 million tonnes. Complete crop failure is rare, but occasional outbreaks can destroy between 60 and 95% of the crop (Yambao et al. 1993).

4.6 Pests of Pulses

In pulses and vegetable crops, Gram pod borer (*Helicoverpa armigera*) is found to be the major insect pest of chickpea, tomato and pigeon pea, whereas white grubs like *Anomala dimidiata*, *Holotrichia seticollis* and *H. longipennis*, cutworm (*Agrotis ipsilon*) and blister beetles (*Mylabris phalerata* and *Epicauna mannerheimmi*) are the main pests of soybean and other kharif crops grown in Kumaon hills. Mustard aphid (*Aphis erysemi*) is found to be regular and major pest of vegetables and mustard in addition to its act as a vector for yellow mosaic virus of black gram. Cabbage aphid (*Brevicornye brassicae*) is also reported as an important pest of cabbage in hilly areas (Garg 1996; Garg and Sachan 1992).

4.7 Pest of Cotton

Cotton, *Gossypium* spp., is the most important fibre crop globally and is grown in almost all tropical and subtropical countries. The most important producers worldwide are PR China, the USA, India, Pakistan and Uzbekistan. For many developing countries, cotton is an essential cash crop. Cotton production is threatened especially by attack from insects (Homoptera, Lepidoptera, Thysanoptera, Coleoptera) and by weed competition during early stages of development (Oerke and Dehne 2004).

The cotton aphid damage starts at the beginning of the crop season and it can cause serious economic losses. The damage can be recognized by the appearance of the leaves which bend down. The upper side of the leaves becomes shiny due to the deposition of a sugary substance secreted by the aphid, called honeydew. During the boll opening phase, the damage affects the fibre quality (Almeida et al. 1999). This insect pest causes foliar alterations and delay of the plant growth and other negative effects that interfere with the physiology of the cotton plants (Calcagnolo and Sauer 1954). In addition, virus transmission can reduce the cotton yield up to 30–40% (Costa 1972).

4.7.1 Control Measures

Cotton aphid population dynamics can be influenced by agronomic and pest management practices (Torrey et al. 2000). Many chemical and nonchemical methods have been used for controlling this pest. But due to high usage of pesticides, some reports have been announced about resistance of pest against organic phosphorus pesticides, carbamates and artificial pyrethroids (Singh and Abrol 2001; Miller et al. 2009). Cotton aphid management has become more difficult in recent years with the implementation of the boll weevil eradication programme throughout the southeastern and mid-southern USA. The insecticide, Malathion, used in boll weevil eradication programmes, provides minimal control of cotton aphids. However, natural enemies in the Coccinellidae family (lady beetles), and the parasitic wasp, *Lysiphlebus testaceipes*, are often reduced after Malathion applications, and cotton aphid densities can increase significantly compared to that in non-treated fields (Abney et al. 2000).

4.8 *Phthorimaea operculella* (Potato Tuber Moth)

It was first recorded by Berthon from Tasmania, Australia. It was suggested that the insect was noticed in New Zealand a year earlier (Berthon 1855). It is an oligophagous insect restricted in host range to a certain of family Solanaceae. It does not require the presence of host plant material for oviposition (Fenemore 1978). It can cause serious damage in tropics and sub-tropics. Losses of up to 50% and more have been recorded in commercial crops in Peru.

4.9 *Hypothenemus hampei* (Ferrari) (Coffee Berry Borer)

It is the most serious pest in commercial coffee countries of the world (Le Pelley 1968). Pest of immature and mature coffee berries bore galleries into the endosperm of coffee seed, causing economic losses such as boring and feeding activities of adults and progeny cause a reduction in yield and quality of the final product (Moore and Prior 1988).

4.10 Major Insect Pests of Stored Products

4.10.1 Cryptolestes ferrugineus *(Stephens) (Red Rust Grain Beetle)*

It usually attacks the germs of broken or cracked grains, thus reducing germination. Other species such as *C. pusillus* (Schonherr) and *C. pusilloides* (Steel and Howe) are common in humid areas of the tropics. This family was formerly included in Cucujidae. It includes two important species: the saw-toothed grain beetle (*Oryzaephilus surinamensis* (L)), recognized by the toothed lateral margins of the pronotum, and the merchant grain beetle (*Oryzaephilus mercator* (Fauvel)), which is found in association with *O. surinamensis.* However, *O. surinamensis* prefers cereal products, while *O. mercator* is more frequent on oil-seed products and more temperature sensitive. They enter damaged grains and feed specially on the germ. Optimum conditions for development are between 30–350 °C and 70–90% relative humidity.

Adults are 3 mm flattened narrow winged beetles, but they rarely fly. Females lay their eggs loosely within the stored products. Larvae are free living and start by feeding on the embryo and the endosperm. They require 60–90% humidity for optimal development, and neither species cannot develop or breed at temperatures less than 190 °C. All stages die in 10 min if exposed to 550 °C (Howe 1956).

4.11 Fungal Contamination

Improper handling of crops during postharvest processes can cause fungal infestation. Any damage to stored products increases their susceptibility to fungal contamination. The fungal diseases spread through transmitting the spores and increase the surface area susceptible to fungal infection, which eventually increases the production of mycotoxins (Tagliaferri et al. 1993).

4.12 Prevention and Control of Pests

Pests of crops can be prevented or controlled through the following methods:

- Physical or mechanical control
- Cultural control
- Biological control
- Chemical control

4.13 Physical or Mechanical Controls

Physical control strategies include methods for excluding pests or limiting their access to crops, disrupting pest behaviour or causing direct mortality. Physical control methods can be categorized as active and passive. Active methods involve the removal of individual pests by hand, pruning out infested plant tissues and rouging out heavily infested plants. Passive methods usually include the use of a device or tool for excluding or removing pests from a crop (Vincent et al. 2009).

4.14 Cultural Control

Cultural controls are management tools and activities that make the crop habitat less favourable for pests to survive and cause damage. Cultural management practices may make the crop or habitat inhospitable to pests directly, for example, by planting cultivars resistant to pest feeding or rotating crops to deny overwintering pests their preferred food source. Cultural control is a key pest management tool available to growers because the crop variety, habitat and selected inputs set the stage for future pest fitness and abundance. Thus, implementing preventive cultural control tactics that slow pest population growth can delay or negate the need for insecticide applications and significant plant damage (Horne and Page 2008).

4.15 Biological Control

Biological control (biocontrol for short) is the use of animals, fungi or other microbes to feed upon, parasitize or otherwise interfere with a targeted pest species. Successful biocontrol programmes usually significantly reduce the abundance of the pest, but in some cases, they simply prevent the damage caused by the pest (e.g. by preventing it from feeding on valued crops) without reducing pest abundance (Lockwood 2000). Basically there are three types of biological control strategies applied in pest control programmes. These are importation (sometimes called classical biological control), augmentation and conservation (Van Driesche et al. 2008).

4.16 Chemical Control

Pesticides include a wide assortment of chemicals with specialized names and functions. They are commonly grouped according to the type of pest they control (IPM Report).

- Pesticides – chemicals to control pests
- Bactericides – chemicals to controls
- Fungicides – chemicals to control fungi
- Miticides (acaricides) – control to mites
- Defoliants –promote drying
- Avicides – control to pest birds
- Disinfectants (antimicrobials) – control microorganisms
- Molluscicides – control snails and slugs
- Insecticides – chemicals to control insects
- Rodenticides – chemicals to control rodents
- Avicides – chemicals to control birds
- Nematicides – chemicals to control nematodes

The practice of pest management by the rational application of pesticide is supremely multidisciplinary, combining many aspects of biology and chemistry with agronomy, engineering, meteorology, socio-economic and public health together with newer disciplines such as biotechnology and information science.

4.17 Integrated Pest Management (IPM) Tips for Managing Pest

Integrated pest management (IPM) is an effective, environmentally sound approach to pest management (Kabir and Rainis 2015). It provides for the protection of beneficial insects, as well as prevention of secondary pest outbreaks, pest resurgence and the spread of disease. IPM strategies aim to protect air, water and soil resources while meeting specific production objectives (Mangan and Mangan 1998). IPM combines the use of a variety of pest control methods in a way that facilitates biological control of pest insects in crops in order to improve economic, public health and environmental outcomes. Key components of effective IPM strategies are monitoring of pest populations, recognizing pest-resistant plant varieties and modifying cultural, mechanical, chemical and biological controls as needed to achieve production goals (Adams 1996).

4.18 Conclusion

Insect pests and diseases caused enormous losses to the potential agricultural production. There are many evidences which indicate rise in the losses due to pest and insects, despite increasing use of chemical pesticides. At the same time, there is a rising public concern about the potential adverse effects of chemical pesticides on the human health, environment and biodiversity. These negative externalities, though, cannot be eliminated altogether; their intensity can be minimized through development, dissemination and promotion of alternative technologies such as biopesticides and bioagents as well as good agronomic practices rather relying solely on chemical pesticides. Awareness is needed while cropping susceptible varieties in agricultural systems. Multiple host plant-resistant varieties having resistance to nematodes, diseases and insects need to be developed. But due to changing selection pressure of the pest, need-based application of biopesticides supplemented with biocontrol agents, cultural practices and cow dung, etc., is also important.

References

Abney MR, Ruberson JR, et al (2000) Natural enemies and population dynamics of the cotton aphid, Aphis gossypii, in Georgia. In: Proceedings 2000 Beltwide cotton conference National Cotton Council, Memphis, TN, pp 1246–1247

Adams RG (1996) Introduction to integrated pest management: northeast sweet corn production and integrated pest management manual in cooperative extension system. In: Adams RA, Clark JC (eds) UniversityofConnecticut, Storrs

Almeida, De, Silva RP, et al. (1999) ManejoIntegrado de pragas do algodoeiro. In: Beltrão, N.B.E. de. O agronegócio do Algodão no Brasil Brasília SPI 2:753–820

Berthon H (1855) On the potato moth Pap Proc Royal Sot VunDiemun's Lund 3–76

Calcagnolo G, Sauer HFG (1954) Ainfluência do ataque dos pulgõesnaproducão do algodão (Aphis gossypii glover 1876, Hom.: Aphididae). Arq Inst Biol 21:85–99

Costa DS (1972) O pulgãoemevidência. Divulgaçãoagronômica 32:19–22

Daamen RA, Stol W (1994) Surveys of cereal diseases and pests in the Netherlands. 6. Occurrence of insect pests in winter wheat. Neth J Plant Pathol 99:51–56

Daily G (1997) Nature's services: societal dependence on natural ecosystems. Island Press, Washington

Dangles O, Carpio C, etal BA (2008) Temperature as a key driver of ecological sorting among invasive pest species in the tropical Andes. Ecol Appl 18:1795–1809

Davidson SA, Peairs FB, Khosla R (2007) Insect pest densities across site-specific management zones of irrigated corn in northeastern Colorado. J Eco Entomol 100:781–789

Fenemore PG (1978) Oviposition of potato tuber moth, Phthorimaea operculella Zell. (Lepidoptera: Gelechiidae); the physical nature of the oviposition substrate. N.Z. J Zool 5:591–599

Garg DK (1996) Certain parasites and predators of crop pests in hilly regions of Uttar Pradesh. J Appl Zoo Res 7(1):55–56
Garg DK, Sachan SK (1992) Seasonal abundance of insect pests associated with soybean under hill conditions. J Appl Zoo Res 3(1):68–70
Horne P, Page J (2008) Integrated pest management for crops and pastures. Landlinks Press, Collingwood. ISBN 0643092579
Howe R (1956) The biology of the two common storage species of *Oryzaephilus* (Coleoptera: Cucujidae). Ann Appl Biol 44(2):341–355
http://www.agriinfo.in
http://www.agtfoods.com/about-pulses/history-of-pulses.html
http://www.infoplease.com/encyclopedia/science/pulse-botany-the-pulses-their-uses.html
Huang HT, Yang P (1987) The ancient cultured citrus ant. Bioscience 37:665–671
IDAHO Department of Lands 2014
Imms AD (1964) Outlines of entomology, 5th edn. Methuen, London, p 224
Kabir MH, Rainis R (2015) Do farmers not widely adopt environmentally friendly technologies lesson from integrated Pest management (IPM). Mod Appl Sci 9:208–215. https://doi.org/10.5539/mas.v9n3p208
Kremen C, Williams NM, Bugg RL, Fay JP, Thorp RW (2004) The area requirement of an ecosystem service: crop pollination by native bee communities in California. Ecol Lett 7:1109–1119
Le Pelley RH (1968) Las plagasdelcafeto. Editorial Labor SA Barcelona, 693 pp
Le Ru BP, Ongamo GO, Moyal P et al (2006) Diversity of lepidopteran stem borers on monocotyledonous plants in eastern Africa and the islands of Madagascar and Zanzibar revisited. Bull Entomol Res 96:555–563
Lockwood JA (2000) Non target effects of biological control: what are we trying to miss? In: Follett PA, Duan JJ (eds) Nontarget effects of biological control. Kluwer Academic Publishers, Boston, pp 15–30
Mangan M, Mangan MS (1998) A comparison of two IPM training strategies in China: the importance of concept of the rice ecosystem for sustainable insect pest management. Agric Hum Values 15:209–222
Masabni J, Dainello et al. (2009) The Texas vegetable growers handbook. Aggi Horticulture Network
Miller GL, Favret C, Carmichael A et al (2009) Is there a cryptic species within *Aulacorthum solani* (Hemiptera: Aphididae). J Eco Entomol 102:398–400
Moore D, Prior C (1988) Present status of biological control of the coffee berry borer Hypothenemus hampei. In: Proceedings of the Brighton crop protection conference – pests and diseases 1119–1124
Oerke EC (2006) Crop losses to pests. J Agri Sci 144:31–43
Oerke EC, Dehne HW (2004) Safeguarding production—losses in major crops and the role of crop protection. Crop Prot 23(4):275–285
Savary S, Elazeguia FA, Moodya K et al (1994) Characterization of rice cropping practices and multiple pest systems in the Philippines. Agri Syst 46:385–408
Singh JB, Abrol DP (2001) Pest complex of brinjal, *Solanum melongena* L. in Jammu. J Ins Environ 6:172–173
Tagliaferri M, Rigo G, Banfi G et al (1993) Prevention and control of contamination in foods derived from wheat. TecnicaMolitoria 44(11):958–965
Tarlochan Singh Thind (2015) Outlooks on pest management. Summarises the state of crop protection in India
Thomas MB (1999) Ecological approaches and the development of "truly integrated" pest management. Proc Natl Acad Sci U S A 96:5944–5951
Torrey K, Fife H, Leonard BR et al (2000) Effects of conservation tillage systems on cotton aphid populations 1208–1209

Van Driesche RG, Lyon S et al (2008) Greenhouse trials of *Aphidius Colemani* (Hymenoptera: Braconidae) banker plants for control of aphids (Hemiptera: Aphididae) in greenhouse spring floral crops. Fla Entomol 91:583–591

Vincent C, Weintraub P, Hallman G (2009) Physical control of insect pests. In: Resh VH, Cardé RT (eds) Encyclopedia of insects, 2nd edn. Academic, San Diego, pp 794–798. ISBN 9780123741448

Williams CB (1947) The field of research in preventive entomology. Ann Appl Biol 34(2):175–185

Yambao EB, Ingram KT, Rubia EG et al (1993) Case study: growth and development of rice in response to artificial stem borer damage. Int Crop Prot SARP Res Proc

Chapter 5
ABA-Mediated Drought Stress Resistance in Crops for Sustainable Agriculture

M. Ramachandran, D. Arulbalachandran, and K. Jothimani

Abstract A century before, the strategies of plant breeding play a crucial role in yield increase in drought environments for many crop plants at certain extent. Meanwhile, research needs further understanding of the physiological and molecular responses of plants grown in water deficit condition, but there is still a large gap between yields in optimal and stress conditions. After a decade, agriculture will face water-deficient environment which leads to insufficient food production toward forthcoming growing population in the future. Under water deficit environment, plants optimize the metabolism and adapt to water stress up to proximal during drought. Drought is the challenge to farmers too, and it is the most important environmental stress in agriculture. Customarily, in arid and semiarid regions of the world, the crop productivity declined due to induction of drought. However, the crop plants have undergone drought stress and various physiological and molecular mechanisms such as transmitting of signals and altering metabolism by which defends the plant from death. The various efforts have been made to improve crop productivity under water-limiting conditions; the strategies to growing plants in drought are risky and could decline productivity. The physiological strategy, such as plant hormones related to drought stress, is the key which is important in understanding and developing drought-resistant crops. Therefore, studies on the mechanism that regulates the ABA level are essential for understanding plant stress responses to develop drought-tolerant crop varieties which are adaptable to the arid environment. In this review, how the crops adapt and mitigate the drought stress through ABA with alternative metabolism and molecular characterization and develop drought-tolerant crops which lead to sustainable agriculture was discussed.

M. Ramachandran (✉)
Department of Botany, School of Life Sciences, Periyar University, Salem, Tamil Nadu, India
e-mail: ramachandran.atr@gmail.com

D. Arulbalachandran • K. Jothimani
Division of Crop Stress and Molecular Biology, Department of Botany, School of Life Sciences, Periyar University, Salem 636011, Tamil Nadu, India

A. Dhanarajan (ed.), *Sustainable Agriculture towards Food Security*,
https://doi.org/10.1007/978-981-10-6647-4_5

5.1 Introduction

Water deficit is one of the leading environmental factors which limits plant growth and productivity and is expected to become increasingly important in many regions because of the ongoing climate change, and it is well established that drought stress impairs numerous physiological and biochemical processes in crops (Sepehri and Modarres Sanavy 2003; Lawlor and Tezara 2009). It is well known that large yield losses in crops all over the world.

Drought is defined as precipitation of rainfall of below average in a given geographical region leading to acute water deficit to survival of plant. It continues to be an important challenge to agricultural researchers and plant breeders, and it is assumed that by the year 2025, around 1.8 billion people will be facing absolute water shortage and 65% of the world's population will live under water stress environment (Ingram and Bartels 1996).

Environmental stresses come in many forms, yet the most prevalent stresses have in common their effect on plant water status. The availability of water for its biological roles as solvent and transport medium and as evaporative coolant is often impaired by environmental conditions. Abiotic stress conditions cause extensive losses to agricultural production worldwide (Boyer 1982; Bray et al. 2000). Individually, stress conditions such as drought, salinity, or heat have been the subject of intense research (Bray et al. 2000; Cushman and Bohnert 2000).

Abscisic acid (ABA) plays a central role in drought responses through regulating developmental and physiological processes including stomata closure (Leung and Giraudat 1998; Umezawa et al. 2009). It is acted as a major signaling molecule involved in response of crops to drought stress, and exogenous ABA triggers stomatal closure to limit water loss through transpiration, as well mobilizes a battery of genes that serve and protect the cells from oxidative damage in prolonged stress (Wasilewska et al. 2008; Liu et al. 2005).

5.2 Drought Stress and Plant Growth

Drought stresses reduce crop yield worldwide. They are characterized by the reduction of water content in plant, closure of stomata, decrease in cell enlargement, diminished growth, decrease of leaf water potential, and turgor loss. Severe water stress may result in the hold of photosynthesis, stoppage of metabolism, and finally the death of plant (Jaleel et al. 2009). Plant growth and development reduces as a consequence of poor root development, with reduced leaf surface traits (form, shape, composition of cuticular wax, leaf pubescence, and leaf color), which affect the radiation load on the leaf canopy, delay in or reduced rate of normal plant senescence as it approaches maturity, and inhibition of stem reserves (Blum 2011).

5.2.1 Physiological Response of Plants Under Drought Stress

Drought stress affects various physiological processes and induces several physiological responses in crop plants which help them to adapt to such limiting environmental conditions. Optimization of these physiological processes is prerequisite for increased water productivity under water stress (Serraj et al. 2009). Drought causes many changes in relation to altered metabolic functions, either loss of or reduced synthesis of photosynthetic pigments. These changes in the amounts of photosynthetic pigments are closely associated to plant biomass and yield (Jaleel et al. 2009). A key factor determining plant improved productivity under drought conditions is water-use efficiency (WUE), and it is mentioned as a strategy to improve crop performance under water-limited conditions (Araus et al. 2002).

5.3 Plant Growth Regulators and Their Role in Environmental Stress

Phytohormones are naturally occurring chemically synthesized secondary metabolites that act as growth regulators and play a key role in the life span of plants, including cell division, enlargement, and differentiation, organ development, seed dormancy and germination, leaf organ senescence, and abscission (Davies 1995). Besides, phytohormones play a crucial role in stress responses and adaptation of plant (Sharma et al. 2005; Shaterian et al. 2005). Plant growth regulators are either naturally occurring or synthetic when applied exogenously which can increase yields of a target plant. They act as chemical messengers regulating the normal progression of developmental changes as well as responses to environmental signals (Morgan 1990). Plant hormones regulate every aspect of plant growth and development and the responses of plants under biotic and abiotic stresses. They have been used interchangeably, particularly when referring to auxins, gibberellins, cytokinins, ethylene, and abscisic acid (Taiz and Zeiger 2006), act as a plant development controller or a plant environment mediator, and can influence crop yield directly or indirectly; therefore, research into plant hormone physiology has become an important target for agriculture development (Davies 1995).

5.3.1 Types of Phytohormones

A number of phytohormones/growth regulators existing in plants naturally; however, auxins are the first class of plant hormones discovered and play a central role on the regulation of germination, plant growth, flower-bud formation, flowering, and other developmental processes. It regulates the following functions such as cell cycling, growth, and development and formation of vascular tissues (Davies 1995).

Cytokinin plays a significant role during several plant growth and developmental processes including cell division, chloroplast biogenesis, apical dominance, leaf senescence, vascular differentiation, nutrient mobilization, stem differentiation, anthocyanin production, and photomorphogenic development (Davies 2004). Indoleacetic acid (IAA) is one of the most multifunctional phytohormones and is vital not only for plant growth and development but also for governing and/or coordinating plant growth under stress conditions (Kazan 2013). Gibberellic acid (GA) accumulates rapidly when plants are exposed to both biotic (McConn et al. 1997) and abiotic stresses (Lehmann et al. 1995). Brassinosteroids are plant hormones with pleiotropic effects, as they influence diverse physiological processes such as growth, seed germination, rhizogenesis, senescence, and leaf abscission (Sasse 1997). Jasmonic acid (JA) is a ubiquitous plant signaling compound. Exogenous application of jasmonates induces the expression of a number of genes, and under stress conditions, the expression of jasmonate-responsive genes is altered (Farmer et al. 2003; Wasternack and Hause 2002). Ethylene is a plant hormone that plays important roles in growth and development. Responses to ethylene include fruit ripening, abscission, senescence, and adaptive responses to a wide range of biotic and abiotic stresses (Abeles et al. 1992; Bleecker and Kende 2000). Salicylic acid (SA) is vital in plant growth and development as there is evidence that this hormone regulates seed germination, vegetative growth, photosynthesis, respiration, thermogenesis, flower formation, seed production, senescence, and a type of cell death that is not associated with the hypersensitive response and regulation of antioxidant enzymes activity (Durner and Klessig 1995, 1996; Slaymaker et al. 2002).

5.4 Abscisic Acid (ABA)

Abscisic acid (ABA), one of the important phytohormones, was discovered in the 1950s affecting leaf abscission and bud dormancy. It plays essential roles in many physiological and developmental processes in higher plants, including seed maturation, dormancy and germination, seedling growth, stomatal closure, and the control of flowering time (Zeevaart and Creelman 1988; Finkelstein et al. 2002; Razem et al. 2006). It plays also a central role in coordinating the various aspects of the plant response to water stress as well as in regulation of plant growth and development (Verslues et al. 2006). A water deficit can also trigger the production of the phytohormone (ABA) which in turn causes stomata to close and induces the expression of drought stress-related genes to modulate plant responses (Seki et al. 2007).

5.5 ABA and Stress Signaling

ABA is also the major internal signal enabling plants to survive adverse environmental conditions (Keskin et al. 2010). In order to control seed germination and developmental processes, a signal is generated by ABA during a plant's life cycle.

Specifically, guard cells can target by the action of ABA for stomatal closure induction, but for modification toward severe water shortage, it may also signal systemically. It is now well-thought-out as a plant stress hormone because of different stresses to induce ABA synthesis (Mahajan and Tuteja 2005). ABA acts as an endogenous messenger in the plant's water regulation status. It plays an important role in regulating water status in plant through guard cells and growth as well as by induction of genes that encode enzymes and other proteins involved in cellular dehydration tolerance (Zhu 2002).

It plays also an important role in the ability of plants to signal to shoots that they are experiencing stressful conditions around the roots, giving rise to water-saving antitranspirant activity such as stomatal closure and reduced leaf/canopy expansion of plants (Davies et al. 2002;Wilkinson and Davies 2002). ABA is involved in adaptive deeper root growth and architectural modification of plant under drought (Spollen et al. 2000; Giuliani et al. 2005). ABA also modulates aquaporin-related root and shoot hydraulic conductivity for improved soil moisture, scavenging free harmful radicals and plant water distribution (Parent et al. 2009). It upregulates the processes which are involved in cell turgor maintenance and desiccation tolerance such as the synthesis of osmotically active solutes, metabolites, and antioxidant enzymes (Chaves et al. 2003).

5.6 Mechanism of ABA Mediation in Plants Under Drought Stress

ABA mediates many stress responses such as stress perception, signal transduction to cellular components, gene expression, and, finally, metabolic changes imparting stress tolerance (Agarwal et al. 2006). It is also involved in the modification of gene expression, and a number of stress-responsive genes are upregulated by ABA during osmotic imbalance (Ingram and Bartels 1996). ABA regulates the expression of various sets of stress-responsive genes including those involved in accumulation of compatible osmolytes and in the synthesis of LEA proteins, dehydrins, and other protective proteins (Ingram and Bartels 1996; Verslues et al. 2006). These protective proteins help in maintaining cellular water status and protect other proteins and cellular organelles from collapsing under water stress (Kishor et al. 2005).

ABA acts as a mediator in plant responses to a range of stresses, including drought stress. It is also the major internal signal enabling plants to survive adverse environmental conditions such as drought and stress (Keskin et al. 2010). Under water stress, plant cells lose water and decrease turgor pressure while ABA increases endogenously.

5.7 ABA Biosynthesis on Drought Stress

Plant hormones, the endogenous concentration of ABA, can increase more than tenfold within a few hours of drought stress and decrease dramatically to normal levels following rehydration (Kushiro et al. 2004). The elevated ABA content is beneficial for plant under stress conditions in spite of the inhibition effect on plant growth. The water loss is reduced because of stomata closure induced by ABA under osmotic stress and because canopy expansion is reduced. Many stress-responsive genes which are favorable for biosynthesis of compatible osmolytes and LEA-like proteins are induced by ABA and, thus, prevent plants from stress damage and increase plant stress tolerance (Bray 2002; Finkelstein et al. 2002).

The increase in ABA reprograms the gene expression pattern to regulate water relations through adjustment of cellular osmotic pressure, the closure of stomata, a reduced leaf canopy, deeper root growth, and changes in root system architecture (Davies et al. 2002). Biosynthesis of ABA has been relatively well characterized in *Arabidopsis*, and some data is available for other species, such as maize, tomato, potato, and barley (Seo and Koshiba 2002; Seiler et al. 2011). ABA biosynthesis occurs mostly in the chloroplast, and last steps take place in the cytoplasm. Under drought-stressed plants, synthesis de novo of ABA rapidly increases and quickly leads to stomatal closure to protect the plant from rapid desiccation of cells (Schwartz and Zeevaart 2010). Beta-carotene is the precursor for ABA synthesis through several enzymatic steps. Abiotic stress-induced activation of many ABA biosynthetic genes such as zeaxanthin oxidase, 9-cis-epoxycarotenoid dioxygenase, ABA-aldehyde oxidase, and molybdenum cofactor sulfurase appeared to be regulated through calcium-dependent phosphorylation pathway (Xiong et al. 2002;Chinnusamy et al. 2004; Zhu 2002). The drought induction of ABA biosynthesis occurs primarily in vascular tissues, and that vascular-derived ABA might trigger stomatal closure via transport to guard cells (Endo et al. 2008). These genes which encode the hydroxylases that are responsible mostly for ABA catabolism have been identified in *Arabidopsis*, rice, barley, wheat, and soybean (Saika et al. 2002; Millar et al. 2006).

5.8 Signal Transduction Endogenous ABA

Growth retardation, perturbation of metabolites including osmotic protective solutes and proteins, reprogramming of gene expression patterns, elevation of reactive oxygen species (ROS) levels, alteration of plant hormone levels, and onset of developmental plasticity compatible to drought conditions have been studied extensively (Hu and Xiong 2014). It is a wide range of physiological induced by drought; plant drought resistance is achieved by complex networks of signal transduction via both ABA-dependent and ABA-independent resistant mechanisms. In drought stress by plant cells, endogenous ABA levels increase rapidly (Zeevaart and Creelman 1988) and initiate stress resistance mechanisms. Because ABA-dependent signal transduction is the main pathway involved in drought-tolerant responses, ABA signaling components are regarded as major targets to modify for improving drought tolerant traits. ABA-independent pathways involve the transcriptional regulation of genes controlled by promoters containing the dehydration-responsive element/C-repeats (DRE/CRTs) and DRE-binding proteins/C-repeat binding factors (DREBs/CBFs) which bind to DRE/CRTs in response to osmotic stress (Yamaguchi-Shinozaki and Shinozaki 2005; Roychoudhury et al. 2013).

5.9 ABA and Stomatal Closure Under Drought Mechanism

ABA-dependent manner is the closure of stomata; the closing or opening of the pore is an osmotic shrinking or swelling of the two surrounding stoma guard cells. ABA acts directly on the guard cells and induces stomata closure via an efflux of potassium and anions from the guard cells (Schroeder et al. 2001). ABA regulation of membrane ion channels is mediated by the increased cytosolic Ca^{2+} resulting from the release of Ca^{2+} from intracellular stores and a Ca^{2+} influx from the extracellular space. It is worth noting that a number of mutations that affect ABA signaling in regard to stomatal action during drought have been characterized (Gosti et al. 1999). Stomatal closure under drought conditions prevents the intracellular water loss, and thus ABA is aptly called as a stress hormone. The main function of ABA seems to be the regulation of plant water balance and osmotic stress tolerance. Several ABA-deficient mutants, namely, *aba1*, *aba2*, and *aba3*, have been reported for *Arabidopsis* (Koornneef et al. 1998).

Various types of genes of insensitive and supersensitive mutants have been described in earlier studies with dominant mutations that encode type 2C phosphatases ABA insensitive 1 and ABA insensitive 2 (Gosti et al. 1999), whereas recessive mutations that lead to supersensitivity to ABA in regard to stomata closure are found in genes that encode farnesyl transferase β-subunit enhanced responsive to ABA1 (Cutler et al. 1996).

5.10 ABA and Drought Resistance Mechanisms

As earlier mentioned, drought tolerance is defined as the ability to grow, flower, and display economic yield under suboptimal water supply, and it affects the water relations of plants at cellular, tissue, and organ levels, causing specific as well as unspecific reactions, damage, and adaptation reactions (Beck et al. 2007). Plants respond and adapt to survive under drought stress by the induction of various morphological, biochemical, and physiological responses. To cope with the drought, tolerant plants initiate defense mechanisms against water deficit (Chaves and Oliveira 2004).

External application of plant growth regulators influences physiological processes of plants at very low concentrations (Morgan 1990). Abscisic acid is a growth inhibitor and produced under a wide variety of environmental stresses, including drought. All plants respond to drought and many other stresses by accumulating abscisic acid. It is ubiquitous in all flowering plants and is generally recognized as a stress hormone that regulates gene expression and acts as a signal for the initiation of processes involved in adaptation to drought and various environmental stresses. When plants undergo wilting, the abscisic acid levels typically rise as a result of increased synthesis endogenously (Taylor 1991).

5.11 Exogenous Application ABA Under Drought

Application of exogenous ABA enhances the recovery of the net photosynthetic rate, stomatal conductance, and transpiration rate under drought, with increased expression of various drought responsive genes (Teng et al. 2014). ABA regulates stomatal movement and thus is an important component of drought tolerance strategy for reduced water loss, by closing stomata (Ahmad et al. 2014). Exogenous application of ABA also induces the expression of many genes whose products are involved in response to drought. These genes are mainly activated by a group of transcription factors, which specifically bind to promoters containing ABA-responsive elements (Rushton et al. 2012). These ABA-induced genes encode proteins involved in stress tolerance while ABA-repressed gene products are associated with growth.

Exogenous ABA concentration depends on its environmental stress condition to the extent that ABA may induce opposing effects on whole plant transpiration which rapidly limits transpiration (Sobeih et al. 2004). It is a well-known ability to induce stomatal closure. ABA modifies its target cells directly or indirectly by altering its biosynthesis (Davies et al. 2005; Wilkinson and Davies 2010) and is known to alter other traits such as plant growth and development, mimicking the effects of water stress and thereby helping plants to better survive stress conditions (Davies and Jones 1991). ABA-induced effects predominate with respect to altering/adapting plant phenotype and are likely to depend on the chronological progression of changes in ABA concentration, ABA distribution, and downstream effects, and stomatal closure may be an initial rapid effect resulting from ABA accumulation in the apoplast in the vicinity of stomatal guard cells, while phenotypic reversion to sustained, high stomatal conductance may be possible days later, after enough time has passed for osmotically active solutes to be synthesized in leaf tissue, as a result of the sustained presence of ABA in the leaf symplast.

5.12 Gene Expression Under Drought Stress and Endogenous ABA Mediation

During drought stress, ABA is considered to be the primary phytohormone that triggers short-term responses such as stomatal closure, and it controls longer-term growth responses through the regulation of gene expression that favors maintenance of root growth, which optimizes water uptake (Zhang et al. 2006). Most of the genes that respond to drought, salt, and cold stresses are also induced by exogenous application of ABA (Shinozaki and Yamaguchi-Shinozaki 1996; Bray 1997). Hence, it is assumed that a large group of these genes is regulated by the stress ABA. The kind of ABA signaling has been under intensive research over the last decade, and considerable progresses have been made so far in identifying these ABA-responsive genes, especially that the use of ABA biosynthesis and ABA-insensitive mutants has shown that signaling of water stress may be understood in two major pathways: the ABA-dependent and ABA-dependent gene expression (Zhu 2002). Previous study shown in *Arabidopsis thaliana,* that ABA mediated induction of the drought responsive rd29B gene possessing two ABREs essential for its expression. The Arabidopsis genes, rd29A and rd29B, are differentially induced under stress conditions and by ABA treatment. The promoter region of rd29A contains at least two cis-acting elements, two DREs and one ABRE, involved in ABA-independent and ABA-responsive gene expression (Yamaguchi-Shinozaki and Shinozaki 1994).

5.13 Exogenous ABA Gene Expression Under Drought

In the ABA-independent gene expression pathways, some genes may be induced by ABA, but ABA is not the essential condition (Chandler Robertson 1994; Ingram and Bartels 1996; Bray 1997). It has been two regulative elements, of which one is ABA responsive and the other non-ABA responsive. The non-ABA-responsive sequence has a conservative 9-BP element, TACCGACAT, also known as DRE/C-repeat element. This element is essential for water stress-induced gene expression but is not under the regulation of ABA (Leung and Giraudat 1998; Ingram and Bartels 1996). Many drought-inducible genes are induced by exogenous ABA treatment. These genes contain potential ABA-responsive elements (ABREs) in their promoter regions (Thomashow 1994). The ABRE functions as a cis-acting DNA element involved in ABA-regulated gene expression (Chandler and Robertson 1994). The G-box resembles the ABRE motif and functions in the regulation of plant genes in a variety of environmental conditions (Chandler and Robertson 1994).

Two *independent transcription systems mediate ABA-responsive gene expression*: one is controlled by ABA directly through ABRE, and the other may require de novo synthesis of ABA-inducible transcription factors in drought-inducible or ABA-inducible transcription factors and may function in the ABA-responsive gene

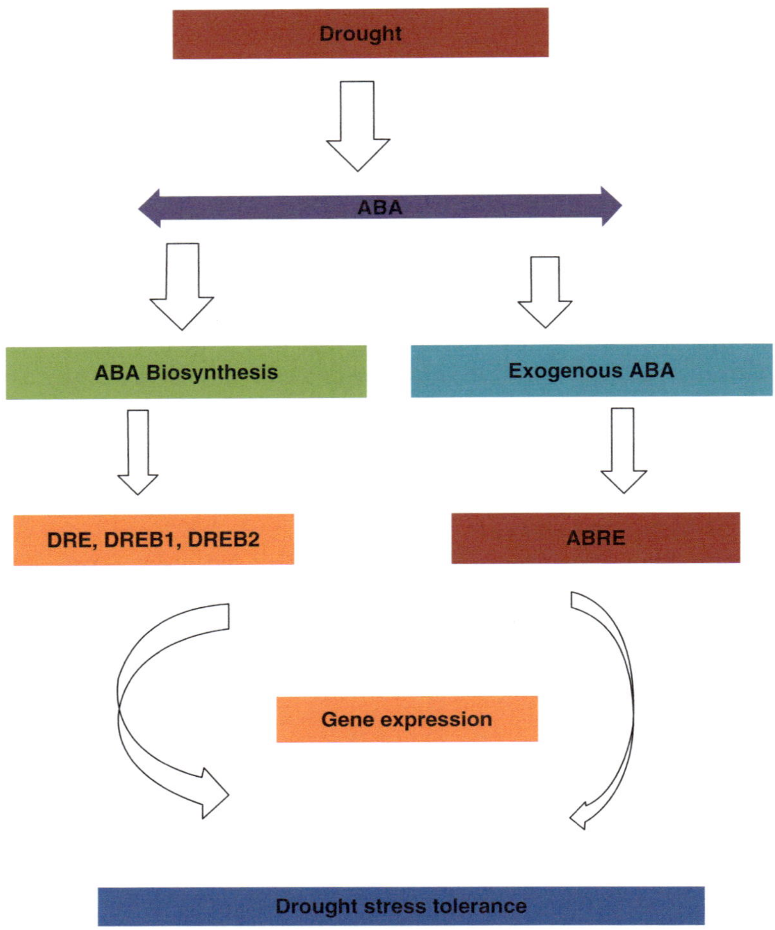

Fig. 5.1 Mechanism of gene responsive in drought

expression under drought stress. It is now well established that ABA induces certain dehydration-responsive genes (Fig. 5.1).

Transcriptional regulatory network of cis-acting elements and transcription factors are involved in ABA and abiotic stress-responsive gene expression. The promoters of the stress-induced genes contain cis-regulatory elements such as DRE/CRT (A/GCCGAC) and ABRE to induce the stress-responsive gene (*RD29B*). Transcription factors like DREB2A and DREB2B transactivate the DRE cis-element of osmotic stress genes and thereby are involved in maintaining the osmotic equilibrium of the cell (Mahajan and Tuteja 2005).

Overexpression of a transcription factor regulating ABA-responsive gene expression conferred multiple stress tolerance. The ABA level goes down during seed imbibition, which allows embryos to germinate and develop into seedlings, while ABA level remains high during abiotic stress conditions, which can arrest the growth and development. Several transcription factors, including abscisic acid insensitive, ABI3 and ABI5, are known to control this developmental checkpoint (Reyes and Chua 2007).

A number of transcription factors encoded by homeodomain-leucine zipper (HD-Zip) genes in *Arabidopsis*, rice, and other plants have been implicated in regulating drought tolerance through either ABA-dependent or ABA-independent pathways (e.g., Soderman et al. 1996, 1999; Gago et al. 2002; Himmelbach et al. 2002; Deng et al. 2006; Agalou et al. 2008; Shan et al. 2012). The HD-Zip genes, however, are an abundant group of transcription factors that are exclusively found in plants (Ruberti et al. 1991; Schena and Davis 1992). These genes are HD-Zip family I and II genes and are regulated by drought in rice (Agalou et al. 2008).

Phylogenetic analysis places Oshox22, Oshox24, Athb-7, and Athb-12 in the same subgroup (Henriksson et al. 2005) of the HD-Zip family I, and Oshox22 is very likely related to Oshox24 via an ancient chromosomal duplication (Agalou et al. 2008).

Drought stress responses by modulation of crop ABA signaling components

Species	Gene	Drought tolerance	References
Rice (*Oryza sativa*)	OsRK1/SAPK6	Overexpression resulted in reduced ABA responses resulted in enhanced ABA responses	Chae et al. (2007) Tao et al. (2011)
Maize (*Zea Mays*)	ZmCPK11	ABA-induced ROS production	Ding et al. (2013)

5.14 Conclusion

In various environmental stresses, drought is most sensitive toward inhibition of plant growth and reduction of crop yield at different geographical regions in the world. Development of sustainable growth of plants and crops is the key player for balancing the environment. As drought is prevalent which limits crop yield, the sustainable agriculture is most urgent to meet the demand for forthcoming growing population. Hence, enhancing the stress tolerance in crop implications is mandatory for sustainable agriculture. In this context, one of the methods for sustainability to improve drought-tolerant crops by mediation of ABA that regulated the plant growth has been reviewed.

References

Abeles FB, Morgan PW, Saltveit ME Jr et al (1992) Regulation of ethylene production by internal, environmental and stress factors. Ethylene in. Plant Biol 2:56–119

Agalou A, Purwantomo S, Overnas E et al (2008) A genome-wide survey of HD-Zip genes in rice and analysis of drought-responsive family members. Plant Mol Biol 66(1–2):87–103

Agarwal PK, Agarwal P, Reddy MK et al (2006) Role of DREB transcription factors in abiotic and biotic stress tolerance in plants. Plant Cell Rep 25(12):1263–1274

Ahmad M, Zaffar G, Razvi SM et al (2014) Resilience of cereal crops to abiotic stress. A review. Afr J Biotechnol 13(29):2908–2921

Araus JL, Slafer GA, Reynolds MP (2002) Plant breeding and drought in C3 cereals: what should we breed for? Ann Bot 89(7):925–940

Beck EH, Fettig S, Knake C et al (2007) Specific and unspecific responses of plants to cold and drought stress. J Biosci 32:501–510

Bleecker AB, Kende H (2000) Ethylene: a gaseous signal molecule in plants. Annu Rev Cell Dev Biol 16:1–18

Blum A (2011) Drought tolerance: is it a complex trait? Funct Plant Biol 38(10):753–757

Boyer JS (1982) Plant productivity and environment. Science 218:443–448

Bray EA (1997) Plant responses to water deficit. Trends Plant Sci 2:48–54

Bray EA (2002) Abscisic acid regulation of gene expression during water deficit stress in the era of the Arabidopsis genome. Plant Cell Environ 25:153–161

Bray EA, Bailey-Serres J, Weretilnyk E (2000) Responses to abiotic stresses. In: Buchanan BB, Gruissem W, Jones RL (eds) Biochemistry and molecular biology of plants. Amercian Society Plant Physiologists, Rockville, pp 1158–1203

Chae MJ, Lee JS, Nam MH et al (2007) A rice dehydration-inducible SNF1-related protein kinase 2 phosphorylates an abscisic acid responsive element-binding factor and associates with ABA signaling. Plant Mol Biol 63(2):151–169

Chandler PM, Robertson M (1994) Gene expression regulated by abscisic acid and its relation to stress tolerance. Annu Rev Plant Biol 45(1):113–141

Chaves MM, Oliveira MM (2004) Mechanisms underlying plant resilience to water deficits: prospects for water-saving agriculture. J Exp Bot 55:2365–2384

Chaves MM, Maroco JP, Pereira JS (2003) Understanding plant responses to drought: from genes to the whole plant. Funct Plant Biol 30:239–264

Chinnusamy V, Schumaker K et al (2004) Molecular genetics perspectives on cross-talk and specificity in abiotic stress signalling in plants. J Exp Bot 55(395):225–236

Cushman JC, Bohnert HJ (2000) Genomic approaches to plant stress tolerance. Curr Opin Plant Biol 3:117–124

Cutler S, Ghassemian M, Bonetta D (1996) A protein farnesyl transferase involved in abscisic acid signal transduction in Arabidopsis. Science 273(5279):1239

Davies PJ (1995) Plant Hormones; Physiology, Biochemistry and Molecular Biology, edn 2 edn. Kluwer Academic Publishers, London

Davies PJ (2004) Plant hormones: biosynthesis, signal transduction, action. Kluwer, Dordrecht

Davies WJ, Jones HG (1991) Abscisic acid biochemistry and physiology. Bios Scientific Publishers, Oxford

Davies WJ, Wilkinson S, Loveys BR (2002) Stomatal control by chemical signalling and the exploitation of this mechanism to hydraulic conductivity in the control of water relations in wheat plants exposed to increased evaporation demand. Planta 233:87–94

Davies WJ, Kudoyarova G, Hartung W (2005) Long-distance ABA signaling and its relation to other signaling pathways in the detection of soil drying and the mediation of the plant's response to drought. Plant Growth Regul 24:285–295

Deng X, Phillips J, Bräutigam A et al (2006) A homeodomain leucine zipper gene from Craterostigma plantagineum regulates abscisic acid responsive gene expression and physiological responses. Plant Mol Biol 61(3):469–489

Ding Y, Cao J, Ni L et al (2013) ZmCPK11 is involved in abscisic acid-induced antioxidant defence and functions upstream of ZmMPK5 in abscisic acid signalling in maize. J Exp Bot 64(4):871–884
Durner J, Klessig DF (1995) Salicylic acid is a modulator of tobacco and mammalian catalases. Biolo Chem 271:28492–28501
Durner J, Klessig DF (1996) Salicylic acid is a modulator of tobacco and mammalian catalases. J Biol Chem 271(45):28492–28501
Endo A, Sawada Y, Takahashi H et al (2008) Drought induction of Arabidopsis 9-cis-epoxycarotenoid dioxygenase occurs in vascular parenchyma cells. Plant Physiol 147(4):1984–1993
Farmer EE, Alme'ras E, Krishnamurthy V et al (2003) Jasmonates and related oxylipins in plant responses to pathogenesis and herbivory. Curr Opin Plant Biol 6:372–378
Finkelstein RR, Gampala SSL, Rock CD (2002) Abscisic acid signallinig in seeds and seedling. Plant Cell 14:S15–S45
Gago GM, Almoguera C, Jordano J et al (2002) Hahb-4, a homeobox-leucine zipper gene potentially involved in abscisic acid-dependent responses to water stress in sunflower. Plant Cell Environ 25(5):633–640
Giuliani S, Sanguineti MC, Tuberosa R et al (2005) Root-ABA1 a major constitutive QTL affects maize root architecture and leaf ABA concentration at different water regimes. J Exp Bot 56:3061–3070
Gosti F, Beaudoin N, Serizet C et al (1999) ABI1 protein phosphatase 2C is a negative regulator of abscisic acid signaling. Plant Cell 11(10):1897–1910
Henriksson E, Olsson ASB, Johannesson H et al (2005) Homeodomainleucine zipper class I genes in Arabidopsis. Expression patterns and phylogenetic relationships. Plant Physiol 139:509–518
Himmelbach A, Hoffmann T, Leube M et al (2002) Homeodomain protein ATHB6 is a target of the protein phosphatase ABI1 and regulates hormone responses in Arabidopsis. EMBO 21(12):3029–3038
Hu H, Xiong L (2014) Genetic engineering and breeding of droughtresistant crops. Annu Rev Plant Biol 65:3.1–3.27
Ingram J, Bartels D (1996) The molecular basis of dehydration tolerance in plants. Annu Rev Plant Biol 47(1):377–403
Jaleel CA, Manivannan P, Wahid A et al (2009) Drought stress in accumulation, proline content and photosynthetic characteristics of rice under drought. Rice Sci 14(1):42–48
Kazan K (2013) Auxin and the integration of environmental signals into plant root development. Ann Bot 112:1655–1665
Keskin BC, Sarikaya AT, Yuksel B et al (2010) Abscisic acid regulated gene expression in bread wheat. Aus Crop Sci 4:617–625
Kishor PB, Kavi et al (2005) Regulation of proline biosynthesis, degradation, uptake and transport in higher plants: its implications in plant growth and abiotic stress tolerance. Curr Sci 88(3):424–438
Koornneef M, Leon-Kloosterziel KM, Schwartz SH et al (1998) The genetic and molecular dissection of abscisic acid biosynthesis and signal transduction in Arabidopsis. Plant Physiol Biochem 36(1–2):83–89
Kushiro T, Okamoto M, Nakabayashi K et al (2004) The Arabidopsis cytochrome P450 CYP707A encodes ABA 8'- hydroxylases: key enzymes in ABA catabolism. EMBO J 23:16471656
Lawlor DW, Tezara W (2009) Causes of decreased photosynthetic rate metabolic capacity in water deficient leaf cells; a critical evaluation of mechanism integration of processes. Ann Bot 103:561–579
Lehmann J, Atzorn R, Brückner C et al (1995) Accumulation of jasmonate, abscisic acid, specific transcripts and proteins in osmotically stressed barley leaf segments. Planta 197(1):156–162
Leung J, Giraudat J (1998) Abscisic acid signal transduction. Annu Rev Plant Biol 49(1):199–222
Liu F, Jensen CR, Andersen MN (2005) A reviews of drought adaptation in crop plants; changes in vegetative and reproductive physiology induced by ABA based chemical signals. Aust Agric Res 56:1245–1252

Mahajan S, Tuteja N (2005) Cold, salinity and drought stresses: an overview. Arch Biochem Biophys 444:139–158
McConn M, Creelman RA, Bell E et al (1997) Jasmonate is essential for insect defense in Arabidopsis. Proc Natl Acad Sci U S A 94(10):5473–5477
Millar AA, Jacobsen JV, Ross JJ et al (2006) Seed dormancy and ABA metabolism in Arabidopsis and barley: the role of ABA 8′-hydroxylase. Plant J 45(6):942–954
Morgan PW (1990) Effects of abiotic stresses on plant hormone systems. In: Stress responses in plants: adaptation and acclimation mechanisms. Wiley-Liss, Inc, pp 113–146
Parent B, Hachez C, Redondo E et al (2009) Drought and abscisic acid effects on aquaporin content translate into changes in hydraulic conductivity and leaf growth rate: a trans-scale approach. Plant Physiol 149:2000–2012
Razem FA, El-Kereamy A, Abrams SR et al (2006) The RNA-binding protein FCA is an abscisic acid receptor. Nature 439(7074):290
Reyes JL, Chua N-H (2007) ABA induction of miR159 controls transcript levels of two MYB factors during Arabidopsis seed germination. Plant 49(4):592–606
Roychoudhury A, Paul S, Basu S (2013) Cross-talk between abscisic acid-dependent and abscisic acid-independent pathways during abiotic stress. Plant Cell Rep 32(7):985–1006
Ruberti I, Sessa G, Lucchetti S et al (1991) A novel class of plant proteins containing a homeodomain with a closely linked leucine zipper motif. EMBO 10(7):1787
Rushton DL, Tripathi P, Rabara RC et al (2012) WRKY transcription factors: key components in abscisic acid signalling. Plant Biotech 10(1):2–11
Saika H, Ohtsu K, Hamanaka S et al (2002) AOX1c, a novel rice gene for alternative oxidase; comparison with rice AOX1a and AOX1b. Genes Genet syst 77(1):31–38
Sasse JM (1997) Recent progress in brassinosteriod research. Physiol Plant 100:696–701
Schena M, Davis RW (1992) HD-Zip protein members of Arabidopsis homeodomain protein superfamily. Proc Natl Acad Sci U S A 89:3894–3898
Schroeder JI, Kwak JM, Allen GJ (2001) mGuard cell abscisic acid signalling and engineering drought hardiness in plants. Nature 410(6826):327
Schwartz SH, Zeevaart JAD (2010) Abscisic acid biosynthesis and metabolism. In: Davies PJ (ed) Plant hormones: biosynthesis, signal transduction and action, 3rd rev edn. Springer, Dordrecht, pp 137–155
Seiler C, Harshavardhan VT, Rajesh K et al (2011) ABA biosynthesis and degradation contributing to ABA homeostasis during barley seed development under control and terminal drought-stress conditions. Exper Bot Era 446
Seki M, Umezawa T, Urano K et al (2007) Regulatory metabolic networks in drought stress responses. Curr Opin Plant Biol 10(3):296–302
Seo M, Koshiba T (2002) Complex regulation of ABA biosynthesis in plants. Trends Plant Sci 7(1):41–48
Sepehri A, Modarres Sanavy SAM (2003) Water and nitrogen stress on maize photosynthesis. Online J Bio Sci 3(6):578–584
Serraj R, Kumar A, McNally KL et al (2009) Improvement of drought resistance in rice. Adv Agron 103:41–99
Shan H, Chen S, Jiang J et al (2012) Heterologous expression of the chrysanthemum R2R3-MYB transcription factor CmMYB2 enhances drought and salinity tolerance, increases hypersensitivity to ABA and delays flowering in Arabidopsis thaliana. Mol Biotech 51(2):160–173
Sharma N, Abrams SR, Waterer DR (2005) Uptake, movement, activity, and persistence of an abscisic acid analog (80 acetylene ABA methyl ester) in marigold and tomato. Plant Growth Regul 24:28–35
Shaterian J, Waterer D, De Jong H et al (2005) Differential stress responses to NaCl salt application in early-and late-maturing diploid potato (Solanum sp.) clones. Environ Exp Bot 54(3):202–212
Shinozaki K, Yamaguchi-Shinozaki K (1996) Molecular responses to drought and cold stress. Curr Opin Biotech 7(2):161–167

Slaymaker DH, Navarre DA, Clark D et al (2002) The tobacco salicylic acid-binding protein 3 (SABP3) is the chloroplast carbonic anhydrase, which exhibits antioxidant activity and plays a role in the hypersensitive defense response. Proceed Nat Acad Sci 99(18):11640–11645

Sobeih WY, Dodd IC, Bacon MA et al (2004) Long-distance signals regulating stomatal conductance and leaf growth in tomato (*Lycopersicon esculentum*) plants subjected to partial root-zone drying. Exp Bot 55(407):2353–2363

Soderman E, Mattsson J, Engstrom P (1996) The Arabidopsis homeobox gene Athb-7 is induced by water deficit and by abscisic acid. Plant J 10:375–381

Soderman E, Hjellstrom M, Fahleson J et al (1999) The HD-Zip gene ATHB6 in Arabidopsis is expressed in developing leaves, roots and carpels and up-regulated by water deficit conditions. Plant Mol Biol 40(6):1073–1083

Spollen WG, LeNoble ME, Samuels TD et al (2000) Abscisic acid accumulation maintains maize primary root elongation at low water potentials by restricting ethylene production. Plant Physiol 122:967–976

Taiz L, Zeiger E (2006) Plant physiology, 4th edn. Sinauer Associates Inc. Publishers, Massachusetts

Tao Z, Kou Y, Liu H et al (2011) OsWRKY45 alleles play different roles in abscisic acid signalling and salt stress tolerance but similar roles in drought and cold tolerance in rice. Exper Bot 62(14):4863–4874

Taylor I (1991) Genetics of ABA synthesis. In: Davies W, Jones H (eds) Abscisic acid, physiology and biochemistry. Bios Scientific, Oxford, pp 23–35

Teng KQ, Li JZ, Liu L et al (2014) Exogenous ABA induces drought tolerance in upland rice: the role of chloroplast and ABA biosynthesis related gene expression on photosystem II during PEG stress

Thomashow MF (1994) Arabidopsis thaliana as a model for studying mechanisms of plant cold tolerance. In: Meyrowitz E, Somerville C (eds) Arabidopsis. Cold Spring Harbor Laboratory Press, Cold Spring Harbor

Umezawa T, Sugiyama N, Mizoguchi M et al (2009) Type 2C protein phosphatases directly regulate abscisic acid-activated protein kinases in Arabidopsis. Proc Nat Acad Sci U S A 106(41):17588–17593

Verslues PE, Agarwal M, Katiyar-Agarwal S et al (2006) Methods and concepts in quantifying resistance to drought, salt and freezing, abiotic stresses that affect plant water status. Plant 45:523–539

Wasilewska A, Vlad F, Sirichandra C et al (2008) An update on abscisic acid signaling in plants and more. Mol Plant 1:198–217

Wasternack C, Hause B (2002) Jasmonates and octadecanoids: signals in plant stress responses and development. Prog Nucleic Acid Res Mol Biol 72:165–221

Wilkinson S, Davies WJ (2002) ABA-based chemical signaling: the co-ordination of responses to stress in plants. Plant Cell Env 25:195–210

Wilkinson S, Davies WJ (2010) ABA-based chemical signalling: the co-ordination of responses to stress in plants. Plant Cell Environ 25:195–210

Xiong L, Lee H, Ishitani M et al (2002) Regulation of osmotic stress-responsive gene expression by theLOS6/ABA1 locus in Arabidopsis. J Biol Chem 277(10):8588–8596

Yamaguchi-Shinozaki K, Shinozaki K (1994) A novel cis-acting element in an Arabidopsis gene is involved in responsiveness to drought, low-temperature, or high-salinity stress Plant Cell 6: 251–264

Yamaguchi-Shinozaki K, Shinozaki K (2005) Organization of cis-acting regulatory elements in osmotic-and cold-stress-responsive promoters. Tren Pl Sci 10(2):88–94

Zeevaart JAD, Creelman RA (1988) Metabolism and physiology of abscisic acid. Annu rev pl physio pl moleculbiol 39(1):439–473

Zhang JW, Yang J et al (2006) Role of ABA in integrating plant responses to drought and salt stresses. Field Crop Res 97(1):111–119

Zhu JK (2002) Salt and drought stress signal transduction in plants. Annu Rev Plant Biol 53:247–273

Chapter 6
Sustainable Power Production from Plant-Mediated Microbial Fuel Cells

Kamaraj Sathish-Kumar, Venkatasamy Vignesh, and Felipe Caballero-Briones

Abstract The diverse capacity of microbial fuel cells (MFCs) lies in the catabolization of complex/simple organic substrates into electricity with the aid of microbial communities and their interactions. One of the most promising types is plant-based MFCs (P-MFCs), whose benefits allow direct generation of electricity while growing the plants. Since a decade, P-MFCs have been intensively researched and developed, leading to an expansion of their functionalities and improvements in their performance, employing cost-effective materials. The power densities have been amplified mainly due to improvements in the setup construction, operation, and materials, which overcome the system restrictions. Moreover, P-MFCs could be operated with a nitrogen removal system incorporated into the cathodic electron acceptor, which would represent some advantages compared with oxygen as the final terminal electron acceptor. Accordingly, P-MFCs might be a future energy-efficient and economical solution for sustainable agriculture processes and wetland-based wastewater treatment methods. This chapter presents the technologies available in MFCs with a summary of their merits and feasible applications in the near future. Plant-mediated bioelectricity will be an alternative source of generating power throughout the world.

K. Sathish-Kumar (✉)
Tecnológico Nacional de México / Unidad Especializada en Energías Renovables, Blvd. San Pedro, 500, Parque Industrial, Ferropuerto Mieleras, C.P. 27400 Torreón, Coahuila, Mexico
e-mail: sathish.bot@gmail.com; sathish@correo.itlalaguna.edu.mx

V. Vignesh
Government of India, Animal Quarantine and Certification Service, Department of Animal Husbandry, Dairying and Fisheries, Ministry of Agriculture & Farmers Welfare, Mumbai, India

F. Caballero-Briones
Materials and Technologies for Energy, Health and Environment (GESMAT), Instituto Politécnico Nacional-CICATA, Altamira, Mexico

A. Dhanarajan (ed.), *Sustainable Agriculture towards Food Security*,
https://doi.org/10.1007/978-981-10-6647-4_6

6.1 Introduction

Due to the constant increase in energy demand and the exponential growth of the world's population over the last 150 years, the availability of energy faces an alarming situation with the diminution of fossil resources and the global warming that has been associated with the use of these.

According to the International Energy Outlook 2016 reference case projections, substantial growth in total world energy consumption from 549 quadrillion British thermal units (Btu) in 2012 to 629 quadrillion Btu in 2020 and 815 quadrillion Btu in 2040 is expected—a 48% increase from 2012 to 2040.

Thus, the current scientific research trend focus is on renewable and alternative energy sources (Fig. 6.1). These can be classified by their mechanisms of energy generation in renewable energy; except for solar energy (a photoelectric effect), the rest of them convert mechanical movement into electrical energy.

Renewable energy is derived from natural processes that are replenished constantly, such as:

- Sunlight
- Wind
- Hydroelectricity
- Geothermal energy
- Biomass

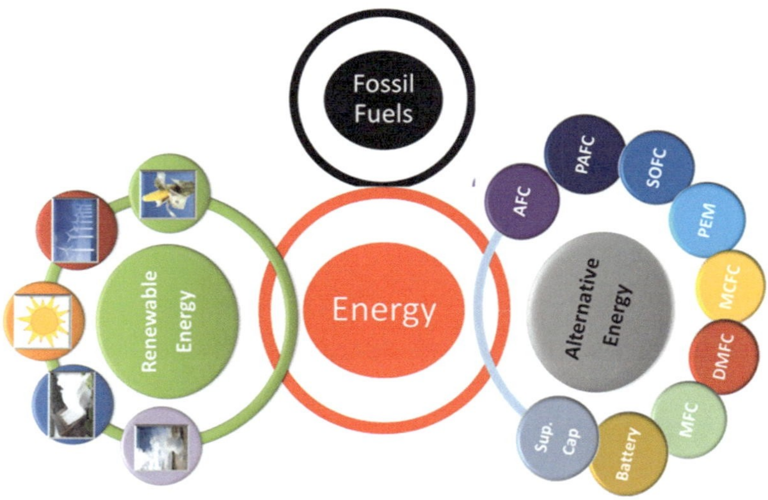

Fig. 6.1 Classification of energy. *AFC* alkaline fuel cell, *DMFC* direct methanol fuel cell, *MFC* microbial fuel cell, *MCFC* molten carbonate fuel cell, *PAFC* phosphoric acid fuel cell, *PEM* polymer electrolyte membrane fuel cell, *SOCF* solid oxide fuel cell, *Sup.Cap* super capacitor

Alternative energy is able to convert chemical energy into electrical energy via faradaic and nonfaradaic reactions, electrochemical power sources (fuel cells), and energy storage devices (batteries and supercapacitors).

In 1839, William R. Groves first reported the gaseous voltaic battery—this invention being considered as the birth of chemical fuel cells. The chemical reaction that takes place in the cell, the temperature range in which it operates, and the kinds of catalysts and fuel required depend on the electrolyte.

Chemical fuel cells are mainly classified by the kind of electrolyte employed, as follows (Larminie et al. 2013):

- Alkaline fuel cells (AFCs)
- Direct methanol fuel cells (DMFCs)
- Molten carbonate fuel cells (MCFCs)
- Phosphoric acid fuel cells (PAFCs)
- Polymer electrolyte membrane (PEM) fuel cells
- Solid oxide fuel cells (SOFCs)
- Microbial fuel cells (MFCs)

The applications for which the above fuel cells are most suitable depend on their characteristics. For example, transportation application requirements are well met by PEM fuel cells, having low sensitivity to orientation, a fast start-up time, and a favorable power-to-weight ratio. These cells are particularly suitable for use in passenger vehicles (cars or buses). Some higher-temperature fuel cells are applied in stationary applications.

The diverse capacity of MFCs lies in the conversion of any kind of organic material into electricity, making possible the sustainability of agriculture processes with the integration of this kind of new technology.

6.1.1 *Principles of Microbial Fuel Cells*

When Potter first discovered the biocatalyst ability to transport electrons directly to an electrode with the aid of an external mediator (Potter 2011), it created the challenge to bring this technology to a more sustainable scale with more economical feasibility. Consequently, MFCs have emerged as an alternative energy system converting organic materials into electricity with the aid of a biocatalyst at an ambient temperature. This can employ a wide range of soluble or dissolved complex organic materials/wastewater and renewable biomass influent/effluent as a substrate that further offers the dual benefits of renewable direct electrical energy generation with simultaneous waste water/remediation, which makes the whole process sustainable.

The biocatalyst found in the anode chamber oxidizes the organic substrates, liberating electrons and protons. The electrons then travel to the cathode side by an external circuit and protons diffuse through the proton exchange membrane, known as PEM. The protons and electrons subsequently are combined at the cathode side with molecular oxygen to produce water (Fig. 6.2).

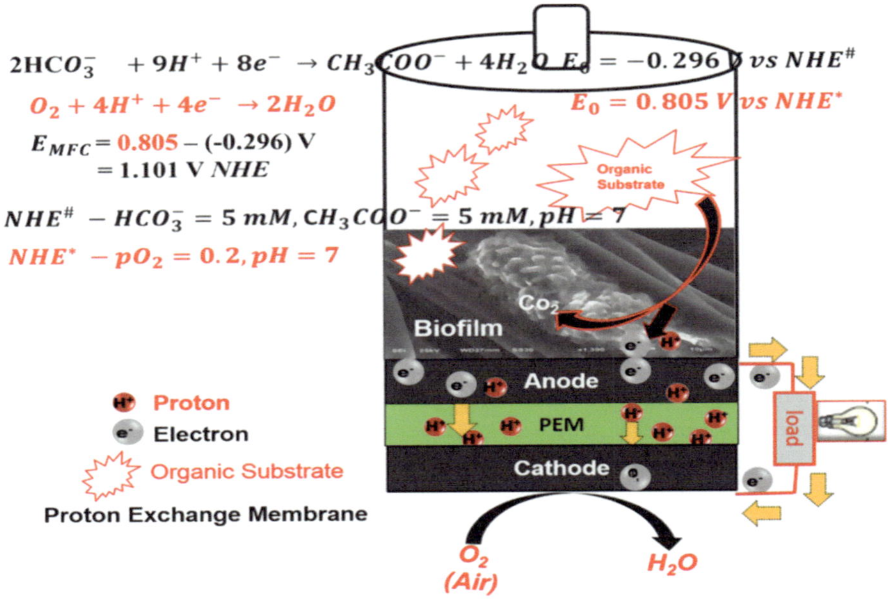

Fig. 6.2 Schematic representation of a single-chamber microbial fuel cell

6.1.2 Microbes for Electricity Generation

The ubiquitous nature of microbes has driven the metal-consuming capability that can adapt biochemical energy metabolism for growth and reproduction. The process of metal conversion is facilitated by four possible mechanisms:

1. Electron-donating organic materials on microbes that use minerals containing metal ions as final terminal electron acceptors for growth (heterotrophy-based respiration)—dissimilatory metal-reducing microorganisms of *Geobacter metallireducens GS-15* (Lovley et al. 1987, 1988) and *Shewanella oneidensis MR-1* (Myers and Nealson 1988 and Lovley et al. 1989b), and oxidation of organic materials or hydrogen, with subsequent transfer of the electrons to redox-active minerals (Fe(III) or Mn(III) or Mn(IV)) in the absence of oxygen and other respiratory terminal electron acceptors (Fig. 6.3a).
2. Electron-donating and/or energy sources such as light, NO_3, and O_2 along with CO_2 for synthesis of organic materials (autotrophic)—use of soluble metal ions as electron and/or energy sources to reduce oxygen, carbon dioxide, and nitrate for growth and reproduction by metal-oxidizing microorganisms such as *Rhodopseudomonas palustris TIE-1* and *Sideroxydans lithotrophicus ES-1* (Jiao et al. 2005; Emerson and Moyer 1997; Shelobolina et al. 2012 and Bose et al. 2014) (Fig. 6.3b).
3. Electron transfer between different microbial species is achieved by semiconductive minerals (hematite and magnetite), which act as conductors of electrons

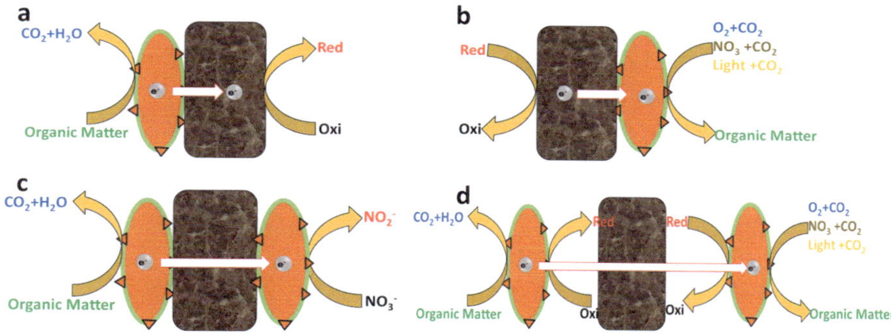

Fig. 6.3 Possible extracellular electron transfer between bacteria and minerals in the environment. (**a**) Microbes use metal ions in the environment as terminal electron acceptors for respiration. (**b**) Energy sources and/or electrons used for growth. (**c**) Electron transfer between microbes of the same and different species with the aid of electrical conductors. (**d**) Microbe metabolism supported by electron storage materials (Adapted from Shi et al. 2016)

9 (Fig. 6.3c). During the oxidation of acetate by *Geobacter sulfurreducens PCA*, the realized electrons transfer to *Thiobacillus denitrificans* and reduce nitrate to nitrite with the aid of hematite and magnetite.

4. Microbial metabolism is supported by electron storage materials (magnetite and clay minerals that contain Fe(II) and Fe(III)). In the absence of other terminal electron acceptors, the electrons released via *G. sulfurreducens PCA* and *S. oneidensis MR-1* are stored in electron storage materials and then donated to *R. palustris TIE-1* and *Pseudogulbenkiania* sp. strain *2002* (Byrne et al. 2015 and Zhao et al. 2015) (Fig. 6.3d). The abovementioned process is facilitated through a cytoplasmic membrane and redox shuttle, transferring the electrons to extracellular minerals; this is referred to as *extracellular electron transfer* (Melton et al. 2014 and Shi et al. 2016). This natural phenomenon is exploited to generate electrical energy while placing an electrode material, which acts as the final terminal electron acceptor. One such approach was reported almost 100 years ago by Potter (2011). Later on, various discoveries prompted interest in the process of energy generation using microbes.

An interesting phenomenon occurs during the transfer of electrons to the external environment. With the focus on electrical energy generation by microbes, the transfer of external electron transport is classified into two processes: (1) a capacitive (non-Faraday) process; and (2) a Faraday process. The processes of adsorption and desorption can occur and charges (i.e., electrons) do not cross the electrode–solution (i.e., electrolyte) interface. Although external current can flow transiently, this depends on the potential, electrode area, and solution composition (Bard and Faulkner 2001). Microbial capacitive processes involve microbes and changes of the double layer capacity on the electrode surface. During the microbe attachment or detachment process, a membrane-bound lipid layer facilitates the displacement of water molecules and ions from the double layer of the electrode surface, leading

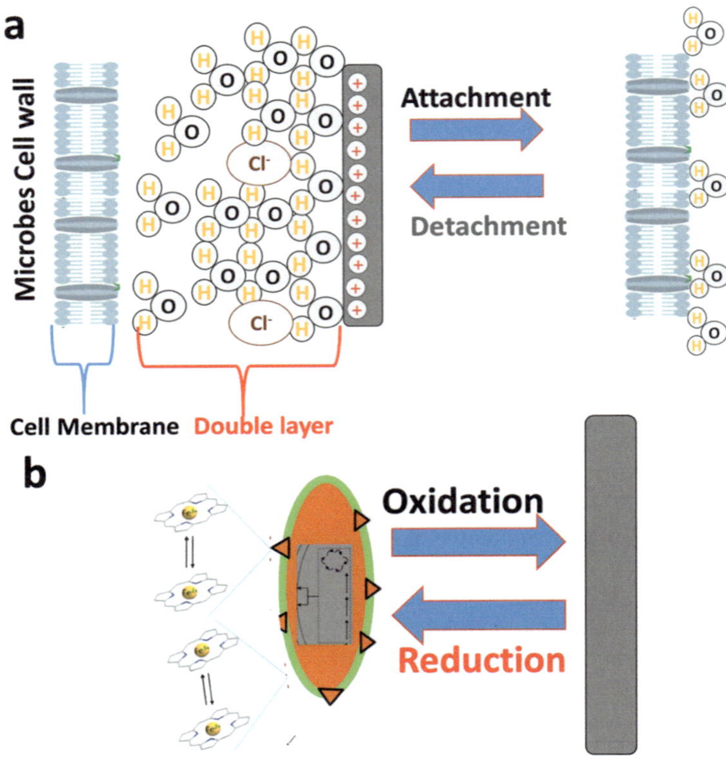

Fig. 6.4 Microbial non-Faraday process (**a**) and Faraday process (**b**)

to a decrease in the double layer capacity of the electrode surface, creating the flow of a charge-balancing electric current (i.e., capacitive) (Schroder et al. 2015) (Fig. 6.4a). In general, use of high-surface-area electrodes for the bioelectrochemical system exhibits a capacitive current in the initial period of operation until biofilm formation on the electrode surface. Transfer of charges (i.e., electrons) occurs at the electrode–solution (i.e., electrolyte) interface. The electron transfer process provokes loss of electrons (oxidative) or receipt of electrons (reduction). Further reactions are governed by the amount of electricity passed, which is directly proportional to the amount of current flow in order to carry out the chemical reaction, called *Faraday's law* (Bard and Faulkner 2001). With regard to microbial Faraday processes, the current exhibits oxidation and reduction of microbes and their molecular species (i.e., redox shuttle) between the electrode surfaces (Fig. 6.4b). These processes are collectively known as *microbial extracellular electron transfer*. Owing to the environmental significance and practical applications of MFC, extracellular electron transfer to minerals, such as Fe(III) and Mn(IV) oxides (Lovley et al. 2004; Gralnick and Newman 2007; Shi et al. 2007), and to electrodes (Lovley 2006 a, b) is probably of the greatest interest at present.

Various reviews have recounted the last century of the study of microbe–electrode interactions (Lovley et al. 2004; Gralnick and Newman 2007) in the Faraday process, demonstrating the mechanisms followed by two kinds of transfer based on the contact of communication between the bacteria and electrode: (1) direct electron transfer; and (2) indirect electron transfer.

Direct electron transfer is facilitated by membrane-bound cytochrome, which can also be involved in various electrocatalysis process. The attachment of microbes or biofilm on the surface of the electrode might also promote the pseudocapacitance charged (oxidatively) and uncharged (reductively) (Malvankar et al. 2012). Long-range electron transfer has been realized with the aid of bacterial nanowires and biofilm-based intraconnected cytochrome of various microbes on biofilm networks (Pfeffer et al. 2012 and Kumar et al. 2017).

Bacteria that have been described as electrochemically active bacterial/electrogenic bacteria are able to transfer electrons directly to an anode surface. A few such notable bacteria are *Desulfuromonas acetoxidans* (Bond et al. 2002), *G. sulfurreducens, G. metallireducens, Rhodoferax ferrireducens, Desulfobulbus propionicus* (Holmes et al. 2004), *Enterococcus gallinarum* (Kim et al. 2005), and *Shewanella putrefaciens* (Kim et al. 1999a, b, c).

Indirect electron transfer occurs either by abiotic oxidation products derived from biological fermentation or through electrochemical mediators, secreted by the bacteria themselves or added to the electrolyte; or directly through the components of the cell membrane. The ability of microbes to switch their metabolism from a soluble electron donor (e.g., hydrogen, glucose, acetate) or acceptor (e.g., oxygen, nitrate, fumarate) to a solid electron donor or acceptor at the surface of a conductive electrode (Fig. 6.5) is the main element of *indirect electron transfer*, which was highlighted by Bond et al. (2002) and Tender et al. (2002).

Bacteria such as *S. putrefaciens* (Kim et al. 1999c), *Clostridium butyricum* (Park et al. 2001), *Aeromonas hydrophila* (Pham et al. 2003), and *S. oneidensis* (Ringeisen et al. 2006) are able to transfer electrons with a combination of direct transfer and indirect transfer through mediators secreted by the bacteria. These processes may occur in three possible ways:

1. *Bacterial metabolic product oxidation*: Bioelectricity generation by oxidation of products derived from fermentation has been found at electrochemically active bacteria (EAB)/conductive material interfaces (Fig. 6.6a). Abiotic oxidation of fermentation products (H_2, alcohols, ammonia, H_2S, or HS^-) occurs at the anodic surface. Bioelectrochemically this produces hydrogen on the anode by the well-known hydrogen-producing strains of *Clostridium beijerinckii, C. butyricum,* and *Escherichia coli K12* (Niessen et al. 2004, 2006). Interestingly, higher current densities related to hydrogen oxidation at the electrode surface have been found. In the marine sediment environment, based on pH dependence, sulfate-reducing bacteria are able to reduce sulfate compounds into H^{S-} or S^{2-}. Further, this has the potential possibility for oxidation directly on the anodic surface into S^0 (Ryckelynck et al. 2005; Lovley 2006a; Reimers et al. 2006).

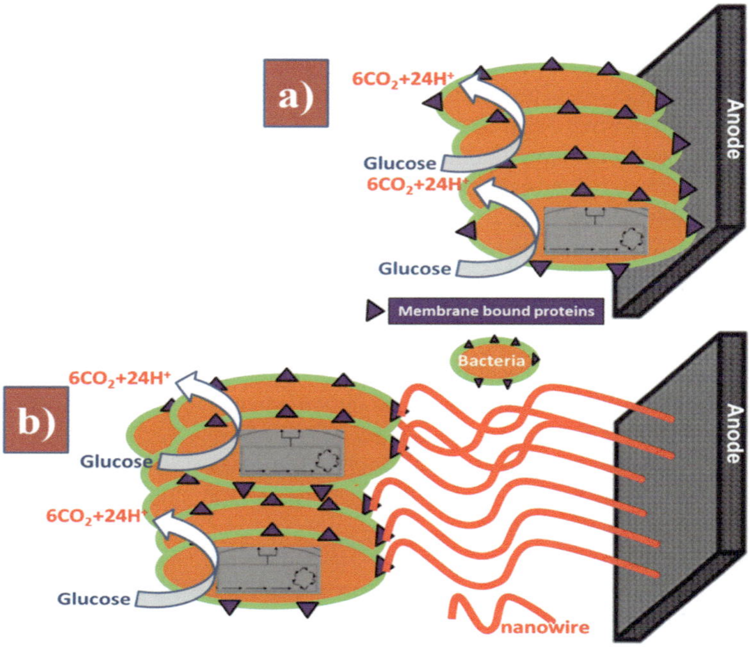

Fig. 6.5 Direct electron transfer. (**a**) Membrane-bound electron transfer. (**b**) Nanowire-mediated electron transfer (Adapted from Rabaey and Verstraete (2005)

2. *Artificial electrochemical mediators*: Electrochemical mediators function as "electron shuttle" molecules that are redox coupling between the electrode surface and the redox-active center in the biological system, mostly involving membrane-bound cytochrome. These mediators are required for electron transfer in the case of nonfermentative bacteria that use the electrode as an electron acceptor to carry out the desired conversion. In general the oxidized forms of mediators cross the cell membrane and receive the electrons from an electron donor within the cell and then come out from the cell in a reduced form, reaching the electrode surface and finally transferring the electron (Shukla et al. 2004) (Fig. 6.6b). *E. coli, Pseudomonas* spp., *Proteus,* and *Bacillus* are required mediators for their internal metabolism outside the cell under the electrochemical energy conversion process (Fig. 6.6b). Commonly used mediators are thionine, neutral red, 2-hydroxy-1,4-naphthoquinone, and varieties of phenazine. However, there is no proper evidence describing microbial growth in the presence of mediators. Furthermore, use of mediators in a biological fuel cell in a continuous batch operation requires the constant presence of mediators that increase the cost of their applications, and mediators are often toxic to the environment, and so should not be discharged without proper treatment.

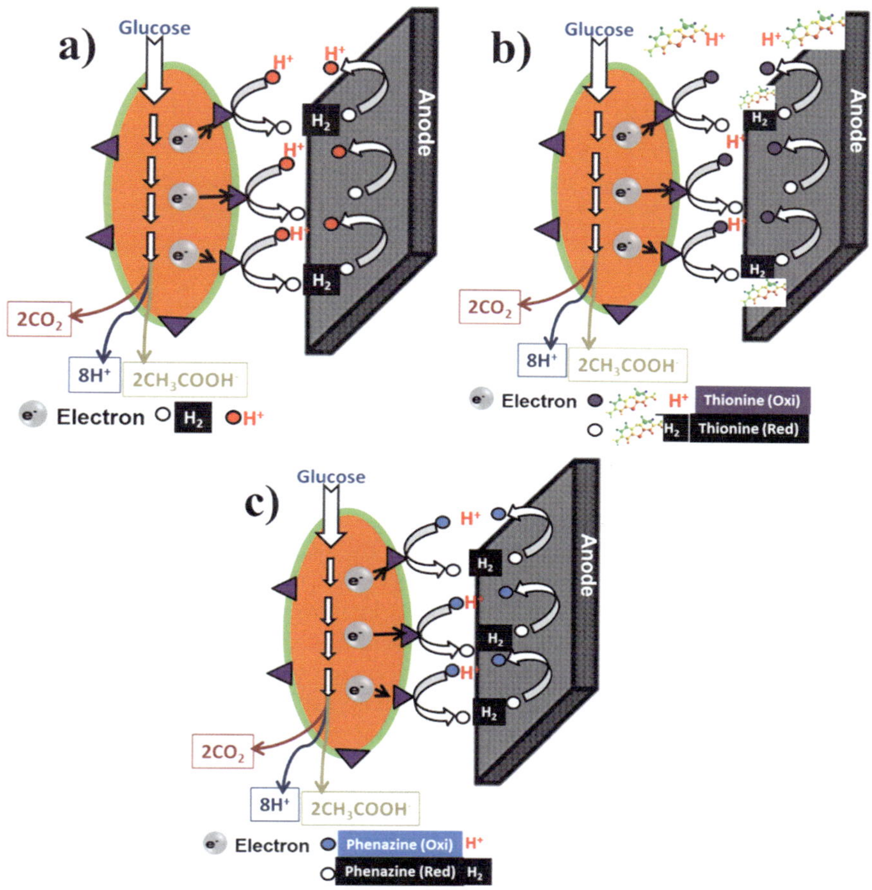

Fig. 6.6 Indirect electron transfer. (**a**) Oxidation of a bacterial metabolism product. (**b**) Electron transfer by artificial electrochemical mediators. (**c**) Bacteria that produce their own mediators (Adapted from Rabaey and Verstraete 2005)

3. *Self-mediators produced by bacteria*: Bacteria are able to produce their own electrochemical mediators in order to facilitate the extracellular electron transfer reaction in bioelectrochemical energy conversion systems. Notable microbes of *Pseudomonas* spp. (Rabaey and Verstraete (2005), *S. putrefaciens* (Kim et al. 1999a, 2002) and *Geothrix fermentans* (Bond and Lovley 2005) can compete/survive in the bioelectrochemical energy conversion system and liberate their own mediators, which increase communication between the electrode and bacteria through an indirect extracellular electron transfer reaction (Hernandez et al. 2004). A self-mediator of phenazine molecules produced from *Pseudomonas aeruginosa* has exhibited improved power output by increased indirect extracellular electron transfer (Rabaey and Verstraete 2005). However, because of inabil-

ity to synthesize a self-mediator, mutant strains of *P. aeruginosa* can produce minimal power output compared with the nondeficient strain. Hence the self-mediator plays a vital role in indirect electron transfer (Rabaey and Verstraete 2005) (Fig. 6.6c).

Integrating these new properties, an opportunity to boost applications of this technology has been opened; in the case of nanotechnology, which has been shown or proposed in areas as diverse as microelectronics, functional materials, and biotechnology, the improvement of technological and commercial developments is notorious (Sapsford et al. 2013).

6.1.3 Types of Microbial Fuel Cells

Microbial fuel cells can be subcategorized, depending on the nature of the biocatalyst, into two complementary classes: enzymatic and whole-cell MFCs (Fig. 6.7). In an enzymatic fuel cell, redox enzymes are used in their isolated forms; this produces electricity from the organic substrate serving as the electron donor. However, it has less stability and is more expensive. A whole-cell MFC exploits the complete microbial system, adapting the bacterial system to facilitate the extracellular electron transfer reaction on the electrode surface (Sathish-Kumar et al. 2012, 2013, 2015). These features allow use of wastewater as fuel, finally combining wastewater treatment and energy recovery. By generation of energy from a "negative-value" waste stream, a greater solution emerges, reducing pollution and costs associated with water and wastewater treatment (Liu et al. 2009).

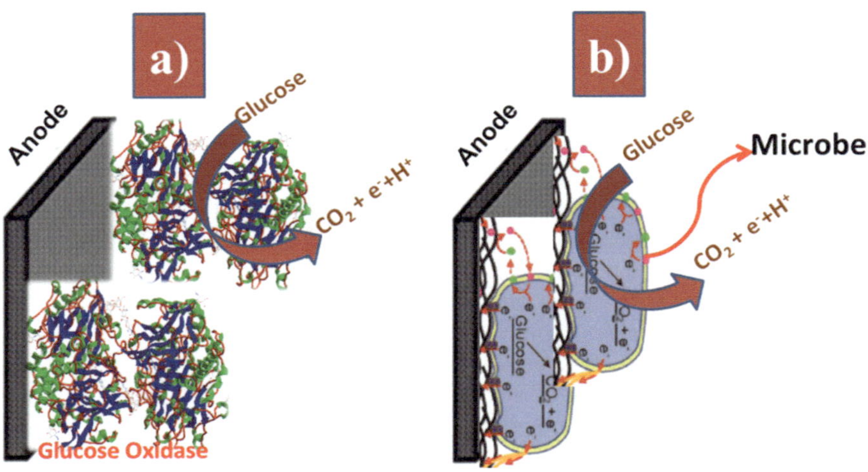

Fig. 6.7 Anodic half-cell reaction of (**a**) an enzymatic fuel cell and (**b**) a whole-cell microbial fuel cell (Adapted from Rabaey and Verstraete 2005)

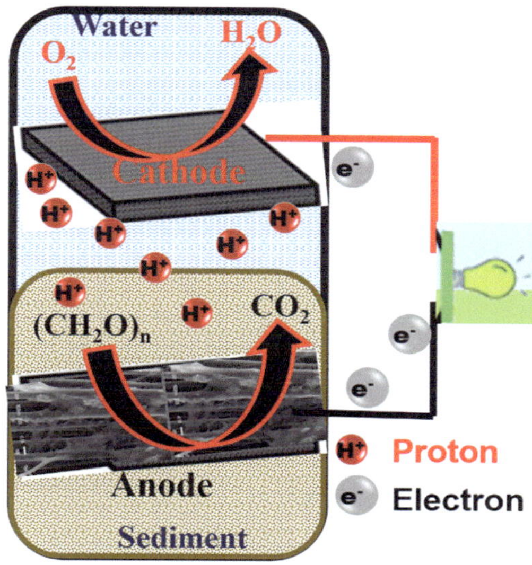

Fig. 6.8 Schematic representation of a sediment-based microbial fuel cell

Implementation of whole-cell (microbe) MFC technology based on organic material source utilization is classified as a sediment-based MFC (S-MFC, also known as a benthic MFC), biomass waste–based MFC, or plant-based MFC (P-MFC).

In some cases, S-MFCs are deployed in a natural system or in simpler engineering structures (e.g., a constructed wetland). Unlike reactor MFCs, which use membranes or separators, creating a clear boundary between the anode and the cathode, S-MFCs rely on a naturally occurring oxygen gradient to separate the anode and the cathode. To achieve that, the anode electrode is embedded in sediment where dissolved oxygen (DO) is depleted, while the cathode electrode is installed in the water phase with relatively higher DO (Fig. 6.8) (Lovley 2006a; Xu et al. 2015).

Biomass influent/effluent–based MFCs systems can be worthwhile in biorefinery processes due to the increased energy recovery from biomass feedstock waste (i.e., biomass discarded after completion of fuel or value-added byproduct production processes) and reduction in waste streams (Borole et al. 2009; Schroder 2008; Borole et al. 2013). Aside from the provided electron donor in the anodic chamber of an MFC, complex organic substrate concentration is important for its influences on the microbial community, power production, and economic viability (Cheng and Logan 2011; Zhi et al. 2014). Several organic substrates have been reported for power generation in MFCs, from simple, pure, or low-molecular sugar to complex organic material (containing carbohydrates, proteins, volatile acids, cellulose, and wastewater) (Chae et al. 2009; Rezaei et al. 2009; Liu et al. 2009). Generally, the chemical composition and the concentrations of the organic substrate that can be converted into products or fuels are potential substrates for the bioelectrochemical systems (Kondaveeti et al. 2014; Lakaniemi et al. 2012; Thygesen et al. 2010). However, there has been a low priority for the utilization of waste/spent feedstock

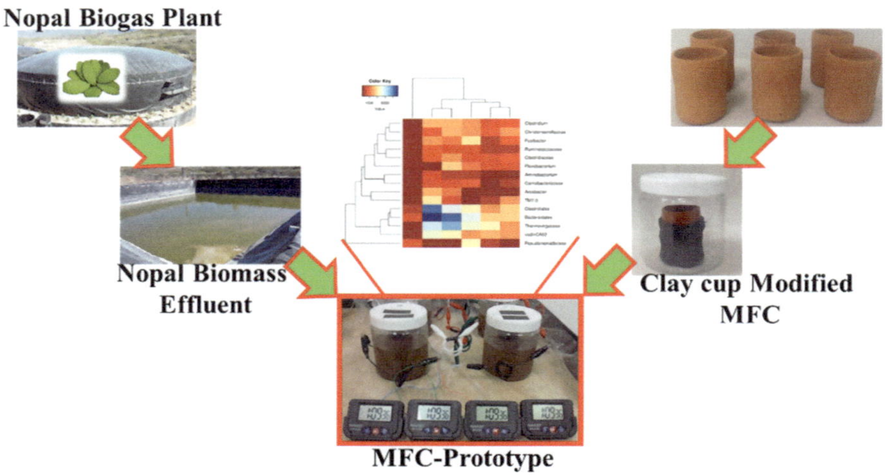

Fig. 6.9 Nopal biomass effluent treatment in a clay-modified microbial fuel cell

from the products or fuels. The spent biomass effluent from the Nopal biogas plant in Calvillo, Aguascalientes, Mexico, was selected in order to improve resource utilization, reduction of the waste stream, and further energy production through the MFC system (Fig. 6.9).

P-MFC systems exploit the organic materials close to the rhizosphere zone of plant, where those organic materials are continuously replenished by the root excretions of the plants. Hence, placing the electrodes close to the rhizosphere zone could deplete the organic materials by the microbes' activity; as a result, receiving the electron could act as an anode and the cathode would be located on the top surface of the soil. In some special cases, the cathode electrode could also be set up in the rhizosphere area to benefit from the oxygen released from the plant roots (Strik et al. 2008).

6.1.4 Principle of Plant-Mediated Microbial Fuel Cells

A promising new technology is the plant–microbial fuel cell (P-MFC), since it could produce sustainable electricity continuously (Strik et al. 2008). In general, plants provide organic matter to microbes at their roots in the form of rhizodeposits, which include excretions, secretions, dead plant material, and *gases* (Neumann 2007). The anodic current collector is placed near the plant's roots, receiving the donated electrons from the organic matter via the electrochemically active bacteria, thus producing electricity (Fig. 6.10).

Even though the technology is solar powered through photosynthesis, it does not rely on direct sunlight. At the same time, excretion of organic matter by the plant continues day and night, referred to as a *diurnal pattern* (Neumann 2007). Since the

Fig. 6.10 Schematic representation of a plant-based microbial fuel cell

first report of a P-MFC (Schamphelaire et al. 2008; Kaku et al. 2008), several experiments with plant systems have been performed. As a result, the theoretical maximum electricity output of a P-MFC is 3.2 W/m^2 plant growth area (PGA) (Strik et al. 2011); currently, a long-term output of 0.155 W/m^2 PGA has been reached (Helder et al. 2012).

The most promising areas in which to incorporate P-MFCs could be natural wetlands. They offer an abundant electricity source without extensive excavation of the soil (Kaku et al. 2008). P-MFCs could also be implemented in rice paddy fields, lowering the competition of food and electricity production (Schamphelaire et al. 2008; Kaku et al. 2008). In addition, P-MFCs could be integrated into green roofs, merging the advantages of insulation, biodiversity, and electricity generation (Helder et al. 2013).

6.1.5 Bioreactor Design

A two-chambered P-MFC has been shown to minimize both inert resistance and oxygen diffusion (Fig. 6.11a). In this configuration, a membrane was fixed at the bottom of the anode with one side in contact with the sediment. The anodic current

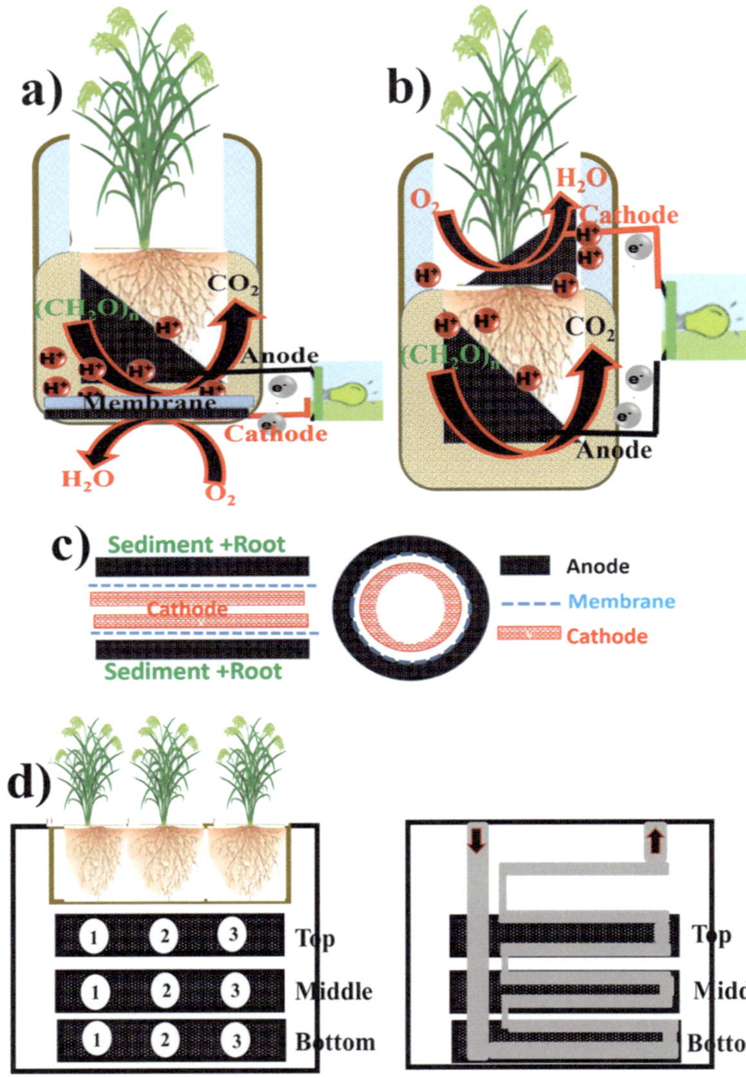

Fig. 6.11 (**a**) Plant-based microbial fuel cell (P-MFC) with a membrane. (**b**) P-MFC without a membrane. (**c**) Novel tubular P-MFC. (**d**) Flat-bed P-MFC

collector was located close to the rhizosphere zone, and the cathode was placed on the other side of the membrane. Use of a membrane on P-MFCs increases the primary cost of this technology (Patra 2008; Strik et al. 2008; Helder et al. 2010), though implementation of single-chamber P-MFCs means that the oxygen concentration declines over the depth of sediment and water, and thus a membrane is not required anymore (Patra 2008). A submerged anode in the support matrix is flooded with soil or other material (vermiculite or graphite granules). A possible problem is proton (H^+) mobility in the soil blocks from the anode to the cathode, with further diffusion of the root deposit to the anode surface; this could result in increasing mass transport and ohmic resistance (Takanezawa et al. 2010). There have been two forms of single-chamber configuration: in the first one, the cathode is situated in overlying water to use oxygen from the air for reduction reactions (Fig. 6.11b); in the second one, the cathode is located in the rhizosphere to use oxygen released from the roots as the electron acceptor. Therefore, the membraneless configuration reduces the capital cost and improves the power densities by decreasing mass transfer and ohmic resistance; hence, it demonstrates potentiality to achieve practical implementation in combination with high performance (Chen et al. 2012) (Fig. 6.11b).

Timmers et al. (2013) have proposed a new tubular P-MFC design where the anode and cathode are integrated into the same unit and generate a current, which makes it possible to apply P-MFC technology with horizontal drilling and thus without excavation of the topsoil felt and graphite granules (Fig. 6.11c). This new design delivers an average power output based on the membrane area, which was 10 mW m^2 for felt and 12 mW m^2 for graphite granules. The corresponding mass and volume power densities were 15 and 69 times greater, respectively, for the felt than for the granules. These findings showed that a decrease in the use of anode electrode material is possible while achieving comparable power outputs per square meter of the membrane, making future applications of P-MFC technology more feasible due to cost reduction per kilowatt hour. Additionally, this P-MFC design does not require excavation of topsoil, which could be likely to apply in the field (Timmers et al. 2013).

A flat-plate P-MFC design was proposed by Helder et al. (2012), showing better performance than the previously used tubular P-MFC. It resulted in lower transport and membrane resistance (Fig. 6.11c). Though the lower total internal resistance was not normalized to the membrane surface area, a lower internal resistance of the flat-plate MFC was achieved comparable to that of the tubular P-MFC by the total internal resistance being a little differently distributed over partial internal resistance components. When using a chemical cathode in a flat-plate P-MFC, slightly higher anodic resistance results. This can be ascribed to either substrate or mass transport limitation (Fig. 6.3d). In order to eliminate substrate limitation issues, the plant should liberate more exudate, otherwise all of the exudate should effectively be converted into either electricity or dead root material (likely other rhizodeposits), which should be used further. During the maturation of the rhizosphere zone, this could liberate a lot of dead root material leading to a higher substrate, which will easily be available to convert electricity in the P-MFC. Plant growth medium adaptation might be stimulated with more exudation in the rhizosphere zone (Blossfeld et al. 2010).

Implementation of an oxygen-reducing biocathode in a P-MFC has shown promising potential viability. It could generate electricity for up to 151 days. In a short period of polarization (10 min), it exhibited maximum power output of 679 mW/m^2 PGA. During further operation for 2 weeks, the average power density was 240 mW/

m^2 PGA produced. The biocathode effectively catalyzed the final terminal electron acceptor of oxygen, resulting in a higher redox potential to achieve the recorded performance. As a result, it showed 127 mV higher cathode potential of the biocathode P-MFC than a P-MFC with a ferricyanide cathode. Moreover, the limitation of substrate availability in the anode is hindering the current generation. The present work shows that the P-MFC can be a completely sustainable biotechnology with enhanced power output, and further research is crucial (Wetser et al. 2015).

The anodic conditions of the P-MFC should be crucial parameters that favor plant growth and electricity generation in return. Hence, care should be taken with the composition of the plant growth medium. So far, a Hoagland medium has been used, with added phosphate buffer to reduce potential losses over the membrane due to the difference in pH between the anode and cathode. Helder et al. (2012) have developed a new and improved plant growth medium, which increases electricity generation, while the plant keeps growing well. This medium comprises a balanced bicarbonate buffer along with a nitrate-less, ammonium-rich medium, with all of the necessary macro- and micronutrients for plant growth. The presence of sulfate in the plant growth medium helps it to retain a low anode potential. As a result, the maximum current generation of the P-MFC has increased from 186 mA/m^2 to 469 mA/m^2 with the new plant growth medium (Helder et al. 2012).

In our study, we proposed a novel P-MFC design from a clay cup (*cantarito*) using the used battery material as an air cathode. Money plant rhizodeposits would serve as an electron donor. This system is able to produce a maximum power density of 61.42 mW/m^3 and a current density of 219.25/m^3 at 0.280 V. (Fig. 6.12a) (Sathish-Kumar 2015). Further, we developed a novel integrated MFC prototype from plants to power digital clocks (Fig. 6.12b) (Mexican patent number: MX/a/2015/001573).

6.1.6 Associated Microbial Communities

A diverse range of microbes is involved in MFCs, which clearly reveals that a single microorganism can develop on the anode. This is probably due to many microorganisms being involved in the production of electricity, like team work in the area of the operating circumstances, electron donors, and acceptors. Meanwhile, a group of microorganisms can be sustained with alternative metabolisms such as fermentation, methanogenesis, and terminal electron acceptors, which are not responsible for electricity production. In MFCs, iron-reducing bacteria such as *Shewanella* and *Geobacter* species are considered to be active electrochemically (Kim et al. 1999a, b, c, d; Bond et al. 2002) but, surprisingly, a diversity of microorganisms are there in the biofilm (Kim et al. 2004; Logan et al. 2005; Rabaey et al. 2004; Phung et al. 2004). In a few MFCs, uncharacterized clones are being used in large Size. Sometimes, wastewater could be an inoculum with dissolved oxygen at the cathode, with a microbial community consisting of 36% unidentified clones, 25% beta- and 20% alphaproteobacteria, and 19% *Cytophaga*, *Flexibacter*, and *Bacteroides* groups (Kim et al. 2004).

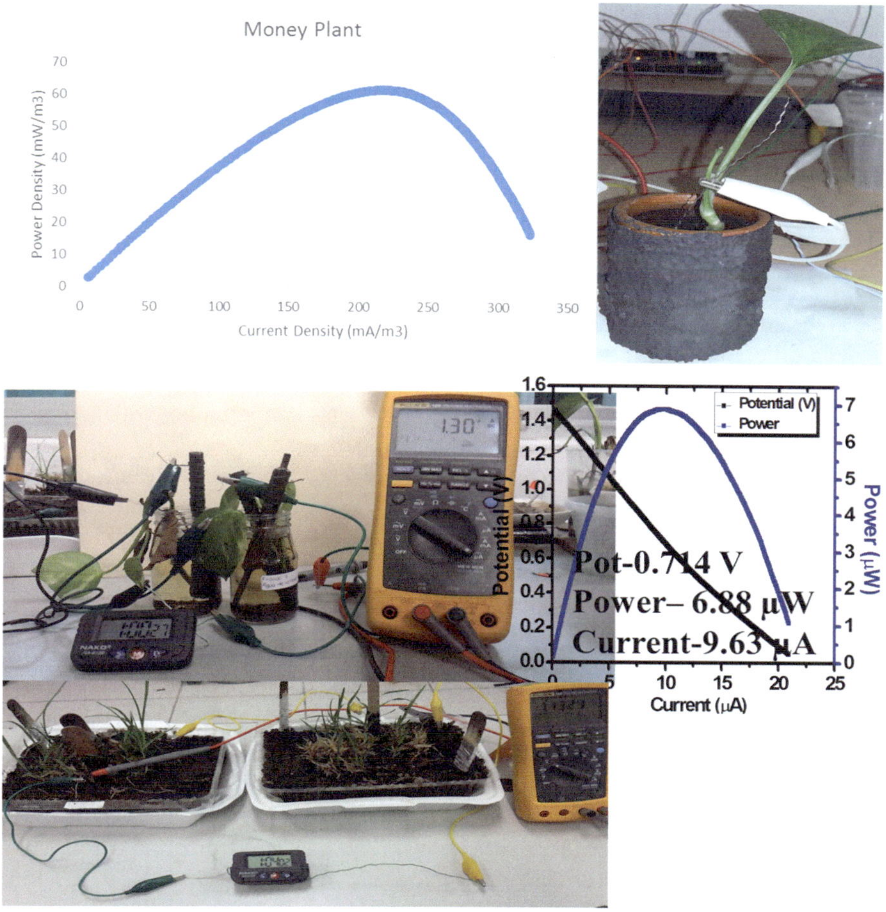

Fig. 6.12 Novel clay cup–modified plant-based microbial fuel cell (P-MFC) and prototype developed from a novel integrated P-MFC (Mexican patent number: MX/a/2015/001573)

6.1.7 Wetland Plant–Microbial Fuel Cells

Natural wastewater treatment systems consist of a series of processes such as physical, chemical, and biological ones (Garcia et al. 2010). Such a system further consists of shallow lined basins with a filter of gravel, which is planted with aquatic plants (macrophytes). In these constructed wetlands, established MFCs may have other uses such as treatment capacity and reduction of both clogging and methane. Further, degradation of organic matter will be brought up upon increasing the ease of use of electron acceptors in such conditions (Schamphelaire et al. 2008). MFCs mobilize organic content in the filter media so that clogging is reduced. Meantime, acetate is used by exoelectrogenic bacteria as a substrate, which results in nonavailability or less availability of carbon to methane-producing bacteria. There is always competition between methane producers and exoelectrogenic bacteria, which makes it a possible way to remarkably decrease methane emissions during treatment of wastewater. It is a wonderful bioelectrochemical tool for assessing treatment performance. It has dual benefits such as optimization of the treatment process and less environmental impact. Implementation of MFCs in marine sediments (Reimers et al. 2001), rice paddy fields (Kaku et al. 2008; Schamphelaire et al. 2008), and planted systems (Strik et al. 2008; Venkata Mohan et al. 2011) has been addressed.

Plants associated with MFCs could be a vast energy provider without polluting the environment. Such systems might be able to produce green electricity or biohydrogen on plantations. P-MFCs focus on solar radiation, which can be converted into green electricity or biohydrogen in a pollution-free environment. The electrochemical system is a relationship between living plants and microbes which, in turn, provide sustainable green electricity or biohydrogen.

A P-MFC is five times as efficient as a bioenergy system. There are multiple ways to implement this technology, from small-scale electricity providers to large-scale energy wetlands and islands, high-tech energy and food-supplying greenhouses, and new biorefineries. In the near future, this technology could be available in all countries.

There are two major processes involved with respect to P-MFCs, such as living plants depositing the organic compounds from which electricity is generated. Photosynthesis is a mechanism for fixing carbon dioxide into carbohydrates, from where 60% of the carbon may be transferred from the leaves to the roots (Lynch and Whipps 1990). Plant root systems turn a diverse variety of organic compounds into soil exudates (sugars, organic acids, etc.), secretions (polymeric carbohydrates and enzymes), lysates (dead cell materials), and gases (ethylene and CO_2) (Uren 2007; Bais et al. 2006). The entire process is known as plant rhizodeposition and the rhizodeposits are used in the P-MFC as a renewable bioenergy substrate. In general, rhizodeposits consist of carbon, of which a portion can be used by microorganisms present in the rhizosphere. This leads to a mutually beneficial relationship between the plants and the microorganisms.

Bacteria interact with plant roots by making biofilms or by producing antibiotics as a biocontrol measure against pathogenic strains (Bais et al. 2006). Rhizodeposits are used by bacteria as major substrates to produce green electricity in MFCs.

MFCs can fulfill the need of the hour and can transform biodegradable substances from wastewater or crops into green electricity (Chaudhuri and Lovley 2003; Logan et al. 2006; Rabaey and Verstraete 2005; Logan and Regan 2006). The active microorganisms in the MFC play a vital role as biocatalysts, using chemical energy for themselves and meantime delivering electrons to the major end like the anode of the fuel cell and the preference final electrode acceptor (anode electrode) is used by the microorganisms (Rabaey and Verstraete 2005).

6.2 Food Crop–Based Microbial Fuel Cells

Agricultural residues contain large amounts of lignocellulosic materials, which are copious and renewable with a low cost for producing energy. However, there are intermediate processes needed to convert complex lignocellulosic materials into simple monosaccharides before introducing microorganisms of MFCs for oxidation (Logan 2004). Two studies were carried out in rice (Kaku et al. 2008; De Schamphelaire et al. 2008) and reed mannagrass (Strik et al. 2008). A pot culture system was employed (Strik et al. 2008; De Schamphelaire et al. 2008). A real rice paddy field was also studied (Kaku et al. 2008). The biological energy conversion process plays a vital role in the case of rice paddy field (RPF) S-MFCs. The inevitable truth is that the most important agronomic plant throughout the world is rice, with 143 billion ha under cultivation, revealing the feasibility of using RPF S-MFCs as a main contributor to meet the electricity needs of the entire globe (Watanabe and Nishio 2010).

6.3 Conclusion

One of the most promising benefits of P-MFCs is that they allow direct generation of electricity while growing plants. In the last 10 years, P-MFCs have been intensively researched and developed, leading to an expansion of their functionalities and improvement in their performance, employing cost-effective materials. Their power densities have been amplified mainly due to improvements in the setup construction, operation, and materials, which overcome the system restrictions.

Moreover, P-MFCs could be operated with a nitrogen removal system incorporated into the cathodic electron acceptor, which represents some advantages compared with oxygen as the final terminal electron acceptor. Accordingly, P-MFCs might be a future energy-efficient and economical solution for sustainable agriculture and wetland-based wastewater treatment. However, the income from electricity generation in such MFCs is doubtful to counterweigh their high capital costs and to build competitive benefit as compared with anaerobic digestion in traditional agricultural processes.

Substantial scaling-up of P-MFCs to cubic meter volumes is essential in order to attain practical operation, with possible integration of the effective bioreactor with an existing infrastructure. Besides, implementation of an effective power-harvesting system needs to provide working power for electronic and electric items. However, the cheap cost of P-MFCs, along with the ecological advantages, would offset the great capital costs of the employed agriculture processes/biorefineries. The productivity and economic aspects of the system could be further enhanced by increasing research on the integration of different industrial processes for the production of value-added materials.

References

Bais HP, Weir TL, Al PLGE (2006) The role of root exudates in rhizosphere interactions with plants and other organisms. Annu Rev Plant Biol 57(1):233–266

Bard AJ, Faulkner LR (2001) Electrochemical methods fundamentals and applications, 2nd edn. Wiley, Hoboken

Blossfeld S, Perriguey J, Sterckeman T et al (2010) Rhizosphere pH dynamics in trace-metal-contaminated soils monitored with planar pH optodes. Plant Soil 330(1):173–184

Bond DR, Lovley DR (2005) Evidence for involvement of an electron shuttle in electricity generation by *Geothrix fermentans*. Appl Environ Microbiol 71:2186–2189

Bond DR, Holmes DE, Tender LM et al (2002) Electrode-reducing microorganisms that harvest energy from marine sediments. Science 295:483–485

Borole AP, Jonathan RM, Tatiana AV et al (2009) Controlling accumulation of fermentation inhibitors in biorefinery recycle water using microbial fuel cells. Biotechnol Biofuels 2(1):1–14

Borole AP, Hamilton CY, Schell DJ (2013) Conversion of residual organics in corn Stover–derived biorefinery stream to bioenergy via a microbial fuel cell. Environ Sci Technol 47:642–648

Bose A, Gardel EJ, Vidoudez C, Parra EA, Girguis PR (2014) Electron uptake by iron-oxidizing phototrophic bacteria. Nat Commun 5:3391

Byrne JM et al (2015) Redox cycling of Fe(II) and Fe(III) in magnetite by Fe-metabolizing bacteria. Science 347:1473–1476

Chae KJ, Choi MJ, Lee JW et al (2009) Effect of different substrates on the performance, bacterial diversity, and bacterial viability in microbial fuel cells. Bioresour Technol 100:3518–3525

Chaudhuri SK, Lovley DR (2003) Electricity generation by direct oxidation of glucose in mediatorless microbial fuel cells. Nat Biotechnol 21(10):1229–1232

Chen Z, Huang YC, Liang JH et al (2012) A novel sediment microbial fuel cell with a biocathode in the rice rhizosphere. Bioresour Technol 108:55–59

Cheng S, Logan BE (2011) Increasing power generation for scaling up single-chamber air cathode microbial fuel cells. Bioresour Technol 102(6):4468–4473

De Schamphelaire L, Rabaey K, Boeckx P et al (2008) Outlook for benefits of sediment microbial fuel cells with two bio-electrodes. Microbiol Biotechnol 1(6):446–462. Energy Information Administration's (EIA's) handling of non-policies in the international energy outlook

Emerson D, Moyer C (1997) Isolation and characterization of novel iron-oxidizing bacteria that grow at circumneutral pH. Appl Environ Microbiol 63:4784–4792

Uren N (2007) Types amounts and possible functions of compounds released into the rhizosphere by soil-grown plants. In: The rhizosphere. CRC Press, pp 1–21

Garcia AJ, Rousseau DPL, Morato J et al (2010) Contaminant removal processes in subsurface-flow constructed wetlands: a review. Crit Rev Environ Sci Technol 40(7):561–661

Gralnick JA, Newman DK (2007) Extracellular respiration. Mol Microbiol 65:1–11

Helder M, Strik DPBTB, Hamelers HVM et al (2010) Concurrent bio-electricity and biomass production in three plant-microbial fuel cells using *Spartina anglica*, *Arundinella anomala* and *Arundo donax*. Bioresour Technol 101(10):3541–3547

Helder M, Strik DPBTB, Hamelers HVM et al (2012) New plant-growth medium for increased power output of the plant-microbial fuel cell. Bioresour Technol 104:417–423

Helder M, Strik DPBTB, Timmers RA et al (2013) Resilience of roof-top plant-microbial fuel cells during Dutch winter. Biomass Bioenergy 51:1–7

Hernandez ME, Kappler A, Newman DK (2004) Phenazines and other redox-active antibiotics promote microbial mineral reduction. Appl Environ Microbiol 70:921–928

Holmes DE, Bond DR, ONeil RA et al (2004) Microbial communities associated with electrodes harvesting electricity from a variety of aquatic sediments. Microb Ecol 48:178–190

Jiao Y, Kappler A, Croal LR, Newman DK (2005) Isolation and characterization of a genetically tractable photoautotrophic Fe(II)-oxidizing bacterium, *Rhodopseudomonas palustris* strain *TIE-1*. Appl Environ Microbiol 71:4487–4496

Kaku N, Yonezawa N, Kodama Y et al (2008) Plant microbe cooperation for electricity generation in a rice paddy field. Appl Microbiol Biotechnol 79(1):43–49

Kim HJ, Hyun MS, Chang IS, Kim BH (1999a) A microbial fuel cell type lactate biosensor using a metal-reducing bacterium, *Shewanella putrefaciens*. J Microbiol Biotechnol 9:365–367

Kim BH, Ikeda T, Park HS et al (1999b) Electrochemical activity of an Fe(III)-reducing bacterium, *Shewanella putrefaciens* IR-1, in the presence of alternative electron acceptors. Biotechnol Tech 13:475–478

Kim BH, Park DH, Shin PK et al (1999c) Mediator-less biofuel cell. Google Patents

Kim HJ, Park HS, Hyun MSA (2002) Mediator-less microbial fuel cell using a metal reducing bacterium, *Shewanella putrefaciens*. Enzym Microb Technol 30:145–152

Kim BH, Park HS, Kim HJ (2004) Enrichment of microbial community generating electricity using a fuel-cell-type electrochemical cell. Appl Microbiol Biotechnol 63(6):672–681

Kim GT, Hyun MS, Chang IS (2005) Dissimilatory Fe(III) reduction by an electrochemically active lactic acid bacterium phylogenetically related to *Enterococcus gallinarum* isolated from submerged soil. J Appl Microbiol 99:978–987

Kondaveeti S, Kwang SC, Ramesh K et al (2014) Microalgae *Scenedesmus obliquus* as renewable biomass feedstock for electricity generation in microbial fuel cells (MFCs). Front Environ Sci Eng 8(5):784–791

Kumar A, Hsu LH-H, Kavanagh P, Barrière F, Lens PNL, Lapinsonnière L, Leech D (2017) The ins and outs of microorganism–electrode electron transfer reactions. Nat Rev Chem 1:0024

Lakaniemi AM, Olli HT, Jaakko AP (2012) Production of electricity and butanol from microalgal biomass in microbial fuel cells. Bio Energy Res 5(2):481–491

Larminie J, Dicks A (2013) Introduction. In: Fuel cell systems explained. Hoboken, Wiley, pp 1–24

Liu Z, Liu J, Zhang S et al (2009) Study of operational performance and electrical response on mediator-less microbial fuel cells fed with carbon- and protein-rich substrates. Biochem Eng J 45:185–191

Logan BE (2004) Extracting hydrogen and electricity from renewable resources. Environ Sci Technol 38:160A–167A

Logan BE, Regan JM (2006) Electricity-producing bacterial communities in microbial fuel cells. Trends Microbiol 14(12):512–518

Logan BE, Murano C, Scott K et al (2005) Electricity generation from cysteine in a microbial fuel cell. Water Res 39(5):942–952

Logan BE, Hamelers B, Rozendal R et al (2006) Microbial fuel cells: methodology and technology. Environ Sci Technol 40(17):5181–5192

Lovley DR (2006a) Bug juice: harvesting electricity with microorganisms. Nat Rev Microbiol 4:497–508

Lovley DR (2006b) Microbial fuel cells: novel microbial physiologies and engineering approaches. Curr Opin Biotechnol 17:327–332

Lovley DR, Phillips EJ (1988) Novel mode of microbial energy metabolism: organic carbon oxidation coupled to dissimilatory reduction of iron or manganese. Appl Environ Microbiol 54:1472–1480

Lovley DR, Stolz JF, Nord GL, Phillips EJP (1987) Anaerobic production of magnetite by a dissimilatory iron-reducing microorganism. Nature 330:252–254

Lovley DR, Baedecker MJ, Lonergan DJ et al (1989a) Oxidation of aromatic contaminants coupled to microbial iron reduction. Nature 339:297–299

Lovley DR, Phillips EJ, Lonergan DJ (1989b) Hydrogen and formate oxidation coupled to dissimilatory reduction of iron or manganese by *Alteromonas putrefaciens*. Appl Environ Microbiol 55:700–706

Lovley DR, Phillips EJP, Gorby YA, Landa ER (1991) Microbial reduction of uranium. Nature 350(6317):413–416

Lovley DR, Holmes DE, Nevin KP (2004) Dissimilatory Fe(III) and Mn(IV) reduction. Adv Microb Physiol 49:219–286

Lynch JM, Whipps JM (1990) Substrate flow in the rhizosphere. Plant Soil 129(1):1–10

Malvankar NS, Mester T, Tuominen MT, Lovley DR (2012) Supercapacitors based on c-type cytochromes using conductive nanostructured networks of living bacteria. ChemPhysChem 13:463–468

Melton ED, Swanner ED, Behrens S, Schmidt C, Kappler A (2014) The interplay of microbially mediated and abiotic reactions in the biogeochemical Fe cycle. Nat Rev Microbiol 12:797–808

Myers CR, Nealson KH (1988) Bacterial manganese reduction and growth with manganese oxide as the sole electron acceptor. Science 240:1319–1321

Neumann G (2007) Root exudates and nutrient cycling. In: Marschner P, Rengel Z (eds) Nutrient cycling in terrestrial ecosystems. Springer, Berlin, pp 123–157

Niessen J, Schroder U, Scholtz F (2004) Exploiting complex carbohydrates for microbial electricity generation—a bacterial fuel cell operating on starch. Elecrochem Commun 6:955–958

Niessen J, Harnisch F, Rosenbaum M et al (2006) Heat treated soil as convenient and versatile source of bacterial communities for microbial electricity generation. Electrochem Commun 8:869–873

Park HS, Kim BH, Kim HS (2001) A novel electrochemically active and Fe(III) reducing bacterium phylogenetically related to *Clostridium butyricum* isolated from a bacterial fuel cell. Anaerobe 7:297–306

Patra A (2008) Low-cost single-chambered microbial fuel cells for harvesting energy and cleansing waste water. J US SJWP 2008, 3:72–85

Pfeffer C, Larsen S, Song J, Dong M, Besenbacher F, Meyer RL, Nielsen LP (2012) Filamentous bacteria transport electrons over centimetre distances. Nature 491(7423):218–221

Pham CA, Jung SJ, Phung NT (2003) A novel electrochemically active and Fe(III)-reducing bacterium phylogenetically related to *Aeromonas hydrophila*, isolated from a microbial fuel cell. FEMS Microbiol Lett 223:129–134

Phung NT, Lee J, Kang KH et al (2004) Analysis of microbial diversity in oligotrophic microbial fuel cells using 16S rDNA sequences. FEMS Microbiol Lett 233(1):77–82

Potter MC (2011) Electrical effects accompanying the decomposition of organic compounds. Proc R Soc Lond B 84(571):260–276

Rabaey K, Verstraete W (2005) Microbial fuel cells: novel biotechnology for energy generation. Trends Biotechnol 23(6):291–298

Rabaey K, Boon N, Siciliano SD (2004) Biofuel cells select for microbial consortia that self-mediate electron transfer. Appl Environ Microbiol 70(9):5373–5382

Reimers CE, Tender LM, Fertig S et al (2001) Harvesting energy from the marine sediment−water interface. Environ Sci Technol 35(1):192–195

Reimers CE, Girguis P, Stecher HA (2006) Microbial fuel cell energy from an ocean cold seep. Geobiology 4:123–136

Rezaei F, Xing D, Wagner R et al (2009) Simultaneous cellulose degradation and electricity production by *Enterobacter cloacae* in a microbial fuel cell. Appl Environ Microbiol 75(11):3673–3678

Ringeisen BR, Henderson E, Wu PK, Jones-Meehan JM (2006) High power density from a miniature microbial fuel cell using *Shewanella oneidensis DSP10*. Environ Sci Technol 40:2629–2634

Ryckelynck N, Stecher HA, Reimers CE (2005) Understanding the anodic mechanism of a seafloor fuel cell: interactions between geochemistry and microbial activity. Bio Geochem 76:113–139

Sapsford KE, Russ AW, Berti L et al (2013) Functionalizing nanoparticles with biological molecules: developing chemistries that facilitate nanotechnology. Chem Rev 113(3):1904–2074

Sathish-Kumar K (2015) Plant based microbial fuel cells, Memorias 1a Reunión de Trabajo, Laboratorio de Materiales Fotovoltaicos, CICATA Altamira. 19 de Junio de

Sathish-Kumar K, Solorza-Feria O, Vázquez-Huerta G et al (2012) Electrical stres–directed evolution of biocatalysts community sampled from a sodic–saline soil for microbial fuel cells. J New Mater Electrochem Syst 15(3):181–186

Sathish-Kumar K, Solorza-Feria O, Tapia-Ramírez J et al (2013) Electrochemical and chemical enrichment methods of a sodic–saline inoculum for microbial fuel cells. Int J Hydrogen Energy 38(28):12600–12609

Sathish-Kumar K, Solorza-Feria O, José T-R et al (2015) Microbial fuel cell with wooden integrated graphite based electrode for waste water treatment applications, Mexican patent number: MX/a/2015/001573
Schamphelaire LD, Bossche LVD, Dang HS et al (2008) Microbial fuel cells generating electricity from rhizodeposits of rice plants. Environ Sci Technol 42(8):3053–3058
Schröder U (2008) From wastewater to hydrogen: biorefineries based on microbial fuel-cell technology. ChemSusChem 1:281–282
Schroder U, Harnisch F, Angenent LT (2015) Microbial electrochemistry and technology: terminology and classification. Energy Environ Sci 8(2):513–519
Shelobolina E et al (2012) Microbial lithotrophic oxidation of structural Fe(II) in biotite. Appl Environ Microbiol 78:5746–5752
Shi L, Squier TC, Zachara JM, Fredrickson JK (2007) Respiration of metal (hydr)oxides by *Shewanella* and *Geobacter*: a key role for multihaem c-type cytochromes. Mol Microbiol 65:12–20
Shi L, Tien M, Fredrickson JK, Zachara JM, Rosso KM (2016) In: Louro R, Diaz-Moreno I (eds) Redox proteins in supercomplexes and signalosomes. CRC Press, pp 187–216
Shukla AK, Suresh P, Berchmans S et al (2004) Biological fuel cells and their applications. Curr Sci 87:455–468
Strik DPBTB, Hamelers HVM, Snel JFH et al (2008) Green electricity production with living plants and bacteria in a fuel cell. Int J Energy Res 32(9):870–876
Strik DPBTB, Timmers RA, Helder M et al (2011) Microbial solar cells: applying photosynthetic and electrochemically active organisms. Trends Biotechnol 29(1):41–49
Takanezawa K, Nishio K, Kato S et al (2010) Factors affecting electric output from rice-paddy microbial fuel cells. Biosci Biotechnol Biochem 74(6):1271–1273
Tender LM, Reimers CE, Stecher HA et al (2002) Harnessing microbially generated power on the seafloor. Nat Biotechnol 20:821–825
Thygesen A, Anne BT, Sam P et al (2010) Integration of microbial electrolysis cells (Mecs) in the biorefinery for production of ethanol, H2 and phenolics. Waste Biomass Valoriz 1(1):9–20
Timmers RA, Strik DPBTB, Hamelers HVM et al (2013) Electricity generation by a novel design tubular plant microbial fuel cell. Biomass Bioenergy 51:60–67
Uren N (2007) Types amounts and possible functions of compounds released into the rhizosphere by soil-grown plants. In: The rhizosphere. CRC Press, pp 1–21
Venkata Mohan S, Mohanakrishna G, Chiranjeevi P (2011) Sustainable power generation from floating macrophytes based ecological microenvironment through embedded fuel cells along with simultaneous wastewater treatment. Bioresour Technol 102(14):7036–7042
Watanabe K, Nishio K (2010) Electric power from rice paddy fields paths to sustainable energy Dr Artie Ng (Ed.). ISBN: 978-953-307-401-6. InTech. Available from: http://www.intechopen.com/books/paths-to-sustainable-energy/electric-power-from-rice-paddy-fields
Wetser K, Sudirjo E, Buisman CJN (2015) Electricity generation by a plant microbial fuel cell with an integrated oxygen reducing biocathode. Appl Energy 137:151–157
Xu B, Ge Z, He Z (2015) Sediment microbial fuel cells for wastewater treatment: challenges and opportunities. Environ Science Water Res Technol 1(3):279–284
Zhao L et al (2015) Biological redox cycling of iron in nontronite and its potential application in nitrate removal. Environ Sci Technol 49:5493–5501
Zhi W, Ge Z, He Z et al (2014) Methods for understanding microbial community structures and functions in microbial fuel cells: a review. Bioresour Technol 171:461–468

Part III
Organic Farming

Chapter 7
Role of Organic Amendments in Sustainable Agriculture

K. Sankar Ganesh, P. Sundaramoorthy, M. Nagarajan, and R. Lawrence Xavier

Abstract Organic farming is one type of sustainable farming. It has potential, especially for farms that still rely on traditional and extensive agricultural methods. Organic farming can provide quality food without adversely affecting the soil's health and the environment; however, a concern is whether large-scale organic farming will produce enough food for India's large population. The past decade's increases in agricultural yields would have been impossible without mineral fertilizer. After the green revolution, the mineral fertilizer usage is progressively increased but yet not to be achieved the expected yield. Until now, insufficient attention has been paid to the adverse effects on the soil and the environment of improper use of mineral fertilizer. Many tropical soils are acidic by nature, and mineral fertilizer speeds up the acidification process. Consequently, soil productivity deteriorates rather than improves in the long term, and the fertilizer cannot have its full effects. Similarly the awareness of organic farming practices increased day by day throughout the world. Now the organic farming practices and organic amendments play a prominent role to mitigate the reclamation of soil fertility.

K. Sankar Ganesh (✉)
Department of Botany, Government Arts College (Autonomous), Kumbakonam, Tamil Nadu, India
e-mail: indsankar@rediffmail.com

P. Sundaramoorthy
Department of Botany, Annamalai University, Annamalainagar, Tamil Nadu, India

M. Nagarajan
Department of Botany, A.V.C. College (Autonomous), Mannampandal, Tamil Nadu, India

R. Lawrence Xavier
Department of Botany, Raja Doraisingam Government Arts College, Sivagangai, Tamil Nadu, India

A. Dhanarajan (ed.), *Sustainable Agriculture towards Food Security*,
https://doi.org/10.1007/978-981-10-6647-4_7

7.1 Introduction

The world's population is poised to reach 9 billion by the middle of this century, and over the next 40 years, 70% more food will be needed to sustain all these people. Most of this additional food will have to be produced where it is needed, namely, in developing countries. These countries will have to double their production to achieve this goal, with implications also for the natural resources that farming depends on and especially water, land for cultivation, and mineral fertilizers. All of these are available in only limited amounts. In many places, the soil has already suffered long-term damage, and water resources are often overused or polluted by fertilizers and pesticides. Agricultural biodiversity has decreased as farming has become industrialized. These negative effects have heightened global awareness of the fact that agriculture does more than simply produce food, animal feed, and energy; it also impacts on the climate and the health of global ecosystems.

Agriculture has been a capital for the development of human civilization, thanks to the agriculturalists that have been the chief managers of terrestrial "usable" lands, shaping and changing the land around us. Before agriculture developing, the hunter-gatherer lifestyle supported about 4 million people globally, whereas today modern agriculture feeds 7,300 million people (Tilman et al. 2002; Scotti et al. 2015) but at the expense of ecosystem health. Agricultural practices, based on large application of fertilizers and pesticides, reduce the ability of ecosystems to provide goods and services.

In fact, the increasing risk of contamination from nutrients and toxic chemical compounds occurs in groundwater and surface waters, incurring in eutrophication and soil quality degradation (Tilman et al. 2002). This incorrect agricultural management can change species composition or reduce biodiversity in ecosystems thus affecting natural abilities of ecosystems (Hector et al. 1999; Loreau et al. 2001). As reported by Tilman et al. (2002), the global use of nitrogen (N) and phosphorus fertilizers increased by 7- and 3.5-fold, respectively, in the past six decades; both fertilizers are expected to increase further threefold by 2050, featuring a conversion of agricultural practices more and more toward intensive agriculture systems. Intensive agriculture is a farming system characterized by a large use of capital and inputs to invest in the acquisition and application of fertilizers and pesticides necessary to crop growing. The abundant use of fertilizers and pesticides also increases the risk that nutrients and pesticides run off into surface and leach into groundwater (Eurostat 2015).

7.2 Sustainable Agriculture

The term "sustainable agriculture" is used in this review with the meaning given by Tilman et al. (2002), as referring to practices that meet current and future society needs for food and feed, ecosystem services, and human health, maximizing the net benefit for people. Namely, sustainability implies both high yields, that can be maintained, and acceptable environmental impact of agricultural management.

It is a difficult task to do well in conventional agriculture; and it is doubly difficult to do well in sustainable agriculture, which is characterized by profitable farming while preserving the environment as much as possible. In fact, switching from conventional to sustainable agriculture is not a simple substitution, such as replacing insecticide with a predator insect or replacing commercial KCl fertilizer with green sand. It is reasonable to say that although more in tune with nature than conventional agriculture, sustainable agriculture is information intensive and requires strong management skills. For example, cultural and biological pest control requires detailed knowledge about a pest's life cycle as well as economic threshold levels of the crop. Similarly, using chicken manure to replace urea as a N source for your crop must take into account the release pattern of organic nitrogen from the manure and the nitrogen demand pattern of your crop, so that the two can be synchronized.

A survey in 1987 shows that 48% of sustainable farmers nationwide have experienced with nutrient deficiencies. Thus, it is no doubt that soil fertility and plant nutrients are important to farming, whether it is conventional or sustainable. (Although sustainable farming is much broader than organic farming, the two usually use the same standards for soil amendments.)

Nitrogen (N_2)

Nitrogen is needed by all plants and usually in large quantities. In fact, nitrogen is so important to plant growth, and thus to food and fiber productions of the world, that the two German scientists who first invented the process of making NH3 from atmospheric nitrogen gas (N2) and natural gas-derived hydrogen (H2) won Nobel prizes for their work in the early of this century (Today's Chemists 1995). For example, average N concentrations needed for normal growth are about 3% for corn and coffee, 4% for tomato, and 2% for macadamia (Tamimi et al. 1994). If plants do not have enough nitrogen, they are stunted. Their leaves are small and pale colored, sometimes even yellow or reddish tinted. The reason is that nitrogen is a component of chlorophyll; less nitrogen results in less chlorophyll, thus less green.

In organic farming systems, you cannot use urea nor NH4NO3 or any synthetic chemicals for that matter. Alternatives must be sought. Table 7.1 lists the total N content of various organic sources.

So on the basis of total N, organic sources would have somewhere between ten and hundred times less N than urea.

Phosphorus (P)

Along with nitrogen, phosphorus is a nutrient that plants need in relatively large quantities for normal growth. In fact, P is a structural component of DNA and RNA, the two genetic entities that are essential for growth and reproduction of living organisms. Living organisms, whether plants or humans, also derive their internal energy from P-containing compounds, mainly adenosine diphosphate (ADP) and adenosine triphosphate (ATP). This means that inadequate P supply will result in a decreased synthesis of RNA, the protein maker, leading to depressed growth.

Table 7.1 Total N concentration in common organic sources as compared with urea (46% N)

Organic source	Total N (%)
Poultry manure	1.5–3.0
Pig, horse, cow manure	0.3–0.6
Green manure	1.5–5.0
Compost	0.5–2.0
Seaweed meal	2.0–3.0
Sewage sludge	1.0–5.0
Fish waste	4.0–10.0
Blood (slaughterhouse)	10.0–12.0
Human urine/night soil	1.0–1.5

Adapted from Caplan (1992), Hue (1995)

Table 7.2 Total P concentration in organically acceptable sources as compared with treble superphosphate (20% P)

Source	Total P (%)
Rock phosphate	17–26
Bone meal	20–30
Fish meal	5–10
Wood ash	2–5
Poultry manure	0.5–1.5
Green manure	0.2–0.5
Compost	0.2–0.5
Sewage sludge	0.4–2.5

Adapted from Nick and Bradley (1994) and Hue (1995)

Table 7.2 shows P content of selected sources that are acceptable to organic farming community in most states. Among these, the first two, rock phosphate and bone meal, have reasonably high total P content: between 20 and 30%. However, P in these two sources is very insoluble, thus much less plant available than P in treble superphosphate. More specifically, P in rock phosphate and bone meal has the formula $Ca_5(OH)(PO_4)_3$ which is hydroxyapatite or appetite for short. This is the same material that our bones and teeth are made of. As our teeth can attest to it, apatite is quite durable and is very hard to dissolve in water, meaning that it provides very little phosphate to your crop in the short term.

Potassium (K)

In plants, potassium is required for maintaining osmotic potential of cell. That is, K makes plants look turgid. Since K regulates the osmotic potential of cells, and the close or open conditions of stomata, it plays an important role in water relations in the plant. Potassium is involved in water uptake from the soil, water retention in the plant tissue, and long-distance transport of water in the xylem and of photosynthates in the phloem. Potassium affects cell extension. With adequate K, cell walls are thicker, thereby improving plant resistance to lodging pests and disease (Table 7.3).

Table 7.3 Total K concentration in selected organically acceptable sources

Source	Total K, %
Sul-Po-Mag [Mg, K, SO_4]	22.0
Polyhalite [Ca, K, SO_4]	10–15
Wood ash	5–10
Green sand	5–7
Green manure	2–5
Seaweed meal	2–3
Compost	0.5–2.0

Adapted from Nick and Bradley (1994) and personal data

7.3 Soil: A Basic Medium for Life on Earth

In recent years, much has been written about soil quality in relation to food security (Lal and Stewart 2010) because of a renewed awareness of the relationship between human population and the Earth's capacity to produce enough food to sustain the world's burgeoning population. In the context of this brief discussion of fertilizers and soil health, it is pertinent to put the global situation with respect to food in perspective. The food balance sheets prepared by the United Nations Food and Agriculture Organization (FAO) show that more than 99.7% of human food (calories) comes from the terrestrial environment, i.e., agricultural land (Pimentel and Wilson 2004). Of the 13 billion ha of land area on Earth, cropland accounts for only 11%. About 78% of the average per capita calorie consumption or energy needs worldwide comes from crops grown directly in soil, and another more than 20% comes from other terrestrial food sources such as meat, eggs, and milk that rely indirectly on soil (Brevik 2013).

Soil is fundamental to crop production and thus constitutes the natural resource that provides mankind the most of its food and nutrients. Because soil is finite and fragile, it is a precious resource that requires special care and conservation so that it can be used indefinitely by future generations. The crucial role of soils in supporting human existence on the planet Earth can be judged from the facts that it takes about 500 years or more for 2.5 cm of topsoil, depending on the weathering environment, to become usable under agricultural conditions; in short, soil formation is a very slow process. With a growing world population and limited possibilities for expansion of cultivated land area, per capita calorie production has consistently decreased in the past decades. For example, the quantity of cereal grains produced per capita has been declining since 1984. The extent of this general trend varies between countries depending on development status, relative population growth, and food diversification. In addition, following decades of significant productivity increases, relative yield gains are declining. One of mankind's greatest challenges is to increase productivity and move off the yield plateau.

7.3.1 Fertilizer Use and Soil Health

Soil is a dynamic natural system that lies at the interface between earth, air, water, and life, providing critical ecosystem service for the sustenance of humanity (Needelman 2013). Preservation of soil quality is among the great challenges and opportunities we have to face in the twenty-first century. Soil quality is usually defined as the capacity of soil to interact with the ecosystem in order to maintain the biological productivity, the quality of other environmental compartments, thus promoting the health of plants and animals, including humans (Doran and Parkin 1994).

Soil quality may quickly deteriorate because of intensive management, stabilize with time under proper management, and improve in the long time by supplying of organic matter. Decline in soil organic matter under intensive farming systems is a major cause of soil fertility loss. Organic matter plays a critical role in soil ecosystem because it provides substrates for decomposing microbes (that in turn supply mineral nutrients to plants), improves soil structure and water holding capacity (Abiven et al. 2009), increases natural suppressiveness against soilborne pathogens (Bonanomi et al. 2010), and reduces heavy metal toxicity (Park et al. 2011). In this scenario, a recovery of depleted soil organic matter and its maintenance to an adequate level is a critical task. It has been shown that application of organic amendments is a reliable and effective tool to ameliorate soil structure and both chemical (Scotti et al. 2013) and biological fertility of soils, as well as to suppress soilborne pathogens (Zaccardelli et al. 2013).

Fertilizer refers to any compound that contains one or more chemical elements, organic or inorganic, natural or synthetic, that is placed on or incorporated into the soil or applied directly on to plants to achieve normal growth. The main supply of plant nutrients includes organic manures, plant residues, biological nitrogen fixation, and chemical fertilizers. In general Inorganic fertilizers or Chemical fertilizers refer to commercially manufactured products containing a substantial amount of one or more plant nutrients. The major impact of inorganic fertilizers on the soil health system and ecosystem functions relates to their effect on primary productivity. Even when fertilizers are applied in somewhat excessive quantities, the effect is on process rates rather than any direct toxic effects.

Initially, the most important indirect consequence of using inorganic fertilizers was a corresponding reduction in the relative amount of organic manure used. Factors that militated against animal manures included limited supplies and energy costs associated with use of manures in cropping systems, e.g., transport and application, in addition to variable quality and low nutrient contents. Subsequently, there was an increased interest in manures due to increasing supplies and their perceived role in soil health as well as nutrient recycling. However, in several developing countries, particularly in Asia, crop production is relying more on fertilizers because of limited availability of animal manures and crop residues.

7.3.2 *Fertilizer Use and Soil Degradation*

Plants must obtain the elements essential for their growth, other than carbon, oxygen, and hydrogen, from the soil. Thirteen elements essential for plant growth have been identified. These essential elements are called nutrients; those needed in the greatest amount are called macronutrients whereas those needed in lesser amounts are called micronutrients. Among the macronutrients are nitrogen, phosphorus, and potassium. These three elements are those most rapidly removed from the soil by plants. Therefore, many commercial plant fertilizers supply these three essential elements. The sources of nitrogen used in fertilizer are many, including ammonia (NH_3), diammonium phosphate ($(NH_4)_2HPO_4$), ammonium nitrate (NH_4NO_3), ammonium sulfate ($(NH_4)_2SO_4$), calcium cyanamide ($CaCN_2$), calcium nitrate ($Ca(NO_3)_2$), sodium nitrate ($NaNO_3$), and urea (N_2H_4CO). Phosphorus is generally supplied as a phosphate, such as diammonium phosphate ($(NH_4)_2HPO_4$) or calcium dihydrogen phosphate ($Ca(H_2PO_4)_2$). Potassium comes from potassium sulfate (K_2SO_4) or potassium chloride (KCl), which is also called muriate of potash.

7.4 Soil Management Strategies for a Sustainable Agriculture

Soil organic matter plays an important role in long-term soil conservation and/or restoration by sustaining its fertility, and hence in sustainable agricultural production, due to the improvement of physical, chemical, and biological properties of soils. The organic matter content is the result of the inputs by plant, animal, and microbial residues and the rate of decomposition through mineralization of both added and existing organic matter. More specifically, the generic term "organic matter" refers to the sum of all organic substances present in the soil. This sum comes from residues at various stages of decomposition, substances synthesized through microbial-chemical reactions, and biomass of soil microorganisms as well as other fauna, along with their metabolic products (Lal 2007). Decomposition of organic matter is chiefly carried out by heterotrophic microorganisms. This process is under the influence of temperature, moisture, and ambient soil conditions and leads to the release and cycling of plant nutrients, especially nitrogen (N), sulfur, and phosphorus (Murphy et al. 2007).

At present, most optimistic estimates show that about 25–30% of nutrient needs of Indian agriculture can be met by various organic sources. Supplementation of entire N through FYM sustains crop productivity at more than use of conventional N fertilizers. Since the estimates of NPK availability from organic sources are based on total nutrient content, efficiency of these sources to meet the nutrient requirement of crops is not as assured as mineral fertilizers, but the joint use of chemical fertilizers along with various organic sources is capable of sustaining higher crop productivity, improving soil quality, and productivity on long-term basis.

In recent years, biofertilizers have emerged as a promising component of integrating nutrient supply system in agriculture. Our whole system of agriculture depends on many important ways, on microbial activities, and there appears to be a tremendous potential for making use of microorganisms in increasing crop production. Microbiological fertilizers are an important part of environment friendly sustainable agricultural practices.

7.4.1 *Physical Fertility*

As widely reported in literature, the use of organic amendments increases soil organic matter (Thangarajan et al. 2013; Khaliq and Abbasi 2015) and as consequence soil aggregate stability, water holding capacity, and soil porosity (Celik et al. 2004; Leroy et al. 2008), thus improving soil quality. Scotti et al. (2013) applied fast field cycling NMR relaxometry as innovative application of spectroscopic technique to soil and highlighted how the combined use of compost and wood scraps under intensive farming system induced an increase of the soil pore size through the formation of organo-mineral aggregates which, in turn, can have positive effects on soil structure and soil aeration.

The application of organic amendments such as sheep manure, cow manure, rice husk, reeds, and wheat straw increased soil aggregate stability and decreased soil bulk density (Karami et al. 2012). In a long-term study in China, Zhao et al. (2009) found that farmyard manure and straw application determined a decrease of soil bulk density (1.21 and 1.18 Mg m-3, respectively) when compared with untreated soils (1.43 Mg m-3) due to increase in soil organic C and porosity. Also organic amendments obtained from manufacturing by-products, such as biochar, can affect particle size distribution and aggregate stability. As reported in Liu et al. (2014), in agricultural soils less than 40 t ha^{-1} biochar, soil water-stable aggregate (>0.25 mm) in the 0–15 cm soil layer had a remarkable increase with respect to other treatments, especially the acroaggregate with particle size larger than >2 mm, suggesting that biochar incorporation into soil improves soil structure. Sometimes, organic amendments can affect in directly soil physical properties. Lucas et al. (2014) demonstrated that organic amendments containing high amount of bioavailable C derived from cellulose can promote fungal proliferation and improve soil structure through stabilization of soil aggregates, suggesting a use of organic amendments to manipulate soil microbial community structure and to promote aggregation in soils.

7.4.2 *Chemical Fertility*

Intensive agriculture, without organic amendments for the restoration of soil organic C stock, negatively affects soil chemical properties producing a reduction in soil C content that, in turn, produces deleterious effects on soil microbial biomass, soil enzymatic activities, and functional and species diversity, besides a drastic increase in soil salinity (Bonanomi et al. 2011).

A large body of empirical studies carried out in different agricultural systems demonstrated that the application of organic amendments in the form of compost is an effective tool to recover soil organic C stock (Hargreaves et al. 2008; Zhang et al. 2015). By contrast, only few studies addressed the capability of compost amendments to recover soil C stock for vegetable cultivations under plastic tunnels (Morra et al. 2010). Iovieno et al. (2009) found no significant organic C recovery after 3 consecutive years of compost amendments (up to 45 t ha^{-1} $year^{-1}$), likely as a result of the rapid compost mineralization due to its relatively high biochemical quality (i.e., C/Nequal to 13). These data highlight how the recovery of soil C stock is challenging because of temperature and water availability, related to plastic cover and irrigation regime. They are not adjustable conditions under this specific farming system, being crucial for producing out-of-season vegetables. Therefore, a valid alternative is to identify organic amendments with specific C biochemical quality able to maximize stable C stock recovery and, at the same time, provide a continuous release of mineral nutrients satisfying crop requirements. The addition of chemical fertilizers generally leads to a rapid mineral N release, while organic amendments induce a slow mineral N release, but extended overtime (Claassen and Carey 2006). Weber et al. (2007) reported that the slow mineralization of N in soils under compost amendment improves not only the soil fertility but also the conditions of organic matter mineralization. In fact, they found an increase of humic acid/fulvic acid ratio in compost-amended soil which might be partly due to the original composition of humic substances in the compost, where humic acids always predominate over fulvic acids.

7.5 Soil Amendments

Soil amendments include all inorganic and organic substances mixed into the soil for achieving better soil constitution regarding plant productivity. The primary role of soil amendments is to provide nutrients for crop growth or to provide materials for soil improvement. A soil amendment can be any material that is mixed into the existing soil to improve its physical properties such as aeration, water retention, and nutrient holding capacity. Misuse of soil amendments can result not only in damage to crops but can also cause negative impacts on the receiving soil, water, air, or habitat environment (Sankar ganesh 2009).

7.5.1 Inorganic Soil Amendments

Inorganic substances are lime, vermiculite, perlite, tire chunks, pea gravel, and sand. In general they must be bought, which makes them more expensive than organic amendments. Therefore these substances do not have the same degree of sustainability as organic amendments.

7.5.2 Organic Soil Amendments

Organic in this context refers to anything that comes from something that is or was alive such as peat, animal manure, municipal biosolids, green manure, compost, grass clippings, wood chips, bone meal, bat guano, and earthworm castings. Organic amendments have an added benefit of providing an energy source for bacteria, fungi, and earthworms that live in the soil.

7.5.2.1 Peat Moss

Peat moss is the decomposed remains of *Sphagnum* moss, a plant native to many regions of the world. Peat moss is a natural, organic soil conditioner that helps aid in moisture retention and helps regulate air around plant roots for ideal growing conditions. Peat moss helps loosen and aerate soil that is high in clay. For soils that are high in alkaline, peat moss will help lower the pH as it is naturally acidic. Peat moss also helps reduce leaching or runoff of nutrients when added to soil.

7.5.2.2 Animal Manure

Animal manures can be an economical and effective source of crop nutrients. Land application of animal manures is also a best management practice for protecting water quality when it is carried out properly. Manure is composed by feces, urine, and animal bedding stacked and turned until a certain level of composting. It derives from beef, dairy, pork, poultry, and turkey, and its composition depends on its origin, the time that urine and feces are excreted and mixed, and the storage time before being applied to soil. Manure supplies nutrients for crops but also organic matter thus improving soil fertility.

7.5.2.3 Municipal Biosolids

Biosolids are a by-product of municipal wastewater treatment. All municipal wastewater treatment plants produce biosolids, which are the stabilized residuals that settle from the water during the various treatment processes (Tuomela et al. 2000). Biosolids are rich in both organic matter and essential plant nutrients and can be utilized in a variety of ways, directly as a soil amendment and fertilizer and indirectly as a feedstock in the fabrication of value-added products. Recycling sludges into biosolids is not a new management concept. For thousands of years, Chinese society returned sewage sludges to farmland in an effort to maintain soil quality and conditions. In parts of Europe and elsewhere, biosolids have been applied on agricultural land for a century and longer.

7.5.2.4 Green Manure and Cover Crops

Green manures, as the name implies, are used primarily for the addition of nutrients and organic matter to the soil, protecting and improving soil quality. Green manure crops are typically grown for a specified period during a rotation when a field is not in use and are then plowed under and incorporated into the soil before the succeeding crop is established.

A cover crop is any living ground cover planted to protect the soil from erosion. It may be planted into or after a main crop and is killed before the next crop is planted. Though the primary benefit of cover crops is the reduction of water runoff and soil erosion, they provide many other services, including weed, insect, and disease suppression; enhanced soil structural stability; water conservation; increased microbial populations; and preservation of soil nutrients. Because of their many services, cover crops can be used to fill multiple niches within a farming system.

7.5.2.5 Compost

Decomposition of organic wastes leads to the formation of the most used soil amendment, the compost. The use of compost represents both an interesting agricultural practice and a waste recycling management (Perez-Piqueres et al. 2006). Indeed, it allows reducing the costs of green/urban waste disposal, recycling nutrient elements for crops, and providing for soil organic matter depletion. Multiple benefits derive from the use of compost as fertilizer, for example, an increase in organic C content and microbial activity (Scotti et al. 2013, 2015), a greater concentration of plant nutrients like N, P, K, and Mg, and a root reinforcement (Szczech and Smolinska 2001; Borrero et al. 2004; Goss et al. 2013).

7.5.2.6 Bone Meal

Bone meal is a mixture of finely and coarsely ground animal bones and slaughterhouse waste products. It is used as an organic fertilizer for plants and as a nutritional supplement for animals. As a slow-release fertilizer, bone meal is primarily used as a source of phosphorus and protein. Finely ground bone meal may provide a quicker release of nutrients than the coarser ground version of bone meal. Meat and bone meal (MBM) contains appreciable amounts of total nitrogen, phosphorus, and calcium.

7.5.2.7 Bat Guano

Bat guano is the feces of bats rich in carbon, nitrogen, vital minerals, and of course beneficial microbes. Chemical properties and the microbes in the guano enrich the soil fertility and the texture, and the microbes help to clear any toxins in the soil and control the fungi and nematodes in the soil. These properties of the guano again depend upon the bat species, location, and age of the guano. Bat guano is known to contain all the macro- and micronutrients that plants require in a natural form and hence ably serve as plant fertilizer, soil builder, soil cleanser, fungicide, nematocide, and compost activator. In ancient times it was used in agricultural practice as manure, but with advent of chemical fertilizers its usage became less popular. Health menace created by chemical fertilizers is again popularizing organic farming, but bat guano is still not popular among the farming community since no explicit work on its plant growth promoting activity has been done.

7.5.2.8 Earthworm Castings

Worm castings are the richest natural fertilizer known to humans. Worm castings contain a highly active biological mixture of bacteria, enzymes, remnants of plant matter and animal manure, as well as earthworm cocoons. The castings are rich in water-soluble plant nutrients and contain more than 50% more humus than what is normally found in topsoil. Worm castings have N-P-K ratio of about 3.2-1.1-1.5. It is packed with minerals that are essential for plant growth, such as concentrated nitrates, phosphorus, magnesium, potassium, and calcium. It also contains manganese, copper, zinc, cobalt, borax, iron, carbon, and nitrogen. However, the best of all is that these minerals are immediately available to the plant, without the risk of ever burning the plant. Remember that animal manure and chemical fertilizers have to be broken down in the soil before the plant can absorb them. The humus in the worm castings extracts toxins and harmful fungi and bacteria from the soil. Worm castings therefore have the ability to fight off plant diseases.

7.6 Conclusion

Sustainable agriculture seeks close harmony with nature. The challenge of sustainable agriculture is that it is information intensive and requires the farmer to have a deep and detailed understanding of natural processes. Because of its special requirements, farmers practicing sustainable agriculture may need to develop more management skills than even modern "conventional" agriculture requires.

References

Abiven S, Menassero S, Chenu C (2009) The effect of organic inputs over time on soil aggregate stability – a literature analysis. Soil Biol Biochem 41:1–12

Bonanomi G, Antignani V, Capodilupo M et al (2010) Identifying the characteristics of organic soil amendments that suppress soilborne plant diseases. Soil Biol Biochem 42:136–144

Bonanomi G, D'Ascoli R, Antignani V et al (2011) Assessing soil quality under intensive cultivation and tree orchards in Southern Italy. Appl Soil Ecol 47:187–194

Borrero C, Trillas M, Ordovás J et al (2004) Predictive factors for the suppression of Fusarium wilt of tomato in plant growth media. Phytopathology 94:1094–1101

Brevik EC (2013) Soils and human health – an overview. In: Brevik EC, Burgess LC, Raton B (eds) Soils and human health. CRC Press, Boca Raton, pp 29–56

Caplan B (1992) Organic gardening. Headline Book Publishing, London

Celik I, Ortas I, Kilic S (2004) Effects of compost, mycorrhiza, manure and fertilizer on some physical properties of a Chromoxerert soil. Soil Tillage Res 78:59–67

Claassen VP, Carey JL (2006) Comparison of slow-release nitrogen yield from organic soil amendments and chemical fertilizers and implications for regeneration of disturbed sites. Land Degrad Dev 18:119–132

Doran JW, Parkin TB (1994) Defining and assessing soil quality. In: Doran JW, Coleman DC, Bezdicek DF, Stewart BA (eds) Defining soil quality for a sustainable environment. Soil Science Society of America, Inc., Madison, pp 3–21

Eurostat (2015) Statistics explained. http://ec.europa.eu/eurostat/statistics-explained. Accessed March 2015

Goss MJ, Tubeileh A, Goorahoo D (2013) A review of the use of organic amendments and the risk to human health. Adv Agron 120:275–379

Hargreaves JC, Adl MS, Warman PR (2008) A review of the use of composted municipal solid waste in agriculture. Agri Ecosys Environ 123:1–14

Hector A, Schmid B, Beierkuhnlein C et al (1999) Plant diversity and productivity experiments in European grasslands. Science 286:1123–1127

Hue NV (1995) Sewage sludge. In: Rechcigl JE (ed) Soil amendment and environmental quality. Lewis Publishers, Boca Raton, pp 193–239

Iovieno P, Morra L, Leone A et al (2009) Effect of organic and mineral fertilizers on soil respiration and enzyme activities of two Mediterranean horticultural soils. Biol Fertil Soils 45:555–561

Karami A, Homaee M, Afzalinia S et al (2012) Organic resource management: impacts on soil aggregate stability and other soil physico-chemical properties. Agri Ecosys Environ 148:22–28

Khaliq A, Abbasi MK (2015) Improvements in the physical and chemical characteristics of degraded soils supplemented with organic-inorganic amendments in the Himalayan region of Kashmir, Pakistan. Catena 126:209–219

Lal R (2007) Anthropogenic influences in world soil and implications for global food security. Adv Agron 93:69–93

Lal R, Stewart BA (2010) Food security and soil quality. CRC Press, Boca Raton

Leroy BLM, Herath HMSK, Sleutel S et al (2008) The quality of exogenous organic matter: short-term effects on soil physical properties and soil organic matter fractions. Soil Use Manag 24:139–147

Liu Z, Chen X, Jing Y (2014) Effects of biochar amendment on rapeseed and sweet potato yields and water stable aggregate in upland red soil. Catena 123:45–51

Loreau M, Naeem S, Inchausti P (2001) Biodiversity and ecosystem functioning: current knowledge and future challenges. Science 294:804–808

Lucas TS, D'Angelo EM, Williams MA (2014) Improving soil structure by promoting fungal abundance with organic soil amendments. Appl Soil Ecol 75:13–23

Morra L, Pagano L, Iovieno P (2010) Soil and vegetable crop response to addition of different levels of municipal waste compost under Mediterranean greenhouse conditions. Agron Sustain Dev 30:701–709

Murphy KM, Campbell KG, Lyon SR, Jones SS (2007) Evidence of varietal adaptation to organic farming systems. Field Crop Res 102:172–177

Needelman BA (2013) What are soils? Nat Educ Knowl 4(3):2

Nick J, Bradley F (1994) Growing fruits and vegetables organically. Rodale Press, Emmaus

Park JH, Lamb D, Paneerselvam P et al (2011) Role of organic amendments on enhanced bioremediation of heavy metal(loid) contaminated soils. J Hazard Mater 185:549–574

Perez-Piqueres A, Edel-Hermann V, Alabouvette C (2006) Response of soil microbial communities to compost amendments. Soil Biol Biochem 38:460–470

Pimentel D, Wilson A (2004) World population, agriculture and malnutrition. World Watch, September/October, 22–25

Sankar Ganesh K (2009) Response of paddy cultivars to chromium pollution. PhD thesis, Annamalai University, Annamalai nagar

Scotti R, Conte P, Berns AE (2013) Effect of organic amendments on the evolution of soil organic matter in soils stressed by intensive agricultural practices. Curr Org Chem 17:2998–3005

Scotti R, D'Ascoli R, Bonanomi G et al (2015) Combined use of compost and wood scraps to increase carbon stock and improve soil quality in intensive farming systems. Eur J Soil Sci 66(3):463–475. https://doi.org/10.1111/ejss.12248

Szczech M, Smolińska U (2001) Comparison ofsuppressiveness of vermicomposts produced from animal manures and sewage sludge against *Phytophthora nicotianae* Breda de Haan var. nicotianae. J Phytopathol 149:77–82

Tamimi Y, Silva JA, Yost RS et al (1994) Adequate nutrient levels in soils and plants in Hawaii. Fact sheet no. 3.CTAHR

Thangarajan R, Bolan NS, Tian G et al (2013) Role of organic amendment application on greenhouse gas emission from soil. Sci Total Environ 465:72–96

Tilman D, Cassman KG, Matson PA et al (2002) Agricultural sustainability and intensive production practices. Nature 418:671–677

Tuomela M, Vikman M, Hatakka A et al (2000) Biodegradation of lignin in a compost environment: a review. Bioresour Technol 72:169–183

Weber J, Karczewska A, Drozd J et al (2007) Agricultural and ecological aspects of a sandy soil as affected by the application of municipal solid waste composts. Soil Biol Biochem 39:1294–1302

Zaccardelli M, De Nicola F, Villecco D et al (2013) The development and suppressive activity of soil microbial communities under compost amendment. J Soil Sci Plant Nutr 13:730–742

Zhang H, Ding W, Yu H et al (2015) Linking organic carbon accumulation to microbial community dynamics in a sandy loam soil: result of 20 years compost and inorganic fertilizers repeated application experiment. Biol Fertil Soils 51:137–150

Zhao Y, Wang P, Li J et al (2009) The effects of two organic manures on soil properties and crop yields on a temperate calcareous soil under a wheat-maize cropping system. Eur J Agron 31:36–42

Chapter 8
Plant Growth–Promoting Microbes: A Boon for Sustainable Agriculture

S. Lalitha

Abstract Biofertilizers are defined as living cells with efficient strains of microorganisms that help crop plant uptake of nutrients by rhizosphere soil when applied through soil. Certain microbial processes in the soil convert nonavailable forms of nutrients to be easily assimilated by plants. In nature, microorganisms are not as efficient in the rhizosphere and its surroundings. Therefore, many efficient cultures have been artificially selected and these microorganisms play a vital role in accelerating microbial processes in soil. Use of biofertilizers is one of the important components of integrated nutrient management, as they are cost effective and renewable sources of plant nutrients to supplement chemical fertilizers for sustainable agriculture. Chemical fertilizers affect and may cause toxicity to both the environment and sustainable agriculture. Moreover, they can cause soil infertility. Biofertilizers add nutrients through the natural processes of nitrogen fixation, solubilizing phosphorus and stimulating plant growth through the synthesis of growth-promoting substances. Biofertilizers can be expected to reduce the soil's natural nutrient cycle and build soil organic matter.

8.1 Introduction

Through the use of biofertilizers, plants can grow healthy and the sustainability and the health of the soil can be enhanced. Since they play several roles, the preferred scientific term for such beneficial bacteria is *plant growth–promoting rhizobacteria* (PGPR) (Mishra 2014). It has been established in agriculture that nitrogen is the most important and major plant nutrient that determines success in crop productivity.

Crops fail to produce enhanced yield if nitrogen becomes a limiting factor. In general, farmers apply a substantial quantity of nitrogenous chemical fertilizers to increase the productivity of crop plants. Unfortunately, the costs of production and distribution of nitrogenous fertilizers are increasing day by day, since they involve

S. Lalitha (✉)
Department of Botany, School of Life Science, Periyar University, Salem, Tamil Nadu, India
e-mail: lara9k@gmail.com

A. Dhanarajan (ed.), *Sustainable Agriculture towards Food Security*,
https://doi.org/10.1007/978-981-10-6647-4_8

substantial quantities of fossil fuels (hydrocarbons) as an energy source. Poor and marginal farmers can ill afford to purchase fertilizers for crops. Therefore, scientists and policy makers have turned their attention to alternative sources of this important plant nutrient. This will hopefully reduce the costs of application of chemical fertilizers and hence the costs of crops. Biological nitrogen fixation (BNF) surely appears to be an alternative technology to circumvent the fertilizer crisis. For India and other developing countries, BNF offers great scope in the agricultural sector (Rao 2002).

Biofertilizers such as *Rhizobium*, *Azotobacter*, *Azospirillum*, and blue green algae (BGA) are widely used. *Rhizobium* inoculants are used for leguminous crops. *Azotobacter* can be used with crops such as wheat, maize, mustard, cotton, potato, and other vegetable crops (Fernandes and Bhalerao 2015). *Azospirillum* inoculants are recommended mainly for sorghum, millets, maize, sugarcane, and wheat.

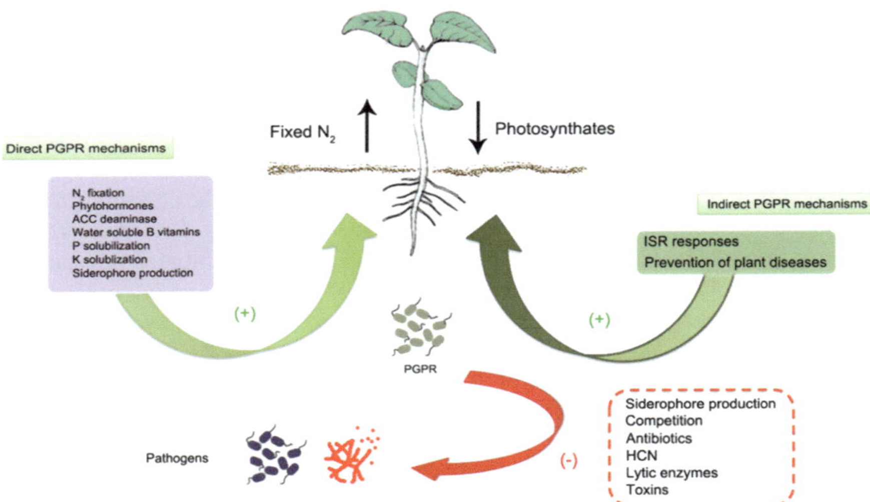

Country:https://www.aimspress.com

Mechanisms of plant growth–promoting rhizobacteria

Biofertilizers are seen as an important alternative technology, given the negative externalities of chemical fertilizers. The use of the latter has led to considerable environmental costs. Biofertilizers do not pollute the soil and do not disrupt the ecological balance, and hence are environmental friendly. An increasing number of farmers are using biofertilizers, and the numbers of biofertilizer-manufacturing units have also grown considerably. BNF benefits not only the legumes themselves but also any intercropped or succeeding crop, reducing or removing the need for nitrogen fertilization. In soils with low mineral nitrogen content, nitrogen-fixing microorganisms provide ammonium to the legume biomass, allowing faster growth than that of other plant competitors. In contrast, if nitrogen is abundant, nitrogen-fixing microorganisms tend to be competitively excluded by nonfixing species because the nitrogen fixation process is bioenergetically costly (Houlton et al. 2008).

In the last century, chemical fertilizers were introduced and this made farmers happy to get increased agricultural yields in the beginning. But, slowly, chemical fertilizers started displaying their ill effects such as leaching, polluting water basins, destroying microorganisms and friendly insects, making the crop more susceptible to attack by disease, reducing the soil fertility, and thus causing irreparable damage to the overall system. One of the other most important effective factors in increasing plant yield is seed inoculation or priming with PGPR (Ashrafi and Seiedi 2011), which also increases the activity of colonization of plant roots and plant growth and yield (Heidari et al. 2011).

The three essential macronutrients that all plants need for growth are nitrogen (N), phosphate (P), and potassium (K). Even if only one of these nutrients is in short supply, plant growth is limited and yield is reduced especially under harsh growth conditions in sandy soil. However, intensive cultivation results in the soil being deficient in these important nutrients. The use of synthetic NPK fertilizers replaces the chemical components that are taken from the soil by growing plants (Datta et al. 2009). They maintain soil productivity and significantly support food security to improve the quality and quantity of the food available today, although their long-term use is debated by environmentalists. With fertilizers, the yield of the plant can often be doubled or even tripled; fertilizers can also be tailored to suit the type of crop that is being grown, creating a better growing environment. Mineral fertilizers are one of the most important tools for agricultural development and will continue to play a decisive role irrespective of which new technologies may yet emerge (FAO and IFA 2000).

In India, the term *biofertilizers* specifies fertilizers used to meet the nutritional requirements of a crop through microbiological means; other countries use the term *microbial inoculants* (Brahmaprakash and Sahu 2012). These biofertilizers are usually carrier-based microbial preparations containing beneficial microorganisms in a viable state intended for seed or soil application, which enhances plant growth through nutrient uptake and/or growth hormone production. Important and popular microbial inoculants in India are those that supplement nitrogen, phosphorus, and PGPR. The soil microorganisms used in biofertilizers are phosphate-solubilizing microorganisms (PSM) such as mycorrhiza, *Azospirillum* spp., *Azotobacter* spp., *Rhizobium* spp., *Sesbania*, blue green algae, *Nitrosomonas* spp., *Nitrobacter* spp., and *Azolla* spp. Application of organic manure, particularly biofertilizer, is the only option to improve soil organic carbon for sustenance of soil quality and future agricultural productivity (Ramesh 2008).

In agriculture, the use of artificial fertilizers still ensures better yields, but soils and the environment become more polluted and depleted of important nutrients. Biofertilizers can contain symbiotic or nonsymbiotic microorganisms that can result in higher resistance of plants to diseases and increase plant growth rate (Kumar and Chandra 2008). Biofertilizers are environmentally friendly fertilizers that not only prevent damage to the natural source but also help to some extent to clear the natural environment of precipitated chemical fertilizers, and can provide better nourishment to plants. Several research studies have revealed that crop yield and growth are increased by 20–30% by replacing chemical nitrogen and phosphorus with biofertilizers. They can also provide protection against drought stress and some soilborne diseases (Vessey 2003).

8.2 *Rhizobium*

Rhizobium is a Gram-negative, free-living organism present in soil and it has the ability to fix atmospheric nitrogen in a symbiotic state only. *Rhizobium* also exists as an endosymbiotic nitrogen-fixing microorganism associated with the roots of legumes. It enters into plants through the root system, then it forms nodules (Devananada 2000). The fixation of nitrogen in root nodules reacts to the available hydrogen molecules present in the form NH_3, which is obtained from the host. The carbohydrates produced by legume plants are transported to nodules are utilized by *Rhizobium* as the sole source of hydrogen in the conversion of nitrogen to ammonia. The root nodules act as a microfermenter for BNF where they can convert atmospheric nitrogen into ammonia. *Rhizobium* is able to induce shoot and root growth in rice plants.

Rhizobium interaction with the legume host is quite specific and it will fix nitrogen in particular host plants only. *Rhizobium* shows host specificity mediated by plant compounds such as flavonoids, which are produced by the host plants. Flavonoids activate the *nod* genes present in *Rhizobium*. This starts the infection on the root surface, after which the root begins curling. After infection, *Rhizobium* starts to move into the root hair; at a certain level the bacterium stops its multiplication and forms bacteroids, which fix nitrogen *nod* and *nif* genes coding for protein called nodules, including leghemoglobin, which is used for nodule development. *Nif* genes are responsible for nitrogen fixation. They are present in both free-living and symbiotic nitrogen-fixing bacteria, including structural genes for nitrogenase and other regulatory enzymes (Choudhry and Kennedy 2004). There are several methods available to introduce *Rhizobium* inoculants in soil; the seed-dipping method is one of the common methods for the application of *Rhizobium*. During seed germination the bacteria infects the root hair and spreads toward the root. When the plant grows, *Rhizobium* will convert nitrogen into ammonia for plant growth.

Rhizobium is a soil habitat bacterium, which can to colonize roots, form a nodule, and fix atmospheric nitrogen symbiotically. The morphology and physiology of *Rhizobium* will vary from the free-living condition to the bacterial nodules. Biological nitrogen fixers are efficient biofertilizers as far as the quantity of nitrogen fixed is concerned. They have seven genera and are highly specific for forming nodules in legumes, referred to as a *cross-inoculation group*. *Rhizobium* inoculant was first made in the USA and commercialized by private enterprise in the 1930s. Use of biofertilizers can also prevent depletion of soil organic matter (Jeyabal and Kupuswamy 2001). Inoculation with bacterial biofertilizers may reduce the application of fertilizers for nitrogen uptake by plants (Mia et al. 2005). But most of this technique is mainly limited between legumes and *Rhizobium* in a symbiotic process, which can fix atmospheric nitrogen.

8.2.1 Role of **Rhizobium** *in Abiotic and Biotic Stress in Plants*

Plant growth in agricultural soils is influenced by many abiotic and biotic factors.

The thin layer of soil surrounding plant roots is extremely important, and that area of root activity and metabolism is known as the *rhizosphere*. The rhizosphere concept was first introduced to describe the narrow zone of soil surrounding the roots where microbe populations are stimulated by root activities. The original concept has now been extended to include the soil surrounding a root in which physical, chemical, and biological properties have been changed by root growth and activity (McCully 2005). A large number of microorganisms such as bacteria, fungi, protozoa, and algae coexist in the rhizosphere. In this microbial ecosystem, bacteria are most abundant and contribute most to their fitness by releasing organic compounds through exudates, creating a very selective environment where diversity is low. It is highly probable that bacteria influence plant physiology to a great extent, especially considering their competitiveness in root colonization (Barriuso et al. 2008).

Rhizobia are aerobic, rod-shaped, motile, Gram-negative, heterotrophic, and nonspore bacteria, which live freely in soil and in the root region of leguminous and nonleguminous plants. Rhizobia fix atmospheric nitrogen and thus not only increase the production of inoculated crops but also leave a fair amount of nitrogen in the soil, which benefits the subsequent crops. Most of these bacteria are very sensitive to soil water deficit, which adversely affects their nitrogen fixation capacity and hence the productivity of the whole legume plant. They are also sensitive to higher salt concentrations and are capable of growing at a 200 mM salt concentration, but their growth is more abundant at lower salt concentrations (Afzal and Bano 2008).

PGPR are a heterogeneous group of bacteria present in the rhizosphere, both at the root surface and in endophytic associations, and they improve the quality of plant growth (Rothballer et al. 2009). PGPR can facilitate plant growth and development in two different ways: directly and indirectly. The direct promotion of plant growth by PGPR generally entails providing the plant with a compound synthesized by the bacterium or facilitating the uptake of nutrients from the environment. The indirect promotion of plant growth occurs when PGPR reduce or prevent the deleterious effects of pathogens on plants by producing inhibitory substances or by increasing the natural resistance of the host (Schuhegger et al. 2006).

The direct growth-promoting mechanisms are as follows: (1) N_2 fixation; (2) phosphate solubilization; (3) complexation of insoluble ferric iron by siderophore production; (4) production of phytohormones such as auxins, cytokines, and gibberellins; and (5) lowering of ethylene concentrations (Mayak et al. 2004). The field of salt-tolerant rhizobacteria was recently reviewed by Egamberdieva and Kucharova (2008). The information on halotolerant diastrophic PGPR is still limited; some examples of salt-tolerant PGPR include *Azospirillum halopraeferens* (Loganathan and Nair 2004) and the *Azospirillum brasilense* strain NH (Nabti et al. 2007). Because soil salinity is increasing in many parts of the world, there is a need for general improvement of plant performance and BNF.

The mechanisms by which PGPRs promote plant growth have been widely described. Several mechanisms have been suggested by which PGPR can promote plant growth, and this include auxins (Egamberdieva 2005), enhancing stress resistance, symbiotic N_2 fixation (Orhan et al. 2006), solubilization of inorganic phosphate, and mineralization of organic phosphate or other nutrients (Jeon et al. 2003); and increasing the supply or availability of primary nutrients to the host plant and antagonism against phytopathogenic microorganisms by production of siderophores, synthesis of antibiotics, enzymes, or fungicidal compounds, and competition with detrimental microorganisms (Ashrafi and Seiedi 2011). Cucumber and spinach seedlings supported more growth of siderophore-producing *Pseudomonas* strains than nonproducers and rootlet elongation was also promoted on cucumber (De Bellis and Ercolani 2001), Moreover, coinoculation of *Pseudomonas* spp. with *Bradyrhizobium/Mesorhizobium* spp. has been found to cause a significant increase in nodule numbers, nodule weight, and plant dry weight of green gram and chickpea when grown under sterilized chillum jar conditions (Sindhu et al. 2002).

PGPR are directly involved in increased uptake of nitrogen, synthesis of phytohormones, solubilization of minerals such as phosphorus, and production of siderophores, which chelate iron and make it available to the plant root. PGPR have been shown to solubilize precipitated phosphates and enhance phosphate availability to cauliflower, which represents a possible mechanism of plant growth promotion under field conditions (Thakuria et al. 2004). It is important to note that several phosphate-solubilizing bacilli occur in soil but their numbers are not usually high enough to complete with other bacteria commonly established in the rhizosphere.

Substantial amounts of potash can be obtained from crop residues if managed to add in soils. Use of biofertilizer in adequate amounts can produce an increase (5–30%) in yield. Vermicomposting of rural waste holds great promise for mitigating nutrient hunger of soils in NE India considering supply of composting earthworms and need-based training in compost technology. Soil amelioration with the use of limestone deposits available in the northeast can be brought into use. Finally, watershed-based technology with proper soil and water conservation measures can be an effective avenue to nurture soil health for sustainable organic food production (Dobbelaere et al. 2003).

PGPR decrease the amount of soil under the plant root system and favor active microbial production. Plants produce metabolically active cells from their roots and 20% of carbon deposits in the rhizosphere, so the plant's and microorganism's relationship will be close. Finally, rhizosphere control the plant disease, with the interaction of, dramatic change. The PGPR strain is used commercially as a biofertilizer. The general mechanisms PGPR use for nitrogen fixation are decreasing ethylene levels, production of siderophores and phytohormones, induced systemic resistance (ISR), solubilization of nutrients, decreasing pollutant toxicity, etc. Agriculture is more dependent on the use of chemical fertilizers and pesticides for good yields but these create a lot of problems, such as environmental pollution, health hazards, and destruction of biological communities. But they support crop production. In this context, PGPR provide benefits to agricultural soils afflicted by biotic (or) abiotic factors (Sivasakthi et al. 2013) and root growth activity. They give lots of report on PGPR to prevent stresses from the environment (Paul and Nair 2008).

8.2.2 The Root Microbiome

The rhizosphere represents a thin layer of soil surrounding plant roots and is influenced by root activities. A soil microbiologist, Lorenz Hiltner, first introduced the term *rhizosphere* in the 1900s (Hartmann and Rothballer 2008). The rhizosphere contains nutrient-rich soil for microorganisms and the root releases compounds such as amino acids, sugars, and other organic compounds, which can be utilized by microorganisms, such as fungi, bacteria, algae, and protozoa (Dobbelaere et al. 2003; Singh and Cameotra 2004; Lambers and Mougel 2009). PGPR are most abundant among all species in the rhizosphere.

8.3 Role of Plant Growth–Promoting Rhizobacteria

PGPR root-colonizing bacteria are beneficial for plant growth to increase seedling emergence, biomass, crop yield, and the root system. PGPR free-living bacteria affect plants directly or indirectly. PGPR, as biocontrol agents, promote plant growth by mechanisms such as production of auxins and phytohormones (Patten and Glick 1996) and decreasing plant ethylene levels (Glick et al. 2007). PGPR stimulate plant growth in two ways, by direct and indirect methods. The direct methods include production of phytohormones such as auxins, gibberellins, cytokines, etc.; BNF, phosphate-solubilizing bacteria (PSB), siderophores, enzymes, and ISR, while indirect stimulation is related to biocontrol, antibiotic production, synthesis of extracellular enzymes, etc. (Zahir et al. 2004; Van Loon 2007). PGPR can be categorized in three ways: biofertilizers, phytostimulators, and biopesticides. PGPR may enhance plant growth in any one of these ways it will be regulate plant growth as experiment alone. Plants produce metabolically active cells from their roots and 20% of carbon deposits in the rhizosphere, so the plant–microorganism relationship will be close.

8.3.1 Traits of Efficient Biocontrol Plant Growth–Promoting Rhizobacteria

The following points should be considered in the selection of PGPR isolates for the management of plant pathogens:

- Screening of candidate bacteria from pathogen-suppressive soil will help in the selection of efficient strains.
- Organisms should have multiple antagonistic mechanisms (e.g., antibiotic, siderophore, and bacteriocin production) and the ability to mediate ISR in plants.

- The in vitro screening assay method for selection of antagonists should be modified in such a way that it simulates natural conditions. Examples include tuber-slice assays and seedling bioassays.
- Selected organisms should have wider environmental adaptability (e.g., pH, temperature, and moisture).
- Organisms must have the ability to colonize plant roots with sufficient population levels.
- Organisms must have good ecological competence and the ability to compete and survive in nature.
- Because the variability in root colonization may bring inconsistent results, organisms should have good rhizosphere competence. If required, crop-specific inoculants should be developed.
- Motile bacteria are hypothesized to colonize roots more efficiently than nonmotile bacteria; however, in-depth research is needed.
- Better rhizosphere colonization at lower temperatures and pH extremes may help bacteria escape competition from native microbes.
- Root colonization by the applied strain may be affected by the host genotype; similarly, the use of crop-specific microbial strains may enhance root colonization.
- PGPR traits in relation to rhizosphere competence have not been clearly described. We do know, however, that selection of bacteria based on their exo-polysaccharide production, fimbriae, flagella, chemotaxis, osmotolerance, and utilization of carbohydrates in root exudates are essential for better colonization.
- Exo-polysaccharide production by bacteria plays an important role in bacteria–plant associations.
- Fimbriae-producing bacteria aid the adhesion of bacterial cells to plant roots.
- The ability to tolerate dry soil and low osmotic potential will improve rhizosphere colonization under adverse conditions. Reports have indicated that proline-overproducing bacteria acquire more osmotolerance.
- PGPR strains should be selected based on the rhizosphere matrix potential of the crop plants. It has been reported that a rhizosphere matrix potential between −0.3 and −0.7 bars is ideal for bacterial cell growth.
- Microbes that possess cellulolytic activity will be more competent than nonproducers in terms of carbohydrate utilization in the rhizosphere.
- Commercial formulations must preserve microbial activity for long periods of time without deterioration of their shelf life.
- Repeated in vitro culturing of PGPR strains may lead to loss of field efficacy.
- The PGPR must include a broad spectrum of activity. If it suppresses only one pathogen, other predominant pathogens will render the treatment ineffective.
- When used along with other compatible bioinoculants and organic waste recyclers, a biocontrol agent may show better disease control than in singular use.

8.3.2 Nitrogen Fixers

Nitrogen fixers, or N_2-fixing organisms, are used in biofertilizers as a living fertilizer composed of microbial inoculants or groups of microorganisms, which are able to fix atmospheric nitrogen. They are grouped into free-living bacteria (*Azotobacter* and *Azospirillum*) and blue green algae and symbionts such as *Rhizobium* and *Frankia* and *Azolla*. *Rhizobium* inoculation is a well-known agronomic practice to ensure adequate nitrogen of legumes instead of nitrogen fertilizer. In root nodules the oxygen level is regulated by special hemoglobin called leghemoglobin. This globin protein is encoded by plant genes but the heme cofactor is made by the symbiotic bacteria. This is only produced when the plant is infected with *Rhizobium* (Asharfuzzaman et al. 2009). The plant root cells convert sugar to organic acids, which they supply to the bacteroids. In exchange, the plant receives amino acids rather than free ammonia. Rhizobia have been reported to significantly inhibit the growth of pathogenic fungi such as *Macrophomina phaseolina*, *Rhizoctonia* spp., *Fusarium* spp., and *Pythium* spp. in both leguminous and nonleguminous plants.

Certain microorganisms such as bacteria and blue green algae have the ability to use atmospheric nitrogen and taxi this nutrient to crop plants. Some of these N_2 fixers· like rhizobia are obligate symbionts in leguminous plants, while others colonize the root zones and fix nitrogen in a loose association with the plants. A very important bacterium in the latter category is *Azospirillum*, which was discovered by a Brazilian scientist and which made headlines in the mid-1970s. The crops that respond to *Azospirillum* inoculation in India are maize, barley, oats, sorghum, pearl millet, and forage and other crops. *Azospirillum* applications increase the grain productivity of cereals by 5–20%, millets by 30%, and fodder by over 50%. The third group includes free-living nitrogen fixers such as blue green algae and *Azotobacter*. Mycorrhizal fungi and PGPR have also been shown to have agronomical implications (Tilak et al. 2003).

Biofertilizers help to improve plant nutrition either by solubilizing these nutrients or by fixing atmospheric N_2. In the case of solubilization, several mechanisms may be involved depending on the nature of the nutrient. For example, phosphate can be released from insoluble organic forms by several microbial enzymes like phytases or nonspecific phosphatases, while inorganic phosphorus stocks are solubilized through the production of organic acids by the beneficial bacteria. Phytostimulation is the direct promotion of plant growth through the modulation of the plant's hormonal balance. Several microorganisms are capable of producing and excreting a variety of plant hormones including auxins, gibberellins, cytokines, etc. Some microbial agents produce enzymes that degrade a precursor of ethylene, thus limiting the levels of this hormone in the plant by increasing plant growth especially under stress conditions (Francis and Holsters 2010).

Soluble nitrogen, phosphorus, and potash, which are the main nutrients in vermiwash, were found to contain an enzyme cocktail of protease, amylase, urease, and phosphatase. Microbiological study of vermiwash revealed that it contains nitrogen-fixing bacteria such as *Azotobacter* spp., *Agrobacterium* spp., and

Rhizobium spp., and some PSB. A laboratory-scale trial showed effectiveness of vermiwash for cowpea plant growth (Zambare and Yadav 2008). Both biofertilization and phytostimulation are important phenomena in the context of the constant need to produce more food on fewer surfaces with the simultaneous wish to reduce chemical fertilizers. Moreover, a microorganism that possesses a combination of these growth-promoting activities and biocontrol potential offers the advantage to supply the crop in one application with both a biopesticide and a biofertilizer. In addition, better nutrition of the plant often enhances its overall resistance against pathogens and other stress factors (Bent 2006).

Nitrogen is one of the most important limiting nutrients, which could alter the development of mangrove vegetation. Inorganic nitrogen in aquatic environments is in a variety of oxidation states ranging from nitrate to ammonia (Ravikumar and Kathiresan 2007). Fixation of nitrogen in marine and terrestrial environments can be influenced by several microbial genera such as *Azospirillum*, *Azotobacter*, *Rhizobium*, *Clostridium*, and *Klebsiella* (Sahoo and Dhal 2009). Among these, *Azospirillum* are free-living, nitrogen-fixing, heterotrophic bacteria (Swedrzynska and Sawicka 2001), which can also mineralize nutrients from the soil, sequester iron, survive in harsh environmental conditions, and favor beneficial mycorrhiza–plant associations (Datta et al. 2009).

Biological N_2 fixation and mineral soil or nitrogen fertilizer are the main sources for meeting the nitrogen requirements of high-yielding soybean (Salvagiottiet et al. 2008). The mineral nutrient of crops can be supplemented with fertilizer application on soils or foliage (Mallarino et al. 2001). Fertilization with nitrogen (N), phosphorus (P), potassium (K), and other nutrients can affect yield and many physiological processes, which in turn could influence grain yield and protein concentration. Microorganisms in mangroves play an important role in the recycling of nutrients and establishment of mangrove forest through mineralization and also by the production of phytohormones (Ravikumar et al. 2002).

Legumes are self-sufficient for nitrogen requirements, derived from symbiotic nitrogen, but high-yielding crops are difficult to sustain solely on biological N_2 fixation. So soybean requires a large amount of nitrogen for seed production and hence its yield may be sensitive to nitrogen fertilization after flowering (Kinugasa et al. 2011). Nitrogen fertilizer applied during the soybean reproductive stage might increase the capacity and duration of the inorganic nitrogen utilization period while maintaining N_2 fixation (Barker and Sawyer 2005). Supplying nitrogen to the soybean plant during peak seed demand supplements existing nitrogen resources, thus preventing premature senescence and increasing seed yield (Freeborn et al. 2001). The availability of carbon energy source are to fix the nitrogen process. Nonsymbiotic nitrogen fixation is giving a great significance of agronomic. Nonsymbiotic nitrogen-fixing bacteria include *Azotobacter*, *Azospirillum*, *Bacillus*, *Pseudomonas*, *Klebsiella*, *Rhodospirillum*, *Rhodopseudomonas*, and *Xanthobacter*.

8.4 Biofertilizers

Biofertilizers or *microbial inoculants* are generally defined as preparations containing live or latent cells of efficient strains for nitrogen fixation or phosphate solubilization, or cellulytic microorganisms used for application to seed, soil, or composting areas with the objective of increasing the numbers of such microorganisms and accelerating certain microbial process to augment the extent of the availability of nutrients in a form that can assimilated by plants. In a large sense, they may be used to include all organic resources (manure) for plant growth which are rendered in an available form for plant absorption through microorganisms or plant associations or interactions (Board NIIR 2004).

Intensive production of food crops to meet the growing demand for food in the world requires increasing amounts of fertilizer. However, indiscriminate use of such agrochemicals affect soil fertility and productivity in addition to causing environmental pollution. Prior to the introduction of chemical fertilizers, agriculture production was dependent on the use of organic amendments, including chicken manure and crop residues, to maintain soil fertility and productivity for more sustainable agriculture. Therefore, organic farming has emerged as an important system that relies on ecosystem management rather than external agricultural inputs, and as a priority area in view of the growing demand for safe and healthy food and long-term sustainability (Karmakar et al. 2007). Application of biofertilizer is highly recommended to limit the use of mineral fertilizers and decrease agricultural costs, maximizing crop yield by providing the necessary nutritive elements and growth-promoting substances.

The most widely used biofertilizer for pulse crops is *Rhizobium*, which colonizes the roots of specific legumes to form tumor-like growths called *root nodules*. The *Rhizobium*-legume association can fix up to 100–200 kg of nitrogen per hectare in one crop season and can leave behind substantial nitrogen for the following crop. Adequate information on seed inoculation procedures and crop responses is available for stem-nodulating legumes such as *Sesbania rostrata*, *Aeschynomene* spp., and *Neptunia oleracea* have become popular in improving soil fertility. The nitrogen-fixing bacteria associated with such stem-nodulating legumes belong to *Azorhizobium* and fast-growing species of *Rhizobium*. The nitrogen-accumulating potential of stem-nodulating legumes under flooded conditions ranges from 41 to 200 kg N/ha.

Biofertilizer or microbial inoculants can be generally defined as preparations containing live or latent cells of potential strains of nitrogen-fixing and PSM used for treatment of seed or soil. They are organic products containing living cells of different types of microorganism, which have the ability to convert nutritionally important elements from unavailable to available forms through biological processes (Vessey 2003). They have emerged as important components of the integrated nutrient supply system and hold great promise to improve crop yield through environmentally better nutrient supplies (Abdel-Hafez and Saber 1993). Buragohain (2000) found that sugarcane yield was significantly higher in crops that were cultivated with *Azotobacter* than in crops that were not. Adaptation is due to microelements and plant growth regulators contained in the fertilizer.

The use of biofertilizer is nowadays known to confer several benefits on soil solubilization of essential minerals, procurement of nutrients, offering micronutrients in more utilizable forms for plants, and taking part in BNF. Microorganisms of this group are generally known as plant growth–promoting microorganisms (PGPM), which include *Azospirillum*, *Azotobacter*, phosphobacteria, rhizobia, and cyanobacteria. PGPM are capable of putting forth advantageous properties to enhance growth and yield characteristics of several cultivable crops in different parts of the world (Abo-Baker and Mostafa 2011). In the laboratory, the biofertilizers are mass multiplied on a large scale using a traditional culture medium and/or a chief source supplemented medium for agricultural purposes (Fahmi et al. 2011; Singh and Pant 2011). Recently, Rajasekar and Karmegam (2010) reported that vermicasts are able to increase the survival rate of biofertilizer organisms for more than a year when used as carrier material.

8.5 *Rhizobium*

Inoculation with *Rhizobium* spp. causes greater increases in plant growth and yield in comparison with plants without *Rhizobium* spp. under field conditions (Akhtar and Siddiqui 2009). In addition to their beneficial N_2-fixing activity with legumes, rhizobia can improved phosphate nutrition by mobilizing inorganic and organic phosphorus. Many rhizobia isolates from different cross-inoculation groups of rhizobia, isolated from soils in Iran, are able to mobilize phosphorus from organic and inorganic sources (Alikhani et al. 2006). Conjunctive use of *Rhizobium* with PSB released synergistic effects on symbiotic parameters and grain yield of mungbean. PSB improve the competitive ability and symbiotic effectiveness of inoculated *Rhizobium* spp. in lentil under field conditions (Kumar and Chandra 2008). Data recorded from a tillage versus no-tillage experiment revealed more nodulation and leghemoglobin content in the no-tillage treatment (Sharma and Singh 2007). A single dual inoculation of *Rhizobium* and PSB with fertilizer (P_2O_5) significantly increases root and shoot weight, plant height, spike length, and grain yield. Seed phosphorus content, leaf protein, and leaf sugar content of the wheat crop in a phosphorus-deficient natural nonsterilized sandy loam soil were 30–40% better than only phosphate fertilizer for improving grain yield (Afzal and Bano 2008). Phosphorus-solubilizing strains have great potential to be formulated and used as biofertilizers (Cakmakci et al. 2007).

Many researchers have reported that inoculation of plants with *Azospirillum*, *Azotobacter*, *Rhizobium*, and *Pseudomonas* singly, in dual combinations, or in different combinations with organic and mineral fertilizers increased the growth parameters, the yield and its components, and chemical constituents in treated plants. The best results were obtained by various mixture inoculations in which the amount of mineral fertilizers used on wheat plants was reduced.

Rhizobia are soil-inhabiting bacteria with the potential for forming specific root structures called *nodules*. In effective nodules, the bacteria fix nitrogen gas (N_2) from the atmosphere into ammonia, which is assimilated by the plant and supports growth particularly in nutrient deficient soils. Nutrients, predominantly decarboxylic acids, are protected inside the nodule structure. In infective nodules, no nitrogen is fixed, yet *Rhizobium* is still supplied with nutrients and in this situation the rhizobia could be considered parasitic.

Rhizobium and *Azotobacter* are important PGPM used as biofertilizer. They fix atmospheric dinitrogen under free-living conditions and promote plant growth activities such as phosphate solubilization and production of plant growth hormones such as auxins, gibberellins, cytokines, vitamins, and amino acids (Denison and Kiers 2004).

8.6 Plant Growth–Promoting Bacteria

PGPR play an important role in phytostimulation, phytoremediation, and biofertilization. The prime beneficial traits of PGPR include production of phytohormones, fixation of atmospheric nitrogen, mineral phosphate solubilization, and antibiotic resistance. They are also known to provide protection to plants against disease by suppressing deleterious and pathogenic microorganisms. PGPR have a significant role in plant growth and development in two different ways: indirectly or directly. The indirect promotion of plant growth occurs when bacteria decrease or prevent some of the deleterious effects of a phytopathogenic organism by one or more mechanisms.

The direct promotion of plant growth by PGPR generally entails providing the plant with a compound that is synthesized by the bacterium or facilitating the uptake of nutrients from the environment. Plant growth benefits due to the addition of PGPR include increases in germination rates, root growth, shoot and root weights, grain yield (Polynskaya et al. 2000), chlorophyll content (Singh and Jaiswal 2003), tolerance of drought and salt stress, and delayed leaf senescence (Lucy et al. 2004).

Pseudomonas makes up a dominant population in the soil and the rhizosphere, and exerts a growth-promoting influence on a variety of plant species on account of their strong competitive behavior, colonization potential, and sustainability. Seed bacterization with these organisms has emerged as a powerful technology to enhance plant growth and yield besides providing protection against diseases. Krishna et al. (2003) reported that seed bacterization with PGPR and phylloplane bacteria promoted growth and increased yield in groundnut.

The worldwide interest in this group of rhizobacteria was sparked by studies reporting that strains of *Pseudomonas fluorescens* and *Pseudomonas putida* applied to seed tubers improved growth of potato. These findings were confirmed and later exemplified in radish, sugar beet, cotton, vegetables, groundnut, lentil, and tomato (Earnapalli 2005). Pal and Chauhan (2003) reported that four PGPR strains belonging to the fluorescent *Pseudomonas* group increased root length, shoot length, and pod yield of groundnut, which was attributed to production of siderophores and indole-3-acetic acid (IAA)–like substances. Two strains of fluorescent *Pseudomonas* isolated from potato epidermis and celery roots significantly increased growth of potato plants up to 50% more than control in greenhouse assays.

It was reported that inoculation of tuber pieces with isolates of *Pseudomonas fluorescens* caused better growth of potato, tomato, cucumber, and lettuce. Potato plants bacterized with a plant growth–promoting *Pseudomonas* strain showed increased root and shoot fresh weight, and simultaneous suppression of deleterious pathogenic microflora was observed. Inoculation of wheat plants with *Pseudomonas fluorescens* in potted soil naturally infested with *Gaeumannomyces graminis* var. *tritici* showed a 29% higher grain yield over untreated controls. Plant growth–promoting strains of *Pseudomonas fluorescens* ANP15 and *Pseudomonas aeruginosa* 7NSK2 were found to protect maize seeds from cold stock damage and significantly increased germination of maize seeds and enhanced the dry matter content of inoculated plants. *Pseudomonas* species isolated from the rhizosphere of fir and spruce plants on the forest floor of a Kashmir valley have been reported to increase plant height, number of leaves, girth, and weight of fir and spruce plants significantly in addition to enhancing the nitrogen, phosphorus, and potassium contents of the plants (Zargar et al. 2005).

It was reported that inoculation of western hemlock (*Tsuga hecterophylla*) with plant growth–promoting *Bacillus polymyxa* strain L6-16R resulted in significant increases in seedling emergence, height, and biomass accumulation. The reported secretion of succinic and lactic acids by the PGPR strain *P. putida* stimulated root growth in *Asparagus* seedlings. Devananda (2000) reported maximum plant growth, yield, and nutrient uptake in pigeonpea with combined inoculation treatment comprising *Rhizobium*, *Azospirillum* and *Pseudomonas striata*.

Isolation and identification of rhizosphere microorganisms is important for proper utilization of their beneficial effects to increase growth of plants in general and teff in particular. In Ethiopia, only a few studies on teff root–associated microorganisms have been undertaken. Accordingly, effects of PGPR on growth and yield of teff were evaluated by, and the effects of *Azospirillum* bacterial isolates on the growth and nitrogen content of teff were studied by Zewdie et al. (2000). Studies carried out in the last century proved that soil enrichment with natural zeolites contributes to increased productivity of various agricultural crops and that simultaneously it is characterized by an after-effect: unlike mineral fertilizers, annual application is not necessary, but positive effects of these minerals on enhancing plant productivity are manifested for several years.

There has been increased interest in BNF in the context of sustainable agriculture as a result of the costs of mineral fertilizers and their possible harm to the environment. There are a wide range of microbes in the soil, which are able to act in symbiotic or nonsymbiotic associations with their host plants (Gray and Smith 2005). Soil microbes are a necessary part of the soil ecosystem and can handle the following important functions in the soil (Abbas-Zadeh et al. 2010): (1) recycling soil nutrients available in organic forms; (2) enhancing soil nutrient availability and hence uptake by plants; (3) improving soil structure by producing different biochemicals; (4) controlling the adverse effect of pathogens on plant growth; and (5) and alleviating soil stresses on plant growth and yield production.

Phosphorus (P) is second only to nitrogen as a mineral nutrient required for plant growth (Ogbo 2010). Most soils in Iran are phosphorus deficient or marginally deficient. In many countries such as Iran, a massive increase in the rate of application of chemical fertilizers has been adopted to ameliorate this deficiency. Current annual consumption of phosphate fertilizers in Iran is approximately 750,000 tons, about 250,000 of which are produced in the country and the rest is imported. A large proportion of the phosphorous content of chemical fertilizers is quickly transformed into insoluble forms such as calcium phosphate, thereby making them unavailable to plants. Moreover, there are global concerns that unbalanced use of chemical fertilizers has a role in environmental degradation and climate change (Gyaneshwar et al. 2002).

Evidence of the involvement of microorganisms in solubilization of inorganic phosphates was reported as early as 1903. Since then, extensive studies on the solubilization of mineral phosphates by microorganisms have been reviewed (Achal et al. 2007 and Aseri et al. 2009). PSM are ubiquitous, and their numbers vary from soil to soil. The populations of PSM and organic matter content of some selected arid soils in Rajasthan, India, were reported as being among the most efficient PSB and important bioinoculants due to their multiple biofertilizing activities, improving soil nutrient status, secretion of plant growth regulators, and suppression of soil-borne pathogens (Trivedi 2008).

A variety of beneficial bacteria colonize the aerial parts of rice (Yanni and Dazzo 2010), which can help in the enhancement of plant growth. Benefits from PGPR have also been reported in cereal crops, including rice (Ashrafuzzaman et al. 2009; Keyeo et al. 2011), which may be highly specific to certain plant species. The beneficial effects of plant growth–promoting bacteria have been attributed to BNF (Keyeo et al. 2011), production of phytohormones (Ashrafuzzaman et al. 2009), root development, and proliferation, resulting in more efficient uptake of water and nutrients.

Nitrogen supply is a key limiting factor in crop production. Bio-N fertilizers have greater amounts of symbiotic and nonsymbiotic bacteria that are responsible for atmospheric nitrogen fixation. The use of microbes and nitrobein as commercial nitrogen biofertilizers gave the same effects as full nitrogen application, and also one third of the recommended nitrogen was saved. A similar conclusion was attained from using commercial microbein and nitrobein (Abdalla 2005). The stimulatory effects of microbein and nitrobein biofertilizers could be attributed to activation of the growth of microflora, which might also furnish the soil with many plant growth stimulators.

Moreover, the influence of nitrogen on plant growth and development is often connected with enhancement of photosynthesis because relatively high nitrogen levels determine the formation and the functional state of the assimilation apparatus of plants. Nitrogen also enhances the production of bioactive substances such as hormones and enzymes, which control soil disease and accelerate decomposition of lignin materials in the soil. Concomitant enhancement of growth parameters would then improve crop productivity. Microorganisms in the biological system also evolve mineral phosphate–solubilizing traits to enhance sufficient phosphorous, where these microorganisms (mainly *Pseudomonas*, *Bacillus*, and mycorrhizal species) are found in all soils (Zarabi and Akbari 2011).

Increased soil fertility and plant growth parameters were recorded with the use of commercial phosphorein in normal conditions (Mostafa and Abo-Baker 2010) and under water stress (Zarabi and Akbari 2011), Phosphorous is involved in cell division and development, photosynthesis, breakdown of sugar, energy transfer, nutrient transfer within the plant, and cell signal transduction, so supply of this element to the plant is essential for achieving optimum growth and crop yield. At present, there is also considerable interest in the potassium-mobilizing symbiotic bacterium *Frateuria aurantia*. Biofertilizer formulations containing this microorganism enable potash mobilization from the soil to the plant. It is recommended to be used with many crops including peanut and sunflower. Overall, the effects of these problems require more concentration on greater access to inexpensive biofertilizer technologies, as they are ecologically sound and their application could help to minimize global warming as well as reducing fertilizer input in farming practices.

If a BNF system could be assembled in nonleguminous plants, it could increase the potential for nitrogen supply because fixed nitrogen would be available to the plants directly, with little or no loss. Thus a significant reduction in the relative use of nitrogen fertilizer can be achieved if atmospheric nitrogen is made available to nonleguminous plants directly through an effective associate system with some of the characteristics of legume symbiosis. Since the rediscovery of *Azospirillum*, *Azospirillum* has gained the reputation of being the most studied plant-associated bacterium as it fixes atmospheric nitrogen and produces phytohormones.

8.6.1 Rhizobacteria as Biocontrol Agents

Soilborne pathogens are to improve the plant growth and increase crop yield. It protects the plants against various diseases and pests. Use of PGPR to induce ISR in plant crops and against different pathogens will be experimentally demonstrated under field conditions. PGPR increase plant resistance to bacterial, viral, and fungal diseases, insects, and nematodes. The plant root surface and surrounding area provide a rich carbon source. PGPR generally contact the root surface by flagella and chemotactic activity (De Weert et al. 2002); it varies from plant to plant. Microorganisms could play an important role in resolving threats to sustainable agriculture and ecosystem stability. The widespread use of pesticides and fungicides has led to environmental concerns and even caused pathogen resistance. Bacteria can be used as plant disease–controlling agents or biocontrol agents as an alternative to chemical pesticides. PGPR plays an important role in defense against soil pathogens through several mechanisms such as antibiosis, competition for nutrients, and induction of resistance factors.

The mechanisms used by PGPR against phytopathogens (Shilev 2013) prevent plant diseases, and a variety of antibiotics have been identified: 2,4-diacetylphloroglucinol (DAPG), oomycin A, phenazine, pyoluteorin, pyrrolnitrin, tensin, tropolone, and cyclic lipopeptides produced by *Pseudomonas* (Looper and Gross 2007) suppress soilborne plant pathogens. It's defined as, to reduction of plant disease by one (or) more organisms against plant pathogens. It occupies a very small place on the map of plant production with soil bacteria acting as antagonists to soilborne plant diseases, for example by antifungal action and directly (or) indirectly supporting plant growth (Haas and Defago 2005). PGPR that produce bacterial metabolites reduce the activity of pathogens among the rhizosphere microflora. It has been reported that *Bacillus subtilis* reduced the harmful effect of *Fusarium oxysporum*, the causative agent of tomato wilt disease. Seeds were treated with *B. subtilis* and sown in soil infested with *F. oxysporum*.

8.6.2 Biological Nitrogen-Fixing Bacteria and Plant Growth–Promoting Rhizobacteria Isolates

A number of bacterial species under the genera *Azospirillum*, *Alcaligenes*, *Arthrobacter*, *Acinetobacter*, *Bacillus*, *Erwinia*, *Pseudomonas*, and *Rhizobium*, and *Flavobacterium serratia* are plant-associated rhizosphere bacteria and are able to beneficially affect plant growth (Tilak et al. 2005).

A variety of bacteria help to more benefit from root surfaces, hence its, interact with the plant and the microorganisms. In this formula, can be done for a long-term sustainable agricultural system. PGPR are commonly used as inoculants to improve the growth of plants and increase the yield of agricultural crops, and provide an alternative to replace chemical fertilizers, pesticides, and supplements (Ashrafuzzuman et al. 2009).

8.6.2.1 *Bacillus* species

Bacillus subtilis more stable contact with higher plants and promotes plant growth. It's most abundant genus, rhizosphere soil, this strain was identified as, many years ago and also more metabolites are released by these strains (Charest et al. 2005). *Bacillus* species are commonly used for biofertilizers and directly improve plant growth through synthesis of hormones. In a micropropagation system, bacterial inoculated, at the beginning stage, it will be observed from the soil microorganisms. In tomato and pepper the best results are obtained from inoculation with *Bacillus licheniformis*, which forms a colonization and can be used for biofertilizer without normal greenhouses (Garcia et al. 2004). *Bacillus magisterium* is very good for root parameters, because it improves root length and dry soil also for mint.

8.6.2.2 *Pseudomonas* species

Pseudomonas species are widely used bacteria in agricultural soils and play an important role in PGPR. A variety of effective strains of *Pseudomonas* have been researched around the world. Strains of *Pseudomonas* increase solubilized phosphorous availability to plants. Pseudomonas fluorescens stimulated the shoot and root length and dry weight of wheat plants. The ability of each *Pseudomonas* isolate to colonize roots was screened. Chickpea roots inoculated previously with *Pseudomonas* isolates were collected 1 month after sowing. A 1 g sample of roots was surface sterilized and crushed in sterile normal saline solution, and 0.1 mL serially diluted extracts were placed on nutrient agar plates and incubated at 37 °C for 24 h. Bacterial colony counts were done on a colony counter and plates containing 30–300 colonies per petri dish were selected for calculation of colony-forming units per gram of root.

8.6.2.3 *Azotobacter*

Azotobacter is a free-living aerobic bacteria. *Azotobacter paspali* was described by Rajasekar and Karmegam (2010). *Azotobacter* and *Azospirillum* improve yield in plants. They are important for germination of seeds and seed growth.

8.7 Production of Phytohormones by Plant Growth–Promoting Rhizobacteria

Hormones are effective at very low concentrations. They are synthesized in one part of the plant and transported to other locations. They interact with specific tissues to cause physiological responses, such as ripening of fruit. Because hormones are able to stimulate or inhibit plant growth, many hormones increase plant growth (shoot length or root length), such as auxins, gibberellins, ethylene, cytokines, abscisic acid (ABA), etc. PGPR is most effective for acetic acid, cytokines, and gibberellins. Azospirillum secretes auxins quantitatively. Large amounts of IAA are produced in stems, young leaves, and seed from transamination and decarboxylation reactions of tryptophan. Tryptophan is the most important precursor for biosynthesis of IAA (Etesami et al. 2009). Direct plant growth promotion can be summarized in three topics: biofertilization, growth and development regulation, and stress abatement. Some mineral nutrients, including nitrogen, phosphorus, and iron, are frequently limited in soil, and this consequently inhibits the growth of land plants. Plant-associated microorganisms can act as biofertilizers by fixing and/or solubilizing mineral nutrients that are unavailable to plants. Among those processes, biological N_2 fixation by rhizobia is well known. Nodulated leguminous plants incorporate carbon and nitrogen into soil, which besides increasing nutrient uptake capacity also improves their tolerance of environmental stresses (Vessey 2003). Moreover, rhizobia have been shown to be a potential tool for remediation of organic and metal contamination by degrading organic contaminants and adsorbing, accumulating, and detoxifying them (Teng et al. 2015).

As 1-aminocyclopropane-1-carboxylic acid (ACC) is the immediate precursor for ethylene, lowering the level of ACC in the plant also lowers the amount of ethylene that can be produced. Indirect mechanisms of plant growth promotion can be summarized as inhibition of the growth and activity of plant pathogens. This inhibition can be induced by various mechanisms including competition for space and nutrients, production of biocontrol agents such as antibiotics and antifungal metabolites, and/or induction of systemic resistance. Ethylene is an important phytohormone to control biological activities, and it affects plant growth and development (cell elongation, root elongation, promotion of fruit ripening) (Glick et al. 2007). High concentrations of ethylene cause defoliation and other cellular processes leading to decreased crop yield.

8.7.1 Indole-3-Acetic Acid

IAA is one member of the family of auxin phytohormones that influence cellular functions in plants. Pattern96 says to IAA synthesis plant-associated bacteria to influence plant growth. IAA stimulates plant growth by nitrogen, phosphorus, potassium, calcium, and magnesium uptake in sweet potato cultivars (Farzana and Radizah 2005). Tryptophan increases the production of IAA. IAA stimulates rapid elongation (increasing cell elongation) and long-term cell division and differentiation. Many PGPR strains act as inducers of ISR, providing resistance against plant pathogens to reduce bacterial disease (Ramamoorthy and Viswanathan 2001; Ping and Boland 2004; Ryu et al. 2004). Reported that the combination of PGPR and systemic acquired resistance (SAR) compounds were more effective to control bacterial leaf spot disease caused by *Xanthomonas campestris* pv. *vesicatoria* in tomato. Many individual bacterial compounds induce systemic resistance such as lipopolysaccharides (LPS), flagella, siderophores, cyclic lipopeptides, 2,4-DAPG, and 2,3-butanediol. Several terms have been used to describe induced resistance such as *systemic acquired resistance*, *translocated resistance*, and *plant immunization*. ISR is defined as plants' "defensive capacity" against pathogens and pests after stimulation. SAR is expressed in maximum level, the inducing organism's causes necrosis.

The production of IAA by *Pseudomonas* isolates was determined by a modified method. Bacterial isolates were grown in nutrient broth supplemented with tryptophan (5 mg/mL). Five milliliters of each bacterial culture was centrifuged at 2816 *g* for 15 min. The supernatant was collected and filtered through filter paper (0.2 mm pore size; Millipore, Cole-Parmer). Two milliliters of supernatant was mixed with two drops of *O*-phosphoric acid and 4 mL of freshly prepared Solawaski's reagent (50 mL of 35% perchloric acid, 1 mL 0.5% $FeCl_3$). IAA development caused production of a pink color, which was measured by reading its absorbance at 530 nm. The level of IAA produced was estimated using a standard IAA graph.

8.7.2 Cytokinins and Gibberellins

Several PGPR bacterial species such as *Azotobacter* spp., *Rhizobium* spp., *Pantoea agglomerans*, *Rhodospirillum rubrum*, and *Bacillus subtilis* can produce cytokines and gibberellins or both for plant growth promotion. PGPR produce a very tiny amount of cytokines compared with phytopathogens, so the promoting rhizobacteria that inhibit pathogen growth, while the effect of cytokines activity (Egamberdieva 2008). Ethylene promotes plant root initiation, inhibition, root elongation, fruit ripening, seed germination, and leaf abscission, which activate plant hormones. Higher concentrations of ethylene lead to defoliation and to reduce plant yield.

8.7.3 Antagonistic Activity

PGPR improve plant growth by preventing phytopathogens and to supporting plant growth. PGPR synthesize antifungal and antibiotic compounds, etc. There are different types of antimicrobial compounds produced by bacteria, which include hydrogen cyanide (HCN), ISR, antibiotics, aldehydes, alcohols, ketones, etc.). Bacterial antagonists suppress plant pathogens by secretion of inhibitory metabolites at low concentrations.

Antagonist	Mechanism
Bacillus spp.	Antibiotic production
Pseudomonas fluorescens PF59	Siderophore production
Pseudomonas fluorescens P.f.G32	Antibiotic and siderophore production
Streptomyces corchorusii	Antibiotic production
Bacillus spp.	Antibiotic production

Plant growth–promoting rhizobacteria antagonize plant pathogenic bacteria by secretion of metabolites and antibiotics. Huang (2013) reported that PGPR isolated from a pathogen-prevalent environment possessed better antagonistic activity. The different types of antimicrobial compounds produced by bacteria include volatiles (HCN, aldehydes, alcohols, ketones, and sulfides), nonvolatile polypeptides (DAPG and mupirocin), heterocyclic nitrogenous compounds (phenazine derivatives: pyocyanin, phenazine-1-carboxylic acid (PCA), phenazine-1-carboxamide (PCN), and hydroxylphenazine) (de Souza and Raaijmakers 2003), phenylpyrrole antibiotic (pyrrolnitrin), and lipopeptide antibiotics (iturins, bacillomycin, surfactin, and zwittermicin A). Fluorescent pseudomonads and *Bacillus* species are also active in the suppression of plant pathogenic microorganisms. These bacterial antagonists enforce suppression of plant pathogens by the secretion of the abovementioned extracellular inhibitory metabolites at low concentrations. For example, black rot caused by *X. campestris* pv. *Campestris* (*Xcc*) causes severe economic losses in all developmental stages of crucifers, but lipopeptide-producing *Bacillus* strains actively suppress *Xcc* during the late growth phase.

8.7.4 Production of Antibiotics

Antibiotic production has been studied for aspects of biocontrol but many studies have found it difficult to distinguish between antibiosis and competition. Molecular tests have been effective, because antibiotic production is easily obtained. In vitro tests have shown that *Streptomyces galbus* produced soluble antibiotics which could inhibit spore germination of *Alternaria solani*, *Aspergillus niger*, and *Helminthosporium curvularia pallescens*. The antibiosis controls the activity by secretion of molecules that kill or reduce the growth of the target pathogen (Whipps 2001).

A number of antibiotics have been isolated from fungal and bacterial strains, which inhibit synthesis of pathogen cell walls, units of ribosomes, and intimation comperes. Antibiotics such as polymyxin, curculin, and colistin are produced by *Bacillus* spp. and are active against Gram-positive and Gram-negative bacteria (Maksimov et al. 2011).

8.7.5 Production of Hydrogen Cyanide

Pseudomonas strains secrete bioactive factors to attack pathogens, e.g., enzymes, antibiotics, or HCN. *Pseudomonas fluorescens* is considered a more effective biocontrol agent against soilborne pathogens, because the formation of plant roots (Lugtenberg and Dekkers 2001). A number of bacterial and fungal species produce siderophore compounds. In maize seeds the siderophore-producing pseudomonads increase iron in stressed conditions (Sharma and Johri 2003b). Hydrogen cyanide produced by many rhizobacteria play a major role in the biological control of pathogens. HCN first inhibits electron transport, thus energy is not supplied to cells, which leads to the death of the organisms. It also inhibits proper functioning of enzymes to inhibit the action of cytochrome oxidase.

8.7.6 Production of Siderophores

Siderophore production has also been shown to be a medium composition-dependent process. It was demonstrated that fluorescence production is dependent primarily upon the concentrations of sulfate, iron, and magnesium in the medium. Siderophores play an important role in the biocontrol of some soilborne disease and iron nutrition. Numerous bacterial and fungal species have been studied to produce siderophore compounds (Carson et al. 2000). Two major types of siderophore compounds produced by microorganisms have been reported: hydroxamate and catechol compounds. Hydroxamate contains N-hydroxy ornithine as a ligand, involved in the chelation of iron. De Bellis and Ercolani (2001) studied rootlets after inoculation of cucumber and spinach seedlings with *Pseudomonas* strains. Iron is an essential element for all living organisms. Siderophores are small high-iron compounds, secreted by microorganisms such as fungi, bacteria, and grasses (Miller and Marvin 2009). Siderophores are also important for pathogenic bacteria of iron. It's the strongest binders to Fe^{3+} with enterobactin of these (Raymond et al. 2003). *Pseudomonas fluorescens* has recently been studied for use as a seed inoculant to increase yield of various crops. Siderophore production is a more important mechanism for certain PGPR (*Bradyrhizobium japonicum*, *Rhizobium leguminaosarum*, and *Sinorhizobium meliloti*) (Carson et al. 2000). Siderophores, which promote iron content of low molecular weight iron-chelating compounds, transport the elements into cells. The siderophores are implicated in both direct and indirect PGPR. Siderophores are low molecular weight extracellular compounds, with high content of ferric iron, secreted by microorganisms to take up iron from the environment. PGPR produce low molecular weight compounds called *siderophores*, which acquire ferric iron (Whipps 2001). Most pathogens need iron, which is taken up from the environment. *Pseudomonas* spp. are highly competitive for iron, so the iron reduces the pathogenicity. Environmental factors such as pH, iron levels, the presence of other trace elements, and supplies of carbon, nitrogen, and phosphorus can also modulate siderophore synthesis by a number of bacterial and fungal species to produce siderophore compounds. In maize seeds, siderophore-producing pseudomonads increase iron in stressed conditions (Sharma and Johri 2003b).

To satisfy nutritional requirements for iron, microorganisms have evolved highly specific pathways that employ low molecular weight iron chelates, termed *siderophores*. Siderophores are secreted to solubilize iron from their surrounding environments, forming a complex ferric siderophore that can move by diffusion and be returned to the cell surface (Andrews and Robinson 2003). The active transport system through the membrane begins with the recognition of the ferric siderophore by specific membrane receptors of Gram-negative and Gram-positive bacteria (Boukhalfa and Crumbliss 2002). Siderophores can chelate ferric ion with high affinity, allowing its solubilization and extraction from most mineral or organic complexes (Wandersman and Delepelaire 2004). In aerobic conditions at physiological pH, the reduced ferrous (Fe^{2+}) form is unstable and is readily oxidized to the oxidized ferric (Fe^{3+}) form, which normally occurs as a poorly soluble iron hydroxide basically unavailable to biological systems (Krewulak and Vogel 2008; Osorio et al. 2008). Siderophores can be defined as small-peptide molecules containing side chains and functional groups that can provide a high-affinity set of ligands to coordinate ferric ions (Crosa and Walsh 2002). Based on their iron-coordinating functional groups, structural features, and types of ligands, bacterial siderophores have been classified into four main classes (carboxylates, hydroxamates, phenol catecholates, and pyoverdines) (Crowley 2006). Hundreds of siderophores have been identified and reported for cultivable microorganisms, some of which are widely recognized and used by different microorganisms, while others are species specific (Crowley 2006; Sandy and Butler 2009). In soil, siderophore production activity plays a central role in determining the ability of different microorganisms to improve plant development. Microbial siderophores enhance iron uptake by plants that are able to recognize the bacterial ferric siderophore complex (Masalha et al. 2000; Katiyar and Goel 2004; Dimkpa et al. 2009) and are also important in iron uptake by plants in the presence of other metals such as nickel and cadmium (Dimkpa et al. 2008). However, it is still unclear if bacterial siderophore complexes can significantly contribute to the iron requirements of the plant.

Siderophore production confers competitive advantages to PGPR that can colonize roots and exclude other microorganisms from this ecological niche (Haas and Defago 2005). Under highly competitive conditions, the ability to acquire iron via siderophores may determine the outcome of competition for different carbon sources that are available as a result of root exudation or rhizome position (Crowley 2006). Among most of the bacterial siderophores studied, those produced by pseudomonads are known for their high affinity to ferric ions. The potent siderophore pyoverdin, for example, can inhibit the growth of bacteria and fungi that present less potent siderophores in iron-depleted media. In vitro a pseudobactin siderophore produced by the *P. putida* B10 strain was also able to suppress *F. oxysporum* in soil deficient in iron; this suppression was lost when the soil was replenished with iron, a condition that represses the production of iron chelates by microorganisms. Recent studies have demonstrated suppression of soilborne fungal pathogens through release of iron-chelating siderophores by fluorescent pseudomonads, rendering it unavailable to other organisms (Dwivedi and Johri 2003).

8.8 Phosphate Solubilization

Phosphorus is also one of the major nutrients for nitrogen-requiring plants. Mostly present in soil in the form of insoluble phosphate (Pradhan and Sukla 2006), it can be utilized by plants. Microorganisms convert the insoluble form of phosphorus into a soluble form of phosphorus in plants. The rhizosphere phosphate-utilizing bacteria provide more phosphorus sources as plant growth–promoting agents in agriculture. PSB increase the uptake of phosphorus content by plants (Chen et al. 2006). PSMs are an alternative biotechnological solution in sustainable agricultural plants or crops. The most important PSMs belong to the genera *Bacillus*, *Rhizobium*, and

Pseudomonas. In rhizobia, two species nodulating chickpea, *Mesorhizobium cicero* and *Mesorhizobium mediterraneum*, are known as good phosphate solubilizers (Rivas et al. 2006).

Phosphorus is second only to nitrogen as a mineral nutrient required for plant growth (Ogbo 2010). Most soils in Iran are phosphorus deficient or marginally deficient. In many countries such as Iran, a massive increase in the rate of application of chemical fertilizers has been adopted to ameliorate this deficiency. Current annual consumption of phosphate fertilizers in Iran is approximately 750,000 tons, about 250,000 of which are produced in the country and the rest are imported. A large proportion of the phosphorous content of chemical fertilizers is quickly transformed into insoluble forms such as calcium phosphate, thereby making them unavailable to plants. Moreover, there are global concerns that the unbalanced use of chemical fertilizers has a role in environmental degradation and climate change (Gyaneshwar et al. 2002).

Evidence of the involvement of microorganisms in solubilization of inorganic phosphates was reported as early as 1903. Since then, extensive studies on the solubilization of mineral phosphates by microorganisms have been reviewed (Achal et al. 2007; Aseri et al. 2009). PSM are ubiquitous, and their numbers vary from soil to soil. Populations of PSM in organic matter content of some selected arid soils in Rajasthan, India, have been reported to be among the most efficient phosphate-solubilizing bacteria and important bioinoculants due to their multiple biofertilizing activities of improving soil nutrient status, secretion of plant growth regulators, and suppression of soilborne pathogens (Trivedi 2008).

A variety of beneficial bacteria colonize the aerial parts of rice (Yanni and Dazzo 2010), which can help in the enhancement of plant growth. Benefits from PGPR have also been reported in cereal crops, including rice (Ashrafuzzaman et al. 2009; Keyeo et al. 2011), which may be highly specific to certain plant species. The beneficial effects of plant growth–promoting bacteria have been attributed to BNF (Keyeo et al. 2011), production of phytohormones (Ashrafuzzaman et al. 2009), and root development and proliferation, resulting in more efficient uptake of water and nutrients.

Nitrogen supply is a key limiting factor in crop production. Bio-N fertilizers have greater amounts of symbiotic and nonsymbiotic bacteria, which are responsible for atmospheric nitrogen fixation. It has been suggested that the use of microbes in commercial nitrogen biofertilizers gave the same effects as full nitrogen application and also one third of the recommended nitrogen was saved. A similar conclusion was attained from using commercial microbein and nitrobein (Abdalla 2005). The stimulatory effects of microbes in nitrogen biofertilizers could be attributed to the activation of the growth of microflora, which might also furnish the soil with many plant growth stimulators.

Moreover, the influence of nitrogen on plant growth and development is often connected with enhancement of photosynthesis because relatively high nitrogen levels determine the formation and the functional state of the assimilation apparatus of plants. Nitrogen also enhances the production of bioactive substances such as hormones and enzymes, which control soil disease and accelerate decomposition of lignin materials in the soil. Concomitant enhancement of growth parameters would then improve crop productivity. Microorganisms in the biological system also evolve mineral phosphate–solubilizing trait(s) to enhance sufficient phosphorous, where these microorganisms (mainly *Pseudomonas*, *Bacillus* and mycorrhizal species) are found in all soils (Zarabi and Akbari 2011).

Increased soil fertility and plant growth parameters were recorded with the use of commercial phosphorein in normal conditions (Mostafa and Abo-Baker 2010) and under water stress (Zarabi and Akbari 2011), Phosphorous is involved in cell division and development, photosynthesis, breakdown of sugar, energy transfer, nutrient transfer within the plant and cell signal transduction, so supply of this element to the plant is essential for achieving optimum growth and crop yield. At present, there is also considerable interest in the potassium-mobilizing symbiotic bacterium *F. aurantia*. Biofertilizer formulations containing this microorganism enable potash mobilization from the soil to the plant. It is recommended to be used with many crops including peanut and sunflower. Overall, the effects of these problems require more concentration on greater access to inexpensive biofertilizer technologies, as they are ecologically sound and their application could help to minimize the global warming as well as reducing fertilizer input in farming practices. If a BNF system could be assembled in nonleguminous plants, it could increase the potential for nitrogen supply because fixed nitrogen would be available to the plants directly, with little or no loss. Thus a significant reduction in the relative use of nitrogen fertilizer can be achieved if atmospheric nitrogen is made available to nonleguminous plants directly through an effective associate system with some of the characteristics of legume symbiosis. Since the rediscovery of *Azospirillum* by *Azospirillum* has gained the reputation of being the most studied plant-associated bacterium as it fixes atmospheric nitrogen and produces phytohormones.

8.9 Induced Systemic Resistance

PGPR produce resistance in plants against fungal, bacterial, and viral disease. Several studies have been carried out on ISR produced by PGPR in plants. Induction of systemic resistance by PGPR against viral diseases have been reported in tobacco and cucumber plants. The system protection against inoculation with tobacco necrosis virus in, tobacco ISR against different pathogens in different crop plants. Strains of *P. fluorescens* protected radish from the fungal root pathogen *F. oxysporum* f. sp. *raphani*, and also against a virulent bacterial leaf pathogen, *Pseudomonas syringae*. The wide spectrum of PGPR-mediated ISR is more rewarding than a narrow spectrum of disease protection (Egamberdieva 2008).

Yet another possible mechanism for biological control of plant pathogens is the use of bacterial metabolites that increase a plant's resistance to pathogens by ISR. Resistance that is elicited in plants by application of chemicals or necrosis-producing pathogens is called *systemic acquired resistance* (SAR). Accordingly, ISR is elicited by rhizobacteria or other nonpathogenic microorganisms, and SAR is elicited by pathogens or chemical compounds. Several PGPR strains can act as inducers of ISR; in fact, PGPR-mediated ISR may be an alternative to chemical inducers. Plants treated with PGPR showed systemic resistance against a broad spectrum of plant pathogens to reduce the incidence of bacterial disease (Ramamoorthy and Viswanathan 2001; Ping and Boland 2004; Ryu et al. 2004a, b). The expression of ISR is dependent upon the combination of the host plant and the bacterial strain (Kilic-Ekici and Yuen 2004). Most reports of PGPR-mediated ISR involve free-living rhizobacteria strains, but ISR activity has also been observed in entophytic bacteria. Volatile organic compounds may be key in this process (Ping and Boland 2004; Ryu et al. 2004)—for example, volatiles released by *B. subtilis* GBO3 and *Bacillus amyloliquefaciens* 937A were able to activate an ISR pathway in *Arabidopsis* seedlings challenged with the soft rot pathogen *Erwinia carotovora* ssp. *carotovora*. Jones et al. (2005) reported that the combined effect of PGPRs and SAR compounds was effective in controlling bacterial leaf spot disease caused by *X. campestris* pv. *vesicatoria* in tomato.

A large number of defense enzymes have been reported to be associated with ISR. These include ascorbate peroxidase (APX), β1,3-glucanase, catalase (CAT), chitins, lipoxygenase (LOX), peroxidase (PO), phenylalanine ammonia lyase (PAL), polyphenol oxidase (PPO), proteinase inhibitors, and superoxide dismutase (SOD). These enzymes also bring about liberation of the molecules that elicit the initial steps in the induction of resistance by PGPR. This action against a broad spectrum of pathogens, including fungi and bacteria, has been noticed in a number of crops including *Arabidopsis*, brinjal, chilli (Ramamoorthy and Samiyappan 2001; Bharathi et al. 2004), carnation, cucumber, mango (Vivekananthan et al. 2004), potato, radish, rice (Nandakumar et al. 2001a), sugarcane, and tomato.

8.10 Accumulation of Cell Wall Components

PGPR inoculation of plants gives strengthening to the cell walls and also is responsible for host physiology and metabolic activity. This leads to synthesis of chemicals upon challenge by pathogens (Ramamoorthy and Viswanathan 2001; Nowak and Shulaev 2003). In PGPR-inoculated plants, the cell wall will be thickened due to deposition of callose and accumulation of phenolic compounds.

8.10.1 Production of Signaling Compounds

Salicylic acid (SA), jasmonic acid (JA), and ethylene (ET) are three important compounds in plant systems. PGPR interact to regulate plants against pathogens. This signaling compound is important for SAR and ISR.

Plant name	Causal agent	Bacteria/actinobacteria	Disease name
Apple	*Erwinia amylovora*	Bacteria	Fire blight disease
Bamboo	*Xanthomonas campestris* pv. *vasculorum*	Bacteria	Gumming disease
Blueberry	Nocardia vaccinii	Actinobacteria	Galles disease
Citrus	*Xanthomonas axonopodis*	Bacteria	Bacterial spot disease
Cotton	*Xanthomonas campestris*	Bacteria	Bacterial blight disease
	Erwinia aroideae	Bacteria	Boll rot disease
Grape	*Agrobacterium tumefaciens*	Bacteria	Crown gall disease
	Xylella fastidiosa	Bacteria	Pierce's disease
Hemp	*Pseudomonas cannabina*	Bacteria	Bacterial blight disease
Jute	*Xanthomonas campestris* pv. *nakataecorchori*	Bacteria	Leaf spot disease
Maize	*Corynebacterium nebraskense*	Bacteria	Bacterial wilt disease
	Erwinia carotovora	Bacteria	Bacterial stalk rot disease
Peanut	*Pseudomonas solanacearum*	Bacteria	Bacterial wilt disease
Pear	*Erwinia amylovora*	Bacteria	Fire blight disease
Potato	*Dickeya dadantii*	Bacteria	Soft rot disease
	Pectobacterium carotovorum	Bacteria	Bacterial soft rot
Rice	*Xanthomonas campestris* pv. *oryzae*	Bacteria	Bacterial blight disease
	Xanthomonas oryzae pv. *oryzicola*	Bacteria	Bacterial leaf stark disease
Sugarcane	*Xanthomonas albilineans*	Bacteria	Leaf scald disease

8.10.2 Traits of Efficient Biocontrol Plant Growth–Promoting Rhizobacteria

Bacterial pathogen, suppressive soil help in the selection of efficient strains of ISR with more ability to mediate in plants, from organisms have multiple antagonistic mechanisms (antibiotics, siderophores, bacteria, PSB, HCN, etc.) These organisms have good ability to colonize plant roots with sufficient population levels and have more ecological competence and ability to survive in nature. They have been shown to have good rhizosphere competence in crop plants. Motile bacteria are more efficient than nonmotile bacteria; however in-depth research is needed.

8.11 Conclusion

PGPR can affect plant growth by different mechanisms, such as their ability to produce various compounds (such as phytohormones, organic acids, and siderophores), fix atmospheric nitrogen, solubilize phosphate, produce antibiotics that suppress deleterious rhizobacteria, and produce biologically active substances or plant growth regulators (PGRs). Significant growth has been achieved in the area of PGPR production. PGPR are potential microbes for enriching soil fertility and increasing agricultural yield. Plant growth promotion is a complex phenomenon. Most PGPR influence plant growth through multiple mechanisms, and in some cases their effect may only occur through interactions with other microbes. Resistance-inducing and antagonistic rhizobacteria might be useful in formulating new inoculants, thus offering an attractive alternative of environmentally friendly biological control of plant disease and improving the cropping systems in which it can be most profitably applied. It requires a systematic approach and design, so as to utilize valuable and useful bacterial properties; even crop yields can be maintained by using combinations of different mechanisms of action and reducing the use of chemical fertilizers. Current and future progress in understanding PGPR diversity, ability to colonize, mechanisms of action, formulation, and application could aid in their development for sustainable agriculture. PGPR are excellent model systems which can provide the biotechnologist with novel genetic constituents and bioactive chemicals with diverse uses in agriculture and environmental sustainability.

References

Abbas-Zadeh P, Saleh-Rastin N, Asadi-Rahmani H et al (2010) Plant growth-promoting activities of fluorescent pseudomonads, isolated from the Iranian soils. Acta Physiol Plant 32(2):281–288

Abdalla AA (2005) Maize yield potentiality response to bio and mineral nitrogen fertilizers under drip irrigation regimes in the newly reclaimed soil. J Agric Sci Mansoura Univ 30:5765–5779

Abdel-Hafez AII, Saber SM (1993) Mycoflora and mycotoxin of hazelnut (Corylus avellana L.) and walnut (Juglans regia L.) seeds in Egypt. Zentralbl Mikrobiol 148(2):137–147

Abo-Baker AA, Mostafa GG (2011) Effect of bio-and chemical fertilizers on growth, sepals yield and chemical composition of Hibiscus sabdariffa at new reclaimed soil of south valley area. Asia J Crop Sci

Achal V, Savant VV, Sudhakara Reddy M (2007) Phosphate solubilization by wide type strain and UV-induced mutants of *Aspergillus tubingensis*. Soil Biol Biochem 39(2):695–699

Afzal A, Bano A (2008) Rhizobium and phosphate solubilizing bacteria improve the yield and phosphorus uptake in wheat (*Triticum aestivum*). Int J Agric Biol 10(1):85–88

Akhtar M, Siddiqui Z (2009) Effects of phosphate solubilizing microorganisms and Rhizobium sp. on the growth, nodulation, yield and root-rot disease complex of chickpea under field condition. Afr J Biotechnol 8(15)

Alikhani HA, Saleh-Rastin N, Antoun H (2006) Phosphate solubilisation activity of rhizobia native to Iranian soils. Plant Soil 287(1–2):35–41

Al-Masalha S (2004) A Common taxonomy for modeling construction operations (Doctoral dissertation)
Andrews SC, Robinson AK (2003) Rodríguez-Quiñones F. Bacterial iron homeostasis. FEMS Microbiol Rev 27:215–237
Aseri GK, Jain N, Meghwal PR (2009) Influence of bio-fertilizers on aonla establishment and production in Indian Thar desert. Indian J Hortic 66(4):449–455
Asharfuzzaman M, Hossen FA, Razi Ismail M et al (2009) Efficiency of plant growth promoting rhizobacteria (PGPR) for the enhancement of rice growth. Afr J Biotechnol 8(7):1247–1252
Ashrafi V, Seiedi MN (2011) Influence of different plant densities and plant growth promoting rhizobacteria (PGPR) on yield attributes of corn (*Zea mays* L). Recent Res Sci Technol 3(1):63–66
Barker DW, Sawyer JE (2005) Nitrogen application to soybean at early reproductive development. Agron J 97(2):615–619
Barriuso J, Solano BR, Gutierrez Manero FJ (2008) Protection against pathogen and salt stress by four plant growth-promoting rhizobacteria isolated from Pinus sp. on Arabidopsis thaliana. Phytopathology 98(6):666–672
Bent E (2006) Induced systemic resistance mediated by plant growth-promoting rhizobacteria (PGPR) and fungi PGPF. In: Tuzun S, Bent E (eds) Multigenic and induced resistance in plants. Springer, Berlin, pp 225–258
Bharathi R, Vivekananthan R, Harish S et al (2004) Rhizobacteria-based bio-formulations for the management of fruit rot infection in chillies. Crop Prot 23(9):835–843
Board NIIR (2004) The complete technology book on bio-fertilizer and organic farming National Institute of Industrial Re
Boukhalfa H, Crumbliss AL (2002) Chemical aspects of siderophore mediated iron transport. BioMetals 15:325–339
Brahmaprakash GP, Sahu PK (2012) Biofertilizer for sustainability. J Indian Inst Sci 92(1):376–362
Buragohain SK (2000) Efficacy of biofertilizers with varying levels of N on growth, yield and quality of sugarcane (Saccharum officinarum L.). Indian Sugar 49(10):847–850
Cakmakci R, Dönmez MF, Ertürk Y et al (2007) The influence of plant growth promoting rhizobacteria on growth and enzyme activities in wheat and spinach plants. J Plant Nutr Soil Sci 170:288–295
Carson KC, Meyer JM, Dilworth MJ (2000) Hydroxamate siderophore of root nodule bacteria. Soil Biol Biochem 32:11–21
Charest MH, Beauchamp CJ, Antoun H (2005) Effects of the humic substances of de-inking paper sludge on the antagonism between two compost bacteria and Pythium ultimum. FEMS Microbiol Ecol 52(2):219–227
Chen YP, Rekha PD, Arun AB (2006) Phosphate solubilizing bacteria from subtropical soil and their tricalcium phosphate solubilizing abilities. Appl soil Ecol 34(1):33–41
Choudhry A, Kennedy IR (2004) Prospects and potentials for system of biological nitrogen fixation in sustainable rice production. Biol Fertil Soil 39:219–227
Crosa JH, Walsh CT (2002) Genetics and assembly line enzymology of siderophore biosynthesis in bacteria. Microbiol Mol Biol Rev 66:223–249
Crowley DE (2006) Microbial siderophores in the plant rhizospheric. In: Barton LL, Abadía J (eds) Iron nutrition in plants and rhizospheric microorganisms. Springer, Dordrecht, pp 169–198
Datta JK, Banerjee A, Sikdar MS et al (2009) Impact of combined exposure of chemical, fertilizer, biofertilizer and compost on growth, physiology and productivity of *Brassica campestris* in old alluvial soil. J Environ Biol 30:797–800
De Bellis P, Ercolani GL (2001) Grown interactions during bacterial colonization of seedling rootlets. Appl Environ Microbiol 67:1975–1948
Denison RF, Kiers ET (2004) Lifestyle alternatives for rhizobia: mutualism, parasitism, and forgoing symbiosis. FEMS Microbiol Lett 237(2):187–193
De Souza JT, Raaijmakers JM (2003) Polymorphisms within the prnD and pltC genes from pyrrolnitrin and pyoluteorin-producing Pseudomonas and Burkholderia spp. FEMS Microbiol Ecol 43(1):21–34

De Weert S et al (2002) Flagella-driven chemotaxis toward exudates components is an important trait for tomato root colonization by *Pseudomonas fluorescens*. Mol Plant Microbe Interact 15:1173–1180
Devananada BJ (2000) Role of plant growth promoting rhizobacteria on growth and yield of pigeonpea (*Cajanus cajan* L.) cultivars. M.Sc. (Agri.) thesis, University of Agricultural Sciences, Dharwad
Dimkpa CO, Svatos A, Dabrowska P (2008) Involvement of siderophores in the reduction of metal-induced inhibition of auxin synthesis in Streptomyces spp. Chemosp 74(1):19–25
Dimkpa CO, Merten D, Svatos A (2009) Metal-induced oxidative stress impacting plant growth in contaminated soil is alleviated by microbial siderophores. Soil Biol Biochem 41(1):154–162
Dobbelaere S, Vanderleyden J, Okon Y (2003) Plant growth-promoting effects of diazotrophs in the rhizosphere. Crit Rev Plant Sci 22:107–149
Dwivedi D, Johri BN (2003) Antifungals from fluorescent pseudomonads: biosynthesis and regulation. Curr Sci 85:1693–1703
Earnapalli VN (2005) Screening of antagonistic microorganisms for biological control of early blight of tomato caused by *Alternaria solani*. M.Sc. (Agri.) thesis, University of Agricultural Sciences, Dharwad
Egamberdiyeva D (2005) Plant-growth-promoting rhizobacteria isolated from a Calcisol in a semi-arid region of Uzbekistan: biochemical characterization and effectiveness. J Plant Nutr Soil Sci 168(1):94–99
Egamberdieva D, Kucharova Z (2008) Cropping effects on microbial population and nitrogenase activity in saline arid soil. Turk J Biol 32(2):85–90
Egamberdieva D (2008) Plant growth promoting properties of rhizobacteria isolated from wheat and pea grown in loamy sand soil. Turkish J Biol 32:9–15
Etesami H, Alikhani HA, Akbari AA (2009) Evaluation of plant growth hormones production (IAA) ability by Iranian soils rhizobial strains and effects of superior strains application on wheat growth indexes. World Appl Sci J 6(11):1576–1584
Fahmi AI et al (2011) Effects of salt stress on some nitrogen fixation parameters in faba bean. Pakistan J Biol Sci 14:385–391
FAO and IFA (2000) Fertilizes and their use, 4th edn. Food and Agriculture Organization of the United Nations, Rome
Farzana Y, Radizah O (2005) Influence of rhizobacterial inoculation on growth of the sweet potato cultivar. Online J Biol Sci 1(Suppl 3):176–179
Fernandes P, Bhalerao SA (2015) Effect of biofertilizer on the growth and biochemical parameters of mungbean *Vigna radiata* (L. Wilczek). Int J Adv Res Biol Sci 2(4):127–130
Francis I, Holsters M (2010) The Gram-positive side of plant microbe interactions. Environ Microbiol 12:1–12
Freeborn JR, Holshouser DL, Alley MM et al (2001) Soybean yield response to reproductive stage soil-applied nitrogen and foliar-applied boron. Agron J 93(6):1200–1209
Garcia JAL, Probanza A, Ramos B et al (2004) Effect of inoculation of Bacillus licheniformis on tomato and pepper. Agron 24(4):169–176
Glick BR, Cheng Z, Czarny J, Duan J (2007) Promotion of plant growth by ACC deaminase-producingsoil bacteria. Eur J Plant Pathol 119:329–339
Gray EJ, Smith DL (2005) Intracellular and extracellular PGPR: commonalities and distinctions in the plant–bacterium signaling process. Soil Biol Biochem 37:395–412
Gyaneshwar P et al (2002) Role of soil microorganisms in improving P nutrition of plants. Plant Soil 245:83–93
Haas D, Defago G (2005) Biological control of soil-borne pathogens by fluorescent pseudomonads. Nat Rev Microbiol 3:307–319
Hartmann A, Rothballer M (2008) Lorenz Hiltner, a pioneer in rhizosphere microbial ecology and soil bacteriology research. Plant Soil 312:7–14
Heidari G, Mohammadi K, Khalesro S et al (2011) A review: beneficial effects of the mycorrhizal fungi for plant growth. J Appl Environ Biol Sci 1(9):310–319

Hossain I, Mohammed D (2002) Seed treatment with biofertilizer in controlling diseases of mungbean. BAU Res Prog 12:34
Houlton BZ, Wang Y-P, Vitousek PM, Field CB (2008) A unifying framework for dinitrogen fixation in the terrestrial biosphere. Nature 454:327–330
Huang JS (2013) Plant Pathogenesis and resistance: biochemistry and physiology of plant-microbe interactions. Spring Sci Busin Med
Jeon JS, Lee SS, Kim HY (2003) Plant growth promotion in soil by some inoculated microorganisms. J Microbiol 41(4):271–276
Jeyabal A, Kupuswamy G (2001) Recycling of organic wastes for the production of vermicomposting and its response in rice–legume cropping system and soil fertility. Eur J. Agron 15:153–170
Karmakar S, Lague C, Agnew J (2007) Integrated decision support system (DSS) for manure management: a review and perspective. Comput Electron Agric 57(2):190–201
Katiyar V, Goel R (2004) Siderophore mediated plant growth promotion at low temperature by mutant of fluorescent pseudomonad. Plant Growth Regul 42(3):239–244
Keyeo F, Ai'shah ON, Amir HG (2011) Diazotroph in promoting growth of rice seedlings. Biotechnology 10(3):267–273
Kilic-Ekici O, Yuen GY (2004) Comparison of strains of Lysobacter enzymogenes and PGPR for induction of resistance against Bipolaris sorokiniana in tall fescue. Biol Control 30(2):446–455
Kinugasa T et al (2011) Demand and supply of N in seed production of soybean (*Glycine max*) at different N fertilization levels after flowering. J Plant Res 125(2):275–281
Krewulak KD, Vogel HJ (2008) Structural biology of bacterial iron uptake. Biochim Biophys Acta 1778:1781–1804
Krishna KG, Pande S, Podile AR (2003) Seed bacterizaiton with PGPR and phylloplane bacteria promotes growth and increases yield in groundnut. In: 6th international PGPR workshop, 5–10 October, Calicut
Kumar R, Chandra R (2008) Influence of PGPR and PSB on *Rhizobium leguminosarum* Bv. *viciae* strain competition and symbiotic performance in lentil. World J Agri Sci 4(3):297–301
Lambers H, Mougel C (2009) Plant–microbe–soil interaction in the rhizosphere: an evolutionary perspective. Plant Soil 321:83–115
Loganathan P, Nair S (2004) *Swaminathania salitolerans* gen. nov., sp. nov., a salt-tolerant, nitrogen-fixing and phosphate-solubilizing bacterium from wild rice (*Porteresia coarctata* Tateoka). Int J Syst Evol Microbiol 54:1185–1190
Looper JE, Gross H (2007) Genomic analysis of antifungal metabolite production by Pseudomonas fluorescence Pf-5. Eur J Plant Pathol 119:265–278
Lucy M, Reed E, Glick BR (2004) Application of free living plant growth promoting rhizobacteria. Antonie Van Leeuwenhoek 86:1–25
Lugtenberg BJ, Dekkers L (2001) Molecular determinants of rhizosphere colonization by pseudomonas. Annu Rev Phytopathol 39:461–490
Maksimov IV, Abizgil'Dina RR, Pusenkova LI (2011) Plant growth promoting rhizobacteria as alternative to chemical crop protectors from pathogens. Appl Biochem Microbiol 47(4):333–345
Mallarino AP, Haq MU, Wittry D et al (2001) Variation in soybean response to early season foliar fertilization among and within fields. Agron J 93(6):1220–1226
Masalha N (2000) Imperial Israel and the Palestinians: the politics of expansion. Pluto Press, London
Masalha J, Kosegarten H, Elmaci Ö et al (2000) The central role of microbial activity for iron acquisition in maize and sunflower. Biol Fertil Soils 30:433–439
Mayak S, Tirosh T, Glick BR (2004) Plant growth-promoting bacteria confer resistance in tomato plants to salt stress. Plant Physiol Biochem 42(6):565–572
McCully M (2005) The rhizosphere: the key functional unit in plant/soil/microbial interactions in the field. Implications for the understanding of allelopathic effects. In Proceed 4th World Cong Allelo 21–26
Mia MAB, Shamsuddin ZH, Wahab Z et al (2005) High-yielding and quality banana production through plant growth-promoting rhizobacterial (PGPR) inoculation. Fruits 60(3):179–185

Miller, Marvin J (2009). Siderophores (iron chelators) an siderophore-drug conjudates (new methods for micribially) selective drug delivery. University of Notre Dame. Dame 4/21/2008
Mishra P (2014) Rejuvenation of biofertilizer for sustainable agriculture and economic development. J Sustain Dep 11(1):41–61
Mostafa GG, Abo-Baker AA (2010) Effect of bio-and chemical fertilization on growth of sunflower (Helianthus annuus L.) at South Valley area. Asian J Crop Sci 2(3):137–146
Nabti E, Sahnoune M, Adjrad S et al (2007) A halophilic and osmotolerant Azospirillum brasilense strain from Algerian soil restores wheat growth under saline conditions. Eng Life Sci 7(4):354–360
Nandakumar R, Babu S, Viswanathan R et al (2001a) A new bio-formulation containing plant growth promoting rhizobacterial mixture for the management of sheath blight and enhanced grain yield in rice. Biocontrol 46(4):493–510
Nandakumar R, Babu S, Viswanathan R et al (2001b) Induction of systemic resistance in rice against sheath blight disease by plant growth promoting rhizobacteira. Soil Biol Biochem 33:603–612
Nowak J, Shulaev V (2003) Priming for transplant stress resistance in vitro propagation. In Vitro Cell Dev Biol Plant 39:107–124
Ogbo FC (2010) Conversion of cassava wastes for biofertilizer production using phosphate solubilizing fungi. Biores Technol 101(11):4120–4124
Orhan E, Esitken A, Ercisli S (2006) Effects of plant growth promoting rhizobacteria (PGPR) on yield, growth and nutrient contents in organically growing raspberry. Sci Hortic 111(1):38–43
Osório H, Martinez V, Nieto PA (2008) Microbial iron management mechanisms in extremely acidic environments: Comparative genomics evidence for diversity and versatility. BMC Microbiol 8:203
Pal KK, Chauhan SM (2003) Application of pseudomonads for enhancing peanut growth, yield and nutrient uptake. In: 6th international PGPR workshop, 5–10 October, 2003, Calicut
Patten CL, Glick BR (1996) Bacterial biosynthesis of indole-3-acetic acid. Can J Microbiol 42(3):207–220
Paul D, Nair S (2008) Stress adaptations in a plant growth promoting rhizobacterium (PGPR) with increasing salinity in the coastal agricultural soils. J Basic Microbiol 48(5):378–384
Ping L, Boland W (2004) Signals from the underground: bacterial volatiles promote growth in Arabidopsis. Trends Plant Sci 9(6):263–266
Polyanskaya LM, Orazova MK, Burkanova OA et al (2000) Microsampling of rhizosphere soil and laboratory artifacts. Microbiology 69(4):486–489
Pradhan N, Sukla LB (2006) Solubilization of inorganic phosphates by fungi isolated from agriculture soil. Afr J Biotechnol 5(10)
Rajasekar K, Karmegam N (2010) Earthworm casts as an alternate carrier material for biofertilizer: assessment of endurance and viability of *Azotobacter chroococcum*, *Bacillus megaterium* and *Rhizobium leguminosarum*. Sci Hortic 124(2):286–289
Ramamoorthy V, Samiyappan R (2001) Induction of defense-related genes in Pseudomonas fluorescens treated chilli plants in response to infection by Colletotrichum capsici. J Mycol Plant Pathol 31:146–155
Ramamoorthy V, Viswanathan R (2001) Induction of systemic resistance by plant growth promoting rhizobacteria in crop plants against pests and diseases. Crop Prot 20(1):1–11
Ramesh P (2008) Organic farming research in M.P. Organic farming in rain fed agricultural. Central Institute for Dry Land Agriculture, Hyderabad, pp 13–17
Rao DLN (2002) Nitrogen fixation by tree legumes. In: Kanniyan S (ed) Biotechnology of biofertilizers. Narosa Publishing House, New Delhi, pp 165–178
Ravikumar S, Kathiresan K (2007) Growth of *Avicennia marina* and *Ceriops decandra* seedlings inoculated with halophilic azotobacters. J Environ Biol 28:601–603
Raymond KN, Dertz EA, Kim SS (2003) Enterobactin: archetypes for microbial iron transport. Proc Natl Acad Sci 100(7):3584–3588

Rivas M, Miliwebsky E, Chinen I (2006) Characterization and epidemiologic subtyping of Shiga toxin–producing Escherichia coli strains isolated from hemolytic uremic syndrome and diarrhea cases in Argentina. Foodborne Pathog Dis 3(1):88–96
Ravikumar S, Anburajan L, Ramanathan G (2002) Screening of seaweed extracts against antibiotic resistant post operative infectious pathogens. Seaw Res Util 24(1):95–99
Rothballer M, Schmid M, Hartmannn A (2009) diazotrophic bacterial entophytes in Gramineae and other plants. In: Pawlowski K (ed) Microbiology monographs, vol.8: Procaryotic endosymbionts in plants. Springer, Berlin, pp 273–302
Ryu CM, Farag MA, Hu CH et al (2004a) Bacterial volatiles induce systemic resistance in Arabidopsis. Plant Physiol 134(3):1017–1026
Ryu CM, Murphy JF, Mysore KS et al (2004b) Plant growth-promoting rhizobacteria systemically protect Arabidopsis thaliana against Cucumber mosaic virus by a salicylic acid and NPR1-independent and jasmonic acid-dependent signaling pathway. Plant J 39(3):381–392
Sahoo K, Dhal NK (2009) Potential microbial diversity in mangrove ecosystems: a review
Salvagiottiet F, Cassman KG, Specht JE, Walters DT (2008) Nitrogen uptake, fixation and response to fertilizer N in soybean: a review. Field Crop Res 108:1–13
Sandy M, Butler A (2009) Microbial iron acquisition: Marine and terrestrial siderophores. Chem Rev 109:4580–4595
Schuhegger R, Ihring A, Gantner S et al (2006) Induction of systemic resistance in tomato by N-acyl-Lhomoserine lactone-producing rhizosphere bacteria. Plant Cell Environ 29(5):909–918
Sharma A, Johri BN (2003a) Combat of iron-deprivation through a plant growth promoting fluorescent Pseudomonas strain GRP3A in mung bean. Microbiol Res 158:77–81
Sharma A, Johri BN (2003b) Growth promoting influence of siderophore-producing Pseudomonas strains GRP3A and PRS9 in maize (Zea mays L.) under iron limiting conditions. Microbiol Res 158(3):243–248
Sharma P, Singh G (2007) Biological nitrogen fixation in mungbean: facts and finding. ISHS Acta Hortic 752:597–601
Shilev S (2013) Soil rhizobacteria regulating the uptake of nutrients and undesirable elements by plants. In: Plant microbe symbiosis: fundamentals and advances. Springer, India, pp 147–167
Sindhu SS, Dadarwal KR (2002) Plant growth promotion effects of *Pseudomonas* sp. on coinoculation with *Mesorhizobium* sp. *Cicer* strain under sterile and "wilt sick" soil conditions. Appl Soil Ecol 19:57–64
Singh RK, Jaiswal HK (2003) Wide occurrence of *Rhizobium leguminosarum* BV *phaseoli* as rice entophyte in Indian soils. In: 6th international PGPR workshop 5–10 October, Calicut
Singh P, Cameotra SS (2004) Potential applications of microbial surfactants in biomedical sciences. Trends Biotechnol 22(3):142–146
Singh G, Kumar D, Marwaha TS et al (2011) Conservation tillage and integrated nitrogen management stimulates soil microbial properties under varying water regimes in maize-wheat cropping system in northern India. Arch Agron Soil Sci 57(5):507–521
Singh AK, Pant S (2011) Optimization and comparative study of the sugar waste for the growth of *Rhizobium* cells along with traditional laboratory media. Res J Microbiol 6(9):715–723
Sivasakthi S, Kanchana D, Usharani G et al (2013) Production of plant growth promoting substance by Pseudomonas fluorescens and Bacillus subtilis isolated from paddy rhizosphere soil of Cuddalore district, Tamil Nadu, India. Int J Microbiol Res 4(3):227–233
Swędrzyńska D, Sawicka A (2001) Effect of inoculation on population numbers of Azospirillum bacteria under winter wheat, oat and maize. Pol J Environ Stud 10:21–25
Teng Y, Wang X, Li L et al (2015) Rhizobia and their bio-partners as novel drivers for functional remediation in contaminated soils. Front Plant Sci 6
Thakuria D, Talukdar NC, Goswami C et al (2004) Characterization and screening of bacteria from rhizosphere of rice grown in acidic soils of Assam. Curr Sci 86:978–985
Tilak KVBR, Ranganayaki N, Manoharachar C (2003) Contribution of plant growth promoting rhizobacteria in crop production-Indian Scenario. In Session II Overview of PGPR work in India. 6th International PGPR Workshop Calicut India 247–248

Tilak KVBR, Ranganayaki N, Pal KK et al (2005) Diversity of plant growth and soil health supporting bacteria. Curr Sci 89:136–150

Trivedi P (2008) *Pseudomonas corrugata* (NRRL B-30409) mutants increased phosphate solubilization, organic acid production and plant growth at lower temperatures. Curr Microbio 156:140–144

van Loon LC (2007) Plant response to plant growth – promoting rhizobacteria. Eur J Plant Pathol 119:243–254

Vivekananthan R, Ravi M, Ramanathan A et al (2004) Lytic enzymes induced by Pseudomonas fluorescens and other biocontrol organisms mediate defence against the anthracnose pathogen in mango. World J Microbiol Biotechnol 20(3):235–244

Vessey JK (2003) Plant growth promoting rhizobacteria as biofertilizer. Plant Soil 255:571–586

Wandersman C, Delepelaire P (2004) Bacterial iron sources: From siderophores to hemophores. Annu Rev Microbiol 58:611–647

Whipps JM (2001) Microbial interactions and biocontrol in the rhizosphere. J Exp Bot 52(1):487–511

Yanni YG, Dazzo FB (2010) Enhancement of rice production using endophytic strains of Rhizobium leguminosarum bv. trifolii in extensive field inoculation trials within the Egypt Nile delta. Plant Soil 336(1–2):129–142

Zahir ZA, Muhammad A, Frankenberger WT (2004) Plant growth promoting rhizobacteria: applications and perspectives in agriculture. Adv Agron 81:97–168

Zambare VP, Yadav AA (2008) Vermiwash: biochemical and microbiological approach as eco-friendly soil conditioner. ARPN J Agri Biol Sci 3(4):1–5

Zargar MY, Khan MA, Parry SN et al (2005) Effect of inoculation of Pseudomonas on the growth and development of fir and spruce seedlings. Indian J Microbiol 45(4):313

Zarabi M, Akbari GA (2011) A study on the effects of different biofertilizer combinations on the yield, its components and growth indices of corn (*Zea mays L.*) under drought stress conditions. Afr J Agri Res 6(3):681–685

Zewdie S, Assefa F, Fetene M (2000) Effect of Azospirillum isolates on the growth and nitrogen content of tef [Eragrostis tef (Zucc.) Trotter]. SINET: Ethiop J Sci 23(2):247–262

Chapter 9
Seaweed: A Fertilizer for Sustainable Agriculture

Thillaigovindhan Nedumaran

Abstract Seaweeds are macroscopic marine algae, which form an important component of the marine living renewable resources. They are utilized for the production of phytochemicals such as agar, carrageenan and alginate which are widely used as gelling, stabilizing and thickening agents in many industries like food, confectionary, pharmaceutical, dairy and textile. Plants which are growing with sufficient intake of all kinds of nutrients become insufficient on frequent cultivation on land; it is necessary to feed the plants with combined nutrients to increase the growth and yield of plants, quantity and quality. Recent researches proved that seaweed fertilizers are better than other fertilizers; they are very economic and cheap. Seaweed fertilizers are preferred not only due to their nitrogen, phosphorus and potash content but also because of the presence of trace elements and metabolites similar to plant growth regulators. Seaweed extracts are utilized to enhance seed germination and plant growth; seaweed extracts have been used to increase crop yield, improve growth and induce resistance to frost fungal and insect attack and increase nutrient uptake from soil.

9.1 Introduction

Seaweeds are macroscopic algae found attached to the bottom in relatively shallow coastal waters. They grow in the intertidal shallow and deep sea areas up to 180 m depth and also in estuaries and backwaters on the solid substrate such as rocks, dead corals, pebbles, shells and other plant materials. They form one of the important living resources, which belong to three different groups, empirically distinguished since the mid-nineteenth century on the basis of thallus colour, brown algae class Phaeophyceae, red algae class Rhodophyceae and green algae class Chlorophyceae. Distinguishing these three classes, however, involves more substantial differences than colour. In addition to the pigmentation, they differ considerably in many

T. Nedumaran (✉)
Plant Biology and Plant Biotechnology, Sir Theyagaraya College (Affiliated to University of Madras), Chennai, Tamil Nadu, India
e-mail: tnmaranbot@gmail.com

A. Dhanarajan (ed.), *Sustainable Agriculture towards Food Security*,
https://doi.org/10.1007/978-981-10-6647-4_9

ultrastructural and biochemical features including photosynthetic pigments, storage compounds, composition of cell walls and presence and absence of flagella. Red and brown algae are almost exclusively marine, whilst green algae are also common in freshwater.

Seaweeds are far more complex organisms; many have specialized tissues and growth forms. Seaweed must produce some amazing adhesives as quite small holdfasts seem to be sufficient for quite large plants, and base of dynamic species counts shows that there are about 10,000 species of seaweeds of which 6500 are red algae, 2000 are brown algae and 1500 are green algae; 800 species of Bryopsidophyceae, 50 species of Cyanophyceae, 400 Siphonocladophyceae and 250 marine Ulvophyceae. The seaweeds are in a polyphyletic group. In addition, some tuft-forming blue-green algae (*Cyanobacteria*) are sometimes considered to be seaweed. "Seaweed" is a colloquial term and lacks a formal definition. Seaweed is the common name for countless species of marine plants and algae that grow in the ocean as well as in river, lakes and other water bodies. The history of Indian seaweed research is not more than 75 years. Hence, seaweeds are important renewable resource in the marine environment and have been a part of human civilization from time immemorial; reports on the uses of seaweeds have been cited as early 2500 years ago in the Chinese literature.

The long history of seaweed utilization for a variety of purposes has led to the gradual realization that some of their constituents are more superior and valuable in comparison to their counterpart on land. Seaweed synthesizes a wide range of chemical, some of which stand only as natural resource of food materials, e.g., agar, carrageen and alginate. Every year about 7.5–8 million tons of wet seaweeds are being produced along the coastal regions worldwide (Mc Hugh 2003).

9.2 Seaweed Habitats and Communities

Virtually any substratum in salt or brackish water is habitat for seaweeds; rocks, wood, sand, shell and other nonliving substrata are most typical, along with the surfaces of other algae and other living submerged plants and the shells of living molluscs. Endolithic algae are important in reef processes. Some of the remarkable seaweed habitats have been reported: the tissues of sea pens, where some *Desmarestia* gametophytes grow (Dube and Ball 1971); the beaks of parrotfish, which provide a moving reef for pioneer species such as *Polysiphonia scopulorum* and *Sphacelaria tribuloides*; a symbiotic association with a sponge (Price et al. 1984); the face and belly of the Hawaiian monk seal (Kenyon and Rice 1959); and the necks of green turtles (Tsuda 1965). Physiological ecology must account for the success of seaweeds wherever they grow, whether in the well-studied habitats such as rocky intertidal zone or the unusual habitats.

9.2.1 Zonation Pattern

Almost all marine shores experience tides although tidal amplitudes vary greatly from place to place. The pattern of high and low waters also varies from place to place, depending on the interaction between the tide waves and the standing waves caused by water slopping back and forth in the ocean basins. More important to seaweed ecology and physiology are the changes at one place. Besides the progression from neap to spring tides twice a month, the time of high and low water changes during the lunar month and often from season to season. Sea levels in the tropical pacific can also change because of El Nino/southern oscillation (ENSO) events. Diurnal tides have one high and one low per day; this is an unusual type. Occurring in parts of the Gulf of Mexico, semidiurnal tides rise and fall twice a day, with successive highs and lows more or less equal in height; this type is common along open coasts of the Atlantic Ocean.

Mixed tides occur twice a day but have clearly unequal highs and lows. Mixed tides are characteristic of Pacific and Indian Ocean coast, as well as in smaller basins such as the Caribbean Sea and the Gulf of St. Lawrence. In addition, there are storm tides, with irregular periods, usually of several days, caused by barometric-pressure changes and winds. Where the tide range is less than 1 m, such as on the Swedish west coast or the Caribbean coast of Panama, atmospheric pressure changes and onshore and offshore winds may combine to produce very irregular and unpredictable changes in water level.

9.2.2 Rocky Intertidal Zone

Some of the habitats are so frequently visited by ecologists as the rocky intertidal zone, for it offers intermittent access to a fascinating variety of organism and must be unique, however, in that it is invariably examined when most of its inhabitants who study the shore at high tide when its residents are active and operational could. This is a pity for it is the shore when underwater that is the shore in action. The term “rocky intertidal zone” may slightly mislead the reader, for shores are rarely composed exclusively of bedrock. Many have pebble-littered gullies or sand-carpeted pools, and below the low-water mark, the rock often gives way to sand or mud. The proximity of such mobile substrata greatly enhances the abrasiveness and therefore the ecological importance of waves.

Even where stable bedrock predominates, the effective substratum may not be rock at all. The midshore region is usually covered with closely packed barnacles, with little rock visible between them. Tide pools are often lined with encrusting pink and purple Corallinaceae, which may also carpet the lowermost levels of the shore. By occupying the rock so comprehensively, these organisms replace it. They become the substratum to which other organisms must attach, and yet little is known about their ecological significance as substrata. Do the propagules of other organisms settle preferentially on some crusts and shun others? The interactions between these little studied substrata and other shore dwellers may be major influences on the patterns of intertidal vegetation.

9.2.3 Coral Reef Habitats

Beneath the vast expanse of warm azure waters, tropical biotic reefs comprise spectacularly complex ecosystems on limestone bases, derived mainly from the fossilized remains of calcareous algae and coelenterate corals, and such reefs occur around the globe within the 22 °C isotherms (north and south). Reef systems have evolved an extremely high level of biological diversity, including many uniquely specialized macroalgae. The calcite ($CaCO_3$) cement produced by coralline algae consolidates calcareous (aragonitic) skeletons of coral animals and other calcifiers, along with terrigenous debris, and leads to reef formation. The non-articulated coralline algae may also form a seaweed intertidal ridge that buffers wave shock, thereby reducing erosion and destruction of the more delicate corals and softer organisms typical of reef-flat habitats. Adverse group of calcified green algae deposits the aragonite form of calcium carbonate, which is responsible for much of the sand and lagoonal sediments within the reef-flat and deeper fore-reef areas. For example, skeletal sand-sized components from some tropical Atlantic reef sediments are composed of up to 77% *Halimeda* fragments. Tropical reefs are remarkable for their development of massive structure in conjunction with high primary productivity.

9.2.4 Seaweeds in Estuaries

Estuaries occur all over the world where seawater mixes with river water. Tidal estuaries in temperature climates are dominated by fringing salt marshes on tropical shores; mangrove swamps are ecological equivalent. Tidal estuaries show a marked longitudinal gradient in the environmental factors from the sea to the river. Generally, salinity and wave action decrease, whereas silt, turbidity, light extinction and nutrient concentrations are all increased. These environmental factors show enormous variations over a tidal cycle and even more over the year. The environmental gradients are reflected in the numbers of macroalgae to be found in the estuary; these numbers are decreasing upstream. Estuaries harbour an impoverished seaweed flora shifting from a marine, epilithic assemblage rich in species to a restricted number of euryhaline species attached to silt and sand farther in land. Near the sea, the perennial zonation pattern of large fucoid algae and subtidal laminarians is evident. This pattern as on rocky shores is shaped by tolerance for desiccation at the upper borders and by biological competition and light extinction at the lower borders further inland; brown and red algae disappear leaving the green and blue-green algae in single layered, in distinct zonation pattern, restricted to the intertidal zone and consisting mainly of annual species. In non-tidal, brackish and sheltered lagoons, where turbidity is low and light penetrates deeper, the sublittoral macroalgal vegetation may be abundant, but the species diversity is always low compared with open seacoasts.

9.3 Salt Marshes and Their Algae

To the superficial observer, salt marshes may look dull and unattractive, places where one sinks ankle-deep in soft, black mud. Connoisseurs know better; for them the salt marsh offers a complex pattern of meandering tidal creeks, elevated creek banks and back marshes, with a delicate zonal vegetation pattern from the lower to the upper marsh. Typically, salt marshes have three zones of flowering plants and their associated algal communities. At the lowest level, within full reach of the tide, a sparse growth of a few species of higher plants forms the pioneer zone or low marsh. This intergrades at higher levels with the richer flora of the mature zone middle marsh. At the highest levels, the species of the mature zone are partially replaced by species from non-saline habitats, which can withstand brief and infrequent submergence in salt water; in many estuaries, salt marshes have been embanked and reclaimed or otherwise exploited by humans, yet in some of the most populous and most heavily industrialized areas of the developed world, stretches of salt marshes are among the few remaining habitats not altered by humans.

Salt marshes are to be found in temperate region around the globe. The east coast of the United States has extensive stretches of diversified salt marsh alternating with sandy beaches, protected shallow waters and barrier islands. Among the studies along this coastline are the 20 years of intensive work on the structure and functioning of Great Sippewissett salt marsh where extensive stretches of intertidal mudflats and salt marshes harbour tremendous prairies of seagrasses and along the coasts of Europe where Netherlands salt marshes once formed a major part many of them to make farmland or industrial areas or dredged them for harbours. There are no trees in temperate marshes comparable to the mangroves of tropical salt swamps, but in Europe, there are micromangroves, shrubs of *Halimione portulacoides*, with perennial wooden stems; and red algal genus *Bostrychia* grows attached to these stems just as it does to the woody prop roots of tropical mangroves.

9.4 Mangrove Habitats for Seaweed

In some mangrove stands the most abundant and characteristic intertidal mangrove community which is often called the Bostrychietum, named after its principal component, the red algae *Bostrychia* sp., and is frequently observed on prop roots and pneumatophores. In the Caribbean, other seaweeds associated with the Bostrychietum are algae *Catenella repens* and *Caloglossa leprieurii*. In other areas, algal communities on intertidal prop roots and pneumatophores are dominated by *Cyanobacteria*, also called blue-green algae.

Littler et al. (1985) found that species richness and biomass differed for wave-exposed and sheltered sites at Twin Cays, Belize, related to sea urchin herbivory and wave turbulence. In areas with abundant herbivorous sea urchins, fleshy algae such as *Acanthophora* sp., *Spyridia* sp. and *Caulerpa* sp. dominate on hanging roots of the red mangrove, whereas calcified algae such as *Halimeda* sp. and *Lithophyllum* sp. are more common on attached root. Apparently, the hanging root dominants are 6–20 times more susceptible to herbivory than the calcified algae and are removed from attached roots by urchins that move from the peats up to the roots. In areas with no or low urchin densities, fleshy algae will out complete calcareous algae on both hanging and attached roots.

In shallow subtidal areas enriched by guano from islands with bird rookeries, entangled mats of the hair-like green alga *Chaetomorpha linum* can be found. These mats can reach diameters of a metre or more and are generally not attached to the substrate. Other indicators of nutrient-rich water are the sheet-like green algae *Enteromorpha* sp. and *Ulva* sp. which are found attached to shell and coral fragments or free floating in sheltered bays. On most mud and sand bottoms, seaweeds are relatively rare; the dominant species are green algae with well-developed basal systems (e.g. stolons, rhizoids) for anchoring in unconsolidated sediments.

9.5 Seagrass Beds as Habitats for Seaweed

The leaves of seagrass affects or are affected by algae, especially those that grow on their surfaces. Seagrasses are not algae; they were out of the domain of phycologists. Seagrass leaves as substrata for algae. Seagrasses are angiosperms, though not in the grass family. Their leaves provide substrata for colonization by other organisms, up to 18 m^2 of surface per square metre of bed area. Except for species of *Syringodium*, seagrass leaves are flat, thus maximizing their photosynthetic surface, diffusion of gases and nutrient uptake. Leaves of *Posidonia oceanica* is attached up to 30 weeks, but the leaves of other species turn over much more rapidly. A sequence of epiphytes can be followed on these ephemeral substrata. Novak (1984) reported coccoid bacteria and diatoms as the first organisms to appear on new leaves. This community is followed by animals and macroalgae at the top. Casola et al. (1987) examined the changes in leaf structure with age, and Cinelli et al. (1984) described the relationship to water depth. The composition of the epiphytic community will be determined by the time at which a leaf becomes available, the presence of propagules and the habitat in which the plant appears.

9.6 Distribution of Seaweeds in Marine Environment

Ocean vegetation is dominated by evolutionarily primitive plants: the algae. No other plant groups such as masses, fern or gymnosperms are found in the oceans, and only a few diverse angiosperms (the seagrasses) occur in marine habitats. The water column is chiefly the domain of the phytoplankton – unicellular or colonial plants, including classes not represented in the benthos, but populations of floating seaweeds are common (Norton and Mathiesan 1983). Rocky shores are abundantly covered with macro vegetation that is almost exclusively seaweeds. On and around the larger plants are many benthic microalgae, including early stages of seaweeds. Muddy and sandy areas have fewer seaweeds, because most species cannot anchor there, though some siphonus greens, e.g. some species of *Halimeda* and *Udotea*, are produced penetrating root-like holdfasts that may also serve in nutrient uptake. In such areas, seagrasses become the dominant vegetation, particularly in tropical and subtropical areas.

9.7 Environmental Factor Interactions

Benthic algae interact with other marine organisms, and all interact with their physicochemical environment. As a rule, they live attached to the seabed between the top of the intertidal zone and the maximum depth to which adequate light for growth can penetrate. Among the major environmental factors affecting seaweeds are light, temperature, salinity, water motion and nutrient availability. Among the biological interactions are relations between seaweeds and their epiphytic bacteria, fungi, algae and sessile animal's interactions between herbivores and plants and the predators, including humans. Individual patterns of growth, morphology and reproduction are overall effects of all these factors combined. An organism's physicochemical environment, consisting of all the external abiotic factor that influences the organism, is very complex and constantly varying. Temperature and salinity affect the density of seawater, hence the mixing of nutrient-rich bottom water with nutrient-depleted surface water. Thermoclines can affect plankton movements, including migration of the larvae of epiphytic animals. Water motion can affect turbidity and siltation as well as nutrient availability. These are examples of one environmental variable affecting another.

There are also examples of two environmental variables acting synergistically on plants, for instance, the combination of low salinity and high temperature can be harmful at levels where each alone would be tolerable. In several seaweeds, the combined effects of temperature and photoperiod regulate development and reproduction. Interactions between physicochemical and biological factors are also the rule rather than the exception. Finally, there is factor interaction through sequential effects. Nitrogen limitation may cause red algae to catabolize some of their phycobiliproteins, which will in turn reduce their light-harvesting ability. It is a well-known fact that the coasts in different climatic zones, and those situated in climatically similar zones but in different oceans or hemispheres, are inhabited by dissimilar seaweed floras. Many seaweed species and genera are consequently more or less similar among seaweed floras inhabiting more or less restricted portions of the world's seacoast. Efforts have been made to relate the resulting patterns to present and past environmental conditions: firstly because generic distribution patterns were thought to possibly reflect events and conditions in the geologic past better than species distributions and secondly because more taxonomic tangles and uncertainties were expected on the species than on the genus level. Furthermore, if worldwide trends do exist on the generic level, they can be also expected on the species level.

9.8 Classification of Seaweeds

Seaweeds are classified based on the presence and type of pigments, external and internal structure and reproduction; seaweeds are divided into four broad groups: green, blue-green, brown and red. Botanist refers to these broad groups as Phaeophyceae, Rhodophyceae, Cyanophyceae and Chlorophyceae, respectively. Brown seaweeds are usually large and range from the giant kelp; red seaweeds are usually smaller, generally ranging a few centimetres to about a metre in length, and they are not always red.

They exhibited sometimes a purple and brownish-red colour. Green seaweeds are also small with a similar size range to the red seaweeds. Besides, pigmentation, they differ considerably in many ultrastructural and biochemical features including photosynthetic pigments, storage compounds, composition of cell walls, presence/absence of flagella and fine structure of the chloroplasts. They originated through different evolutionary processes.

9.9 Seaweed Resources of India

India lies to the north of the Equator between latitude 8°4 and 37°6N and longitudes 68°7 and 97°25E and is bounded to the south by the Indian Ocean, to the west by the Arabian Sea, to the east by the Bay of Bengal, to the north. India being an entirely tropical monsoon country has pronounced wet and dry season which characterize the climate. The entire rainfall of the country depends on the frequency and duration of monsoons. India has two monsoons, the southwest and the northeast; the southwest monsoon is well established and blows with great regularity from June to September. The monsoon winds blow up from the Equator and on reaching the southern tip of India.

India is a tropical country in South Asia and has an exclusive economic zone (EEZ) of 2 million km and over 8071 km of coastline interspersed with sandy and rocky beaches. The intertidal and shallow subtidal waters with rocky and coralline substrata are harbour for luxuriant growth of a diversified seagrass species and comprise mostly of tropical species (Subba Rao et al. 2006).

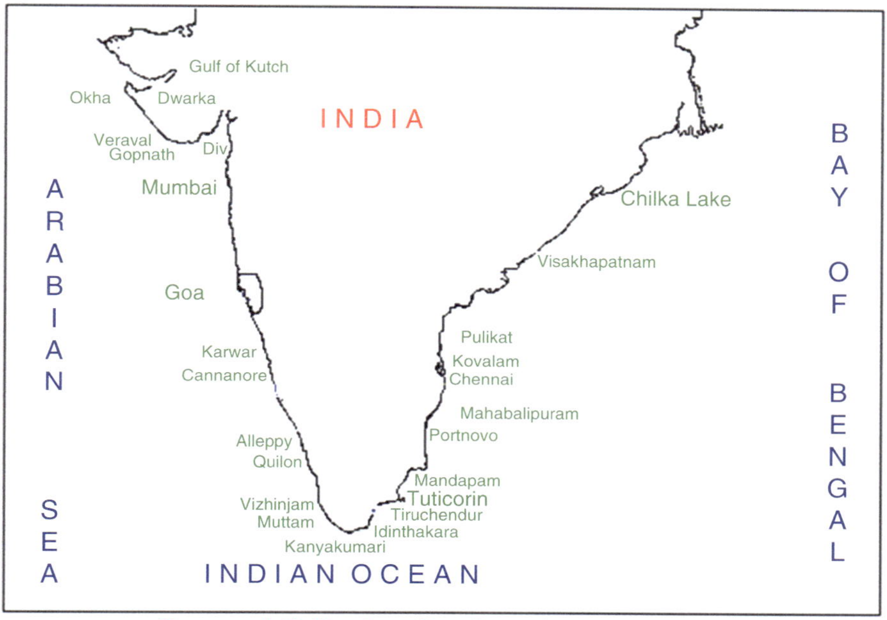

Seaweed Collection site along the Indian coast

India has abundant growth of seaweeds which are distributed along the southeast coast of Tamil Nadu from Mandapam to Kanyakumari, Gujarat coast, Lakshadweep and Andaman and Nicobar islands. Fairly rich seaweed beds were reported in the vicinity of Bombay, Ratnagiri, Goa, Karwar, Varkala, Kovalam, Visakhapatnam and few other places such as Chilka and Pulicat Lake (Thivy 1960; Rao 1969; Kaliaperumal et al. 1987; Rath and Adhikary 2005).

9.10 Utilization of Seaweed

The earliest record of use of seaweeds dates back to 2700 BC in the compilation on" Chinese Herbs" by Emperor Shen Nong. Seaweeds have been a part of the Japanese diet since 3000 BC. Seaweeds are mainly eaten in the oriental countries like Japan, China, Korea and more recently in the United States, Europe, Philippines, Indonesia, Chile, Taiwan, Vietnam, Russia, Italy and India. The Republic of Korea has the highest per capita consumption of seaweeds in the world.

In several industries, seaweeds are utilized as raw materials which have the most commercial value such as phycocolloids like agar, alginate and carrageenan which are used in several industries. The current phycocolloid (seaweed gels) industry stands at over USD 6.2 billion. The world production of commercial seaweeds has grown by 119 species since 1984, and presently, 221 species of seaweeds are utilized commercially including 145 species for food and 110 species for phycocolloid production; these have been used as food for human beings, feed for animals, manure for plants and source of various chemicals. In the recent past, seaweeds have also been gaining new systems for biologist seaweed liquid fertilizer.

Recently, seaweed figure prominently debates about the energy and biofuel production. They have been discussed as a potential source against global warming where seaweeds and algae are very efficient carbon sinks which means that they absorb more carbon that their rate emission.

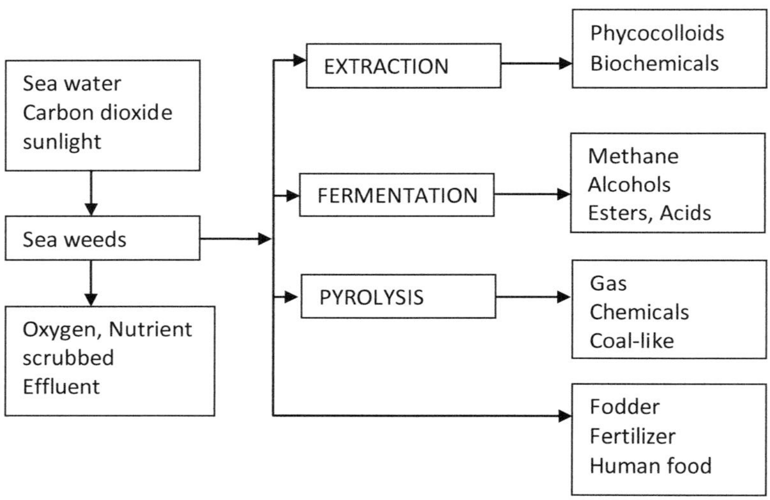

Uses of Seaweeds: Present and Future (Indergaard 1983)

9.11 Seaweeds as Alternative Fertilizer for Sustainable Agriculture

Seaweeds are used in many maritime countries as a source of food, for industrial applications and as fertilizer. The present uses of seaweeds are as human foods, cosmetics, fertilizers and extraction of industrial gums and chemicals. They have the potential to be used as source of long- and short-chain chemicals with medicinal and industrial uses. Marine algae may also be used as energy collectors, and potentially useful substances may be extracted by fermentation and pyrolysis. Seaweeds as an agricultural fertilizer have been demonstrated, especially by coastal farmers with ready access to seaweed (Booth 1965). However, such applications were probably practised in antiquity, the first reference to such use appears in Roman writings from the second century AD where seaweed collected from beach drift is harvested for agricultural use. The use of seaweed as manure is a common practice among the Romans and also practiced in Britain, France, Spain, Japan and China. The use of marine macroalgae as fertilizer in crop production has a long tradition in coastal areas all over the world. Seaweed coast continued to be valuable to farmers even in the early 1900s. In many countries, seaweed and beach cast are still used in both agriculture and horticulture (Verkleij 1992). More than 15 million metric tonnes of seaweed products are produced annually, a considerable portion of which is used for nutrient supplements and as biostimulants or biofertilizers in agricultural and horticultural crop production (FAO 2006). Seaweed extract is a new generation of natural organic fertilizers highly nutritious and provides faster germination of seeds and increases yield and resistant ability of many crops. Ocean is the storehouse of chemicals, and from time immemorial the nutrients in the land are being leached into the sea through rain, rivers, etc. Seaweeds, the marine algae, are known to concentrate high amounts of nutrients from the seawater. Utilization of seaweed manure or seaweed liquid fertilizer (SLFs) for the enhancement of growth and yielded of terrestrial plants have been recorded last few decades. Marine algae contain more than 60 trace elements in a concentration much higher than in terrestrial plants. Historical records show that the use of seaweeds in agriculture is ancient and widespread wherever there are abundant resources in the coastal regions of Iceland, Norway, Great Britain, Ireland and France.

9.12 Why Use Seaweed Fertilizers

As the population is growing at a fast rate, the agro-based products should also increase. With the agro-based products, fertilizer industry is bound to grow, as it is one of the major components for increasing food production. At present there is a shortage of about 3–4 million tonnes of fertilizer (Jeswani 1999 and Zodape 2001). The use of seaweeds as manure is a common practice throughout the world especially in areas adjoining the sea. Seaweeds can be directly applied as organic manure or can be composted. The importance of seaweed fertilizer depends on two factors. Foremost is the presence in seaweeds of appreciable amounts of trace elements besides Ca, P, Mg, Fe, Na and KCl.

Oligo elements such as, Zn, Cu, I, Mo, Cr, etc. and growth hormones like auxin, gibberellin and cytokinin are also present, A second characteristic feature about seaweeds is the presence in them of mucopolysaccharides of various kinds which help in soil conditioning and carbohydrate and other organic matter after the nature of the soil and improve its moisture-retaining capacity. Due to the presence of abundant nutritional value towards plant growth, large quantities of seaweeds can be used as manure in all parts of the country, either directly in the form of compost. Appreciable results were also achieved with vegetables, fruits, flowers and cereals, when manure with seaweed compost, and crotons also grew well with seaweed treatment; application of seaweed manure can maintain a high level of nitrogen availability in soil.

9.13 Why Seaweed Fertilizer Important

Seaweed manure has been recognized for a long time in other countries; however, in India very little information is available on the beneficial effects of seaweed manure to improve seed germination and increase crop growth. Not like other plant fertilizers, seaweed manure is a slow but long active fertilizer, and its application is well suited to light sandy soils, which are generally deficient in potash. Physical condition of these light soils also improves soil structure and is invariably associated with better aeration, enhanced nitrogen fixation and generalized raised proliferation of soil organisms. Capillary action is also increased, and as a result, root system of plant is stimulated into further growth. Humic acid and to a greater extent soluble alginate in seaweed bind particle of clay into larger aggregates or create crumb structure by combining chemically with metallic radicals present in the soil. They bring about aggregation of soil particles; in the case of soluble alginates, each metallic radical combines with two or more alginate molecules to form a polymer or large molecule with branched chains. The polymers are responsible for the formation of crumb structure.

Seaweed extracts are used extensively in agriculture as plant growth supplements, and seaweed meal takes months to become fully effective in the soil as a plant nutrient. Seaweed manure is particularly valuable because of the trace elements that it contains and the total quantity of trace metals removed annually by a crop. The slow loss of water as a result of SLF amendment in the soil which formed a thin layer over the surface prevents water evaporation and increases the soil solution. Seaweed fertilizer application improves the fertility of soils in cultivated fields particularly the brown seaweeds because of their algin content, which helps in conditioning the soil, facilitating aeration, moisture retention and absorption of nutrient elements.

9.14 Methods and Preparation of Seaweed Liquid Fertilizer (SLF)

Seaweed extract is organic manure and is beneficial due to the presence of trace elements and other organic substances such as amino acids, antibiotics, auxins, gibberellins and vitamins in it. Some of the substances are decomposed by heat, and hence it is essential that they should be conserved in the area to benefit the crops. Many seaweed constituents are known to undergo marked seasonal variations, which are being considered, in both commercial seaweed extract production and in the evaluation of inconsistent field trial results that has been reported (Blunden 1972 and Zodape 2001). Many species of seaweeds have been reported to be used for the preparation of SLF such as *Durvillaea potatorum*, *Sargassum wightii*, *Padina pavonica*, *Padina boergesenii*, *Sargassum*, *Champia*, *Turbinaria*, *Helminthocladia*, *Laminaria saccharina*, *Fucus serratus*, *Fucus vesiculata*, *Fucusvesiculata*, *Furcellaria fastigiata*, *Hypnea musciformis*, *Sargassum polyphyllum*, *Ascophyllum nodosum*, *Ulva lactuca*, *Pterocladia* and *Ecklonia radiata*, *Padina tetrastromatica*, *Sargassum tenerrimum*, etc.

9.14.1 First Method (Bhosle et al. 1975)

In this method, seaweeds (seaweeds already used such as *Sargassum tenerrimum* or *Padina tetrastromatica*) chopped must be boiled with distilled water and then filtered. The filtrate should be taken as 100% concentration of the seaweed extract, and from this, different concentrations have to be prepared using distilled water.

9.14.2 Second Method (Challen and Hemingway 1966)

He has described the method there in two samples of commercial seaweed meal; one derived from *Ascophyllum nodosum* and another derived from *Fucus vesiculosus* were used to prepare extract according to the following method. In this method, the powder should be mixed with distilled water and allowed to stand. The mixture should be boiled, allowed to stand for some time and then passed through a fine sieve to remove the solids, and the liquor obtained should be centrifuged. The solid from the sieve and centrifuge should be pressed, and the liquor obtained should be mixed with the main liquor. The combined liquors should be then concentrated under reduced pressure to yield a brown fluid. The percentage of total solids should be determined, and the extract diluted with sufficient water to contain the same percentage of total solids as the commercial seaweed extract of dry seaweed should be further diluted when required.

9.14.3 Third Method (Rama Rao 1990)

In the third method, the seaweed should be washed thoroughly with running tap water to remove epiphytes and sand particles. They should be then shade dried for 5 days at room temperature and stored in polythene bags. The algae should be kept in oven at 60–65 °C for 24 h and hand-crushed and made as coarse powder using a mixer grinder. The coarse powder should be soaked in water for 1 h in the ratio of 1:20 (w/v) without adjusting the pH and then autoclaved at 121 °C, 20 Ibsin2 for

60 min. The hot extracts of the seaweeds should be filtered through a double-layered cheese cloth and allowed to cool at room temperature. The filtrate should be then centrifuged at 10,000 rpm for 30 min at 48 h for drying. The dried powder was considered at 100% seaweed extract and stored at 4 °C for further studies.

9.14.4 Methods of Application: Seaweeds as Fertilizer

Seaweed extracts should be applied in low concentrations/dosages because it is clear that the active ingredients in seaweed extracts are effective in low concentrations, and the different methods of applications are as follows.

9.15 Seaweeds as Manure (or) Compost

The manurial value of seaweed compost has long been recognized but has not been fully exploited. In India, it is used for coconut plantations especially in coastal Tamil Nadu and Kerala, and meal of seaweeds should take months to become fully effective in soil as plant nutrients because of its carbohydrate material which has to be broken down by bacteria (before it can be used by the plant).

9.16 Seaweed Liquid Fertilizer (SLF)

In this method, fertilizer application is by the methods of foliar spray, soaking of seeds (before seed sowing), seedling root tip treatment and by soil/root drench application; presently, the SLFs are widely used in the agricultural and horticulture sections, in many countries.

Liquid fertilizer are obtained from seaweeds contains polysaccharide content in broken down form and hence becomes effective at once. The seaweed extracts can be applied as foliar spray for growth promotion in crop and fruiting plants. This has been largely practiced in countries like the United States, United Kingdom and Norway. In case of seedling root tip method, the tip of the seedling is suspended for 20–30 min in 2% SLF (before transplantation), whereas the seeds were soaked in different concentrations of SLF (before seed sowing). The drench method involves the fertilizer dibbling at root zones and application in furrows near root zone depending on the type of crops for best result.

9.17 Present Status of Seaweed Fertilizer Usage

Seaweed fertilizers have been popularly used in many countries, viz. France, Ireland, Scotland, England, the United States, the United Kingdom, Canada, New Zealand, Australia and Spain. In South Africa, the extracts of green, brown and red seaweeds are sold as soil conditions. The *Sargassum* is used in China, the *Hypnea* in West Indies, whereas the *Hypnea*, *Ulva* and *Enteromorpha* are commonly used in Brazil.

9.18 Some Commercially Available Seaweed Liquid Fertilizer

Seaweed extracts are now commercially available in trade names of Algifert, Algistin, Agral, Algit, Cytex, DSWE, Geomar, GA14, Kelpak 56, Maxicrop, Sea crop 16, Seasol, Seaspray, Seamac, Seamagic3, Organic six, Biozyme, Redicrop, AgriFin, Bio organic, Ecolife, Maize spl, H-8, Bio-Energy, Bio-mol, Sowato, En-zym+, Azooba, Pearl of sea, Saosis, Flower care, Ocean, Maxizyme, Maxigrow, Maxi O-Viral, Maxi 80, Kelpak, Seagro, Seasol, Cytokinin, etc.

9.19 Conclusion

Seaweed is one of the important alternative fertilizers which can be utilized for plant growth in terms of sustainable agriculture. Nowadays, using chemical fertilizers is inevitable for agriculture for increasing crop production which can visibly pollute the environment in all the ways. Although other natural fertilizers are used from various waste sources such as animal dung, sludge and using vermicompost, etc., the plant growth is not effectively grown as used by chemical fertilizers. Seaweed fertilizers which obviously more effective for plant growth are proved, and the source of seaweeds is harbour and abundant to sustain the fertilizer use without environmental pollution.

References

Bhosle NB, Untawale AG, Dhargalkar VK (1975) Effect of seaweeds extract on the growth of Phaseolus vulgaris. Indian Mar Sci 4:208–210

Blunden G. (1972) The effects of aqueous seaweed extracts as a fertilizer additive. In: Proc. Int. Seaweed Symp, 7 : 584-589.

Booth E (1965) The manorial value of seaweed. Bot Mar 8:138–143

Casola E, Scardi M, Mazzella L, Fresi E (1987) Structure of the epiphytic community of *Posidonia oceanica* leaves in a shallow meadow. PSZN: Mar Ecol 8:285–296

Challen S. B., Hemingway J.C. (1966) Growth of higher plants in response to feeding with seaweed extracts. Proc. Fifth Int. Sea. Symp. 359–367.

Cinelli F., Cormaci M., Furnari G., Mazzella L (1984) In: Boudouresque, C.F., de Grissac, A.J., Olivier,J.(Eds.), Epiphytic Macroflora of *Posidonia oceanica* (L.) Delile Leaves Around the Island of Ischia (Gulf of Naples). International Workshop on *Posidonia oceanica* Beds, vol 1. GIS Posidonie Publ, Marseilles, pp. 91–99

Dube MA, Ball E (1971) *Desmarestia* sp. associated with the sea pen *Ptilosarcus gurneyi* (Gray). J Phycol 7:218–220

FAO (2006) State of world aquaculture 2006. Food and Agriculture Organization of the United Nations, FAO Fisheries Technical paper 500, Rome. http://www.Fao.org/docrep/009/ao874e00.htm.2

Indergaard M (1983) The aquatic resource. In: Cote WA (ed) Biomass utilization. Plenum Press, New York, pp 137–168

Jeswani KD (1999) We want to provide farmers updated technology in the field of agriculture. Chem Eng World 34:23–25

Kaliaperumal N, Chennubhotla VSK, Kalimuthu S (1987) Seaweed resource of India. Seaweed Res Utiln 41:51–99

Kenyon C, Rice DW (1959) Life history of the Hawaiian monk seal. Pac Sci 23:215–252

Little MM, Taylor RR, Littler DS, Sims RH, Norris JN (1985) The distribution, abundance and primary productivity of submerged macrophytes in a Belize barrier-reef mangrove system. Atoll Res Bull 289:1–20

Mc Hugh DJ (2003) A guide to the seaweed industry. FAO fisheries technical paper, vol 441. FAO, Rome, p 105

Norton TA, Mathieson AC (1983) The biology of unattached seaweeds. Prog Phycol Res 2:333–386

Novak R (1984) A study in ultra-ecology: microorganisms on the seagrass Posidonia oceanis (L) Delile. PSZNI: Mar Ecol 5:143–190

Price JF, Warner JO, Hey EN, Turner MW, Soothill JF (1984) A controlled trial of hyposensitization with adsorbed tyrosine *Dermatophagoides pteronyssinus* antigen in childhood asthma: in vivo aspects. Clin Exp Allergy 14(3):209–219

Rama Rao K. (1990) Preparation of liquid Seaweed fertilizer from Sargassum. In: Seaweed Research and Utilization Association Workshop on Algal Products and Seminar on Phaeophyceae in India. 4th – 7th June at Madras p16.

Rao U (1969) A catalogue of marine algae in the reference collection of the Central Marine Fisheries Research Institute. Bull Cent Mari Fish Res Inst 9:37–48

Rath J, Ahikary SP (2005) A check list of algae from Chilika lake, Orissa. Bull Bot Surv India 47:101–114

Subba Rao P.V., Vaibhav A., Mantri (2006) Indian seaweed resources and sustainable utilization: Scenario at the dawn of a new century 91, NO. 2, 25: 164–174

Taylor BR, Littler MM, Littler DS (1986) Escapes from herbivory in relation to the structure of mangrove island macroalgal communities. Oecologia 69:481–490

Thivy F (1960) Seaweed utilization in India. In: Proceedings of the symposium on Algology. Indian council of Agricultural Research, New Delhi

Tsuda RT (1965) Marine algae from Laysan Island with additional notes on the vascular flora. Atoll Res Bull 110:1–31

Verkleij FN (1992) Seaweed extract in agriculture and horticulture a review. Biol Agric Hortic 8:309–334

Zodape ST (2001) Seaweeds as a biofertilizer. J Sci Indus Res 60:378–382

Part IV
Alternative Food Sources

Chapter 10
Algal Resource for Sustainable Food Security

M. Ayyappan

Abstract Human population at current scenario faces food demand particularly in undeveloped and developing countries due to unpredictable climate change and other environmental factors. The unsustainable food production and harvest leads to scarcity of food in the future due to overpopulation. However, the world is abundantly covered by oceans, seas, rivers, and other water resources which harbor most of the living organisms. Algae are a group of plants, most of which are capable of performing photosynthesis. There are 80,000 to 100,000 different algae species with widely varying characteristics, and globally, there is growing interest in algae as production organisms. There is substantial evidence for the health benefits of algal-derived food products, but there remain considerable challenges in quantifying these benefits, as well as possible adverse effects. They are primary producers which are a source of many nutrients, and it has high protein content. Not only food but also nutritive ingredient and medicinal value also exist in marine algal source. Hence, algal source is the major initiative to sustainable agriculture to meet the food supply in global population.

10.1 Introduction

Algae, a diverse group of autotrophic organisms, have the ability to grow rapidly, efficiently use light energy, fix atmospheric CO_2, and produce more biomass per acre than vascular plants (Shay 1993). They are abundant in salt water, fresh and stagnant lakes, and humid places and are also found in soil. The presence of algae in surface water has been a long-standing issue all over the world because of their adverse effects on the treatment process and quality of drinking water (Joseph Bruchac 1997). The utilization of algae as food source and for treatment of various ailments has been recorded since over 2000 years (Richmond 1990). Biochar is a natural fertilizer that is useful in cultivation, in particular restoring and maintaining the organic matter in the soil. Algae can be used not only for food and pharmaceutical

M. Ayyappan (✉)
Department of Botany, Government Arts College, Thiruvalluvar University,
Thiruvannamalai, Tamil Nadu 606 603, India
e-mail: ayyap_pn@yahoo.co.in

A. Dhanarajan (ed.), *Sustainable Agriculture towards Food Security*,
https://doi.org/10.1007/978-981-10-6647-4_10

purposes but also for the capture of CO_2 from exhaust gases, in particular from plants for the production of energy, heat, or electricity by microalgae. It is estimated that, for every kg of biomass produced, 2 kg of CO_2 is captured by photosynthetic microalgae (Chisti 2007).

10.2 Green Algae

The green algae are a large, informal group of algae comprising Chlorophyta which are now placed in separate divisions (Jeffrey et al. 2004). Their chloroplasts contain chlorophylls a and b, giving them a bright green color, as well as the accessory pigments beta-carotene and xanthophylls in stacked thylakoids (Hoek et al. 1995). Their cell walls usually contain cellulose and store carbohydrate in the form of starch (Judd et al. 2002). Green algae include large embryophytes belonging to the paraphyletic group; unicellular and colonial flagellates, most of which have two flagella per cell; various colonial, coccoid, and filamentous forms; and macroscopic seaweeds (Perosa et al. 2005).

10.2.1 Blue-Green Algae

Cyanobacteria, also known as blue-green algae, belong to phylum Cyanophyta, which obtain their energy through photosynthesis. They represent some of the most ancient life forms on Earth and are the most studied organisms worldwide. They are characterized as unusual prokaryotic microorganisms that can perform oxygenic photosynthesis (Schopf et al. 1965; Margulis 1975). They live in terrestrial and aquatic habitats, from oceans to freshwater and from bare rock to soil. They are sometimes found in rocks in the desert (Perosa et al. 2005).

10.2.2 Red Algae

Rhodophyta or red algae represent a division that is characterized by chloroplasts that have no external endoplasmic reticulum and unstacked thylakoids, phycobiliprotein pigments, and floridean starch and lack flagella in all stages. They are predominantly marine in distribution with fewer than 3% of more than 6500 species occurring in truly freshwater habitats (Guiry and Guiry 2014; Guiry et al. 2014).

10.2.3 Brown Algae

The brown algae are an important assemblage of plants that are classified in about 265 genera with more than 1500 species (Bold and Wynne 1985). They derive their characteristic color from the large amounts of the carotenoid fucoxanthin (which yields a brown color) contained in their chloroplasts and the presence of various phaeophycean tannins. They occur mainly in the marine environment, where they appear as an intertidal component. Some marine forms penetrate into brackish environments and can be an important part of the salt marsh fauna (Lee 1989).

10.2.4 Yellow Algae

Xanthophycean chloroplasts are yellow green because of the dominance of vaucheriaxanthin and diatoxanthin; the brown accessory pigments are absent. Most of the 500 taxa are unicellular or colonial algae (Mischococcales), some are filamentous (Tribonematales), and others are multinucleate siphonal (Vaucheriales). The cell walls often are divided in two overlapping parts. Recent investigations showed that from the coccoid freshwater species (e.g., members of the genera *Goniochloris* and *Tetraedriella*), a considerable number of species are to be transferred to the class of Eustigmatophyceae (Serge et al. 2013).

10.3 Algae Are Harbor of Natural Resources for Human Society

Interestingly, algae are the elite natural source of new compounds with biological activity that could be used as functional ingredients and sustainable adapted organisms that live in complex habitats submitted to extreme conditions; therefore, they must adapt rapidly to the new environmental conditions to survive, producing a great variety of secondary metabolites (Carlucci et al. 1999).

Moreover, algae are easily cultivated because of their rapid growth and their potential to control the production of some bioactive compounds by manipulating the cultivation conditions. Moreover, one of the least studied aspects is the development of more appropriate, fast, cost-effective, and environmentally friendly extraction procedures able to isolate the compounds or compounds of interest from these natural sources (Merichel Plaza et al. 2008).

10.4 Nutritive and Pharmacological Value of Algae for Sustainable Agriculture

Algae such as *Sargassum vulgare*, *Undaria pinnatifida*, *Himanthalia elongata*, *Chondrus crispus*, and *Porphyra* sp. are containing functional foods and functional ingredients together with their possible effect on human health (Hasler 2002). The species of algae described in this work are macroalgae (viz., *Sargassum vulgare*, *Undaria pinnatifida*, *Himanthalia elongata*, *Chondrus crispus*, *Porphyra* sp., *Cystoseira* spp., and *Ulva* spp.).

The species of algae described in this work are macroalgae *S. vulgare* that present good nutritional values as sources of proteins, carbohydrates, minerals, and vitamins (Marinho-Soriano et al. 2006). *S. vulgare* is the major source of carbohydrates (67.80%), low lipids (0.45%), high percentage of fiber (7.73%), and proteins (15.76%) (Dietrich et al. 1995).

Cyanobacteria produce a wide variety of bioactive compounds, which include 40% lipopeptides, 5.6% amino acids, 4.2% fatty acids, 4.2% macrolides, and 9% amides. Cyanobacterial lipopeptides include different compounds like cytotoxic (41%), antitumor (13%), antiviral (4%), and antibiotics (12%), and the remaining 18% activities include antimalarial, antimycotics, multidrug resistance reversers, antifeedant, herbicides, and immunosuppressive agents (Burja et al. 2001). Other compounds of great importance that can be found in most of the algae described in this work are sterols. Diverse clinical studies have demonstrated that diets with sterols (from plants) might help to reduce cholesterol levels in blood. Additionally, they have anti-inflammatory, antibacterial, antifungicidal, antiulcerative, and antitumoral activity (Dunford and King 2000). In summary, in this work, a bibliographical revision has been carried out on the composition and biological activity of some algae discussing their possibilities as natural sources of functional ingredients. Thus, it is possible to conclude that these organisms show a high potential as natural sources of ingredients with many different biological activities.

10.5 Amino Acid Composition

The amino acid composition of seaweeds has been frequently studied and compared to that of other foods such as eggs or soybean. Most seaweeds and aspartic and glutamic acids constitute together a large part of the amino acid fraction. In *Fucus* sp. brown seaweeds, these two amino acids can represent between 22 and 44% of the total amino acids (Munda 1977).

Rhodophyceae contain a particular protein called phycoerythrin (PE) which is already used in biotechnology applications. PE used for this particular application is obtained from the microalgae *Porphyridium cruentum* in which the pigment can represent up to 50% of the protein fraction (Gantt and Lipschutz 1974).

10.6 Sustainable Food Sources of Algae

The concept of functional food as a mean to protect consumer's health was developed at the beginning of the 1980s in Japan, as a way to reduce the high health costs derived from a population with high life expectations (Arai 1996). In recent years, macroalgae have been increasingly used as animal fodder and in the production of alginate feeds for farm animals. It is alternative protein sources for farmed fish because of their high protein content and productivity. Microalgae and macroalgae are also used as components in polyculture systems and in remediation (Zhou et al. 2006; Marinho-Soriano 2007).

10.7 Algal Food Sources for Fish Farming

Microalgae are rich in β-carotene, astaxanthin, and lutein. It can improve the animal's color and effectively improve the body color of fish. The floating bait that contains*Spirulina* for the goldfish, koi, tropical fish, etc. of ornamental value has been produced in Japan (Zhiqiong Li et al. 2001). Algae can be added to feeds of many farm animals; the main applications of microalgae for aquaculture are associated with nutrition, being used fresh (as sole component or as food additive to basic nutrients) for coloring the flesh of salmonids and for inducing other biological activities (Muller-Feuga 2004). Feeding trials were carried out with many fish species, most commonly red sea bream (*Pagrus major*), ayu (*Plecoglossus altivelis*), nibbler (*Girella punctata*), and striped jack (*Pseudocaranx dentex*); the most extensively studied ones have been the blue-green algae *Spirulina* and *Chlorella*; the brown algae *Ascophyllum*, *Laminaria*, and *Undaria*; the red alga *Porphyra*; and the green alga *Ulva* (Fagbenro 1990). As the main biochemical constituents and digestibility are different among algae, the effect of dietary algae varies with the algae and fish species (Mustafa and Nakagava 1995). Henson (1990) reported that *Spirulina* improved the performances of ayu, cherry salmon, sea bream, mackerel, yellowtail, and koi carp. The levels of supplementation used by Japanese farmers are 0.5–2.5%.

10.8 Algae as Animal Feed

The principal applications of microalgae for aquaculture are associated with nutrition, being used fresh for coloring the flesh of salmonids and for inducing other biological activities (Muller-Feuga 2000). Several algae are additives in fish feed. Feeding trials have been carried out with many fish species, most commonly red sea bream (*Pagrus major*), ayu (*Plecoglossus altivelis*), nibbler (*Girella punctata*), striped jack (*Pseudocaranx dentex*), cherry salmon (*Oncorhynchus masou*), yellowtail (*Seriola quinqueradiata*), black sea bream (*Acanthopagrus schlegelii*), rainbow trout (*Oncorhynchus mykiss*), rockfish (*Sebastes schlegelii*), and Japanese flounder (*Paralichthys olivaceus*) (Spolaore et al. 2006).

10.9 Algal Biofuels and Bioenergy

Microalgae have additional advantages over terrestrial plants. Since they are single-celled organisms that duplicate by division, high-throughput technologies can be used to rapidly evolve strains. This can reduce processes that take years in crop plants, down to a few months in algae. Algae have a reduced impact on the environment compared with terrestrial sources of biomass used for biofuels (Dismukes et al. 2008). The recovery of microalgal biomass generally requires one or more solid-liquid separation steps and is a challenging phase of the algal biomass production process (Wang et al. 2008). Biofuels are heralded by environmentalists and government leaders as the most promising renewable alternatives to achieve the goals of reducing our dependence on fossil fuels and lowering CO_2 emissions (Ragauskas et al. 2006). CO_2 can be harvested from commercial or industrial flue and flare gases produced at electrical generating units, petroleum refineries, and cement, ammonia, hydrogen, and ethanol plants or retrieved through natural underground reservoirs; however, without CO_2 separation processes in place, NOx, SOx, and other contaminants found in emission gases have the potential to stress and inhibit biomass growth and lipid content (Mata et al. 2010). A biodiesel production process that avoids biomass drying and organic solvents for oil extraction could lead to significant energy and cost savings (Singh and Olsen 2011). Algal biomass is rich in nutrients especially nitrogen and phosphorus, for which the use and potential loss may not be environmentally and economically sustainable (Sialve et al. 2009). Ethanol fermentations using flocculating yeast strains were reported in the 1980s and 1990s. There were a variety of fermenters developed with different configurations. Among these configurations are airlift, single packed column, two-stage packed column (with or without settlers), and CO_2 suspended bed (with or without baffle plates) (Bai et al. 2008). Another option for deriving biofuels from botanical biomass is to harness the oils naturally produced by plant seeds and algae. These oils contain triacylglycerol (TAG) fatty acid esters almost exclusively, with each TAG containing three fatty acids of chain lengths most commonly between 16 and 18 carbons (C16–C18) (de la Piscina and Homs 2008).

10.10 Conclusions

World human population is under food scarcity and security in the present scenario of lifestyles. The attempts have been made to sustain the food security since the past decade. Moreover, policies have been governed in major countries in order to protect the environment and find alternative sources of energy to sustain the quality of human life. Source of marine forms especially algal diversity are the huge natural source and boon for human society to meet all the demand such as food and others in sustainable manner. However, no much attention is in marine source to harvest food ingredients toward food demand. Research is still in need to cultivate the marine algae in land. Various culture methods can be applied when cultivating microalgae, and these methods depend on the specific requirements of the species being cultivated. The research is needed to acquire the food production from the marine source to sustain food meets world demand is essential.

References

Arai S (1996) Studies of functional foods in Japan - State of the art. Biosci Biotechnol Biochem 60:9–15

Bai et al (2008) Ethanol fermentation technologies from sugar and starch feedstocks. Biotechnol Adv 26:89–105

Bold HC, Wynne MJ (1985) Introduction to the algae. Prentice-Hall, Englewood Cliffs, p 516

Bruchac J (1997) Keepers of life: discovering plants through native American stories and earth activities for children. Fulcrum Publishing, Golden, p 60

Burja AM, Banaigs B, Abou-Mansour E et al (2001) Marine cyanobacteria-a profilic source of natural products. Tetrahedron 57:9347–9377

Carlucci MJ et al (1999) Inhibitory action of natural carrageenans on Herpes simplex virus infection of mouse astrocytes. Chemotherapy 45(6):429–436

Chisti Y (2007) Biodiesel from microalgae. Biotech Adv 25(3):294–306

de la Piscina PR, Homs N (2008) Use of biofuels to produce hydrogen (reformation process). Chem Soc Rev 2008(37):2459–2467

Dietrich CP, Farias GGM, Deabreu LRD et al (1995) A new approach for the characterization of polysaccharides from algae: presence of four main acidic polysaccharides in three species of the class Phaeophycea. Plant Sci 108:143–153

Dismukes et al (2008) Aquatic phototrophs: efficient alternatives to land-based crops for biofuels. Curr Opin Biotechnol 19(3):235–240

Dunford NT, King JW (2000) Phytosterol enrichment of rice bran oil by a supercritical carbon dioxide fractionation technique. J Food Sci 65(8):1395–1399

Fagbenro OA (1990) Food composition and digestive enzymes in the gut of pond- cultured *Clarias isheriensis* (Siluriformes: Clariidae). J Appl Ichthyol 6(2):91–98

Gantt E, Lipschutz CA (1974) Phycobilisomes of Porhyridium cruentum: pigment analysis. Biochemistry 13:2960–2966

Guiry MD, Guiry GM (2014) Algae base. World-wide electronic publication, National University of Ireland, Galway

Guiry et al (2014) Algae base: an on-line resource for algae. Cryptogam Algol 35:105–115

Hasler CM (2002) Functional foods: benefits concerns and challenges e a position paper from the American Council on Science and Health. J Nut 132:3772–3781

Henson RH (1990) Spirulina algae improves Japanese fish feeds. Aquac Mag 6(6):38–43. Safety and health in agriculture. Inter Lab Organ, p 77

Hoek et al (1995) Algae an introduction to phycology. Cambridge University Press, Cambridge

Jeffrey et al (2004) The plant tree of life: an overview and some points of view. Am J Bot 91(10):1437–1445

Judd et al (2002) Plant systematics, a phylogenetic approach. Associates Inc, Sunderland, p 156

Lee RE (1989) Phycology. Cambridge University Press, Cambridge, p 645

Li Z et al (2001) Feed Research Forum 1:15–18

Margulis (1975) Symbiotic theory of the origin of eukaryotic organelles; criteria for proof. Symp Soc Exp Biol 29:21–38

Marinho-Soriano E (2007) Seaweeds biofilters: Eficiencia de absorçao da macroalga Gracilaria cervicornis Turner Revista Brasileira de Biociencias. Porto Alegre 5(2):807–809

Marinho-Soriano E, Fonseca PC, Carneiro MAA et al (2006) Seasonal variation in the chemical composition of two tropical seaweeds. Bioresour Technol 97(18):2402–2406

Mata et al (2010) Microalgae for biodiesel production and other applications: a review. Renew Sust Ener Rev 14(1):217–232

Merichel Plaza, Cifuentes A et al (2008) In the search of new functional food ingredients from algae. Trends Food Sci Technol 19(1):31–39

Muller-Feuga A (2000) The role of microalgae in aquaculture: situation and trends. J Appl Phys 12(3–5):527–534

Muller-Feuga A (2004) Microalgae for aquaculture. The current global situation and future trends. In: Richmond A (ed) Handbook of microalgal culture. Oxford, Blackwell, pp 352–364

Munda IM (1977) Differences in amino acid composition of estuarine and marine fucoids. Aquc Bot 3:273–280

Mustafa et al (1995) A review: dietary benefits of algae as an additive in fish feed. Israel J Aqua Bamid 47(3–4):155–162

Perosa A et al (2005) Algae as a potential source of food and energy in developing countries sustainability technology and selected case studies Edizioni Ca' Foscari – Digital Publishing per la presente edizione

Ragauskas et al (2006) The path forward for biofuels and biomaterials. Science 311:484–489

Richmond A (1990) Handbook of micoalgal mass culture. CRC Press, Boca Raton

Schopf et al (1965) Electron microscopy of fossil bacteria two billion years old. Science 149:1365–1367

Serge N, Vinogradov SN, Hoogewijs D (2013) Microbial Globins Status and Opportunities. Adv Microb Physiol 63:1–493

Shay EG (1993) Diesel fuel from vegetable oils: status and opportunities. Biomass Bioenergy 4:227–242

Sialve B et al (2009) Anaerobic digestion of microalgae as a necessary step to make microalgal biodiesel sustainable. Biotechnol Adv 27(4):409–416

Singh A, Olsen SI (2011) A critical review of biochemical conversion, sustainability and life cycle assessment of algal biofuels. Appl Energy 88 (10): (3) 548–555

Spolaore P, Joannis-Cassan C, Duran E, Isambert A (2006) Commercial applications of microalgae. J Biosci Bioeng 101:87–96

Wang et al (2008) CO2 bio-mitigation using microalgae. Appl Microbiol Biotechnol 79(5):707–718. https://doi.org/10.1007/s00253-008-1518-y

Zhou C et al (2006) Steroid and xenobiotic receptor and vitamin D receptor crosstalk mediates CYP24 expression and drug-induced osteomalacia. J Clin Invest 116:1703–1712

Chapter 11
Sustainable Food Security: Edible and Medicinal Mushroom

S. Murugesan

Abstract Mushrooms are technically reproductive structures of edible fungi that belong to *Basidiomycotina*, and growth may be epigeal or hypogeal. Like any other fungus, the vegetative part of mushroom consists of thread like thin mycelia, which under suitable conditions form fruit bodies (sporocarps). Mushroom cultivation is large-scale production in industrial of agribusiness in China and India. Early man did not consume edible mushrooms for its nutritional composition but because of the taste and flavor of it. In India it described the important historical developments with year wise for the utilization as food and others. Not only for food but also they have medicinal properties and capable of agro-waste degradation. Edible mushroom contains high value of nutrients such as protein, amino acids, vitamins, fiber, and minerals. Amino acid compositions are analyzed from *A. bisporus*. Different types of edible mushrooms are observed dry weight and proximate analysis. To sustain the food production and security, mushrooms may play an important role to meet the demand of food scarcity and have valuable nutrient compounds that are in mushrooms as in common vegetables.

11.1 Introduction

11.1.1 Mushroom

Mycologically, the word "mycology" (study of fungi) is derived from the Greek word "mykes" = mushrooms and "logos" = discourse. The mycology (Gr. mykes = seta + logos = study), etymologically, is the study of set. Actually, mycology started a long time ago, because the set is part of the biggest mushrooms, which attracted the attention of naturalists before the invention of microscopes by Van Leeuwenhoek in the seventeenth century, starting the systematic study of mushrooms (Alexopoulos and Mims 1985).

S. Murugesan (✉)
Department of Botany, School of Life Science, Periyar University, Salem, Tamil Nadu, India
e-mail: drsmbtpu@gmail.com

A. Dhanarajan (ed.), *Sustainable Agriculture towards Food Security*,
https://doi.org/10.1007/978-981-10-6647-4_11

Many mushroom species are bioluminescent and they emit light in the dark. A mushroom affords a variety of flavor and taste. Eating mushroom is much common in most parts of Europe, America, China, Japan, and other countries. Nowadays the interest of people in mushrooms is increasing greatly and steadily due to the awareness about the usefulness of mushrooms. *Auricularia* sp. is considered as the earlier cultivated mushrooms in the world. It is known to be cultivated even 1000 years ago in China. *Agaricus bisporus* is another edible mushrooms reported to be first cultivated in France during 1650 (Panneerselvam et al. 2009).

About 180 million tons of fresh mushrooms can be grown if half the agricultural waste discarded in the country is used for the purpose (Panneerselvam et al. 2009). Fibrous crop residues such those of rice are major component of livestock feed in many Asian countries including India (Punia 1992). However their use is limited due to low digestibility, very low protein content, fairly high lignin composition, occurrence of silica, and poor palatability. Consequently on the limited use of rice straw (only as feed for cattle), a large quantity of the same remains unutilized or underutilized (Panneerselvam et al. 2009). The spawn produced from the mushroom can be used even for decomposition of coir pith. A large number of lingo-cellulosic degradable enzymes are observed in the spawn which helps for the removal of various toxic and color pollutant in the ecosystem (Panneerselvam et al. 2009).

It is now known that many of the mushrooms under cultivation ranks above vegetables in protein content. Few of the edible mushrooms reduce serum cholesterol, inhibit tumor, stimulate interferon production, and possess antiviral properties (Stamets and Chilton 1983), anti-oxidative property, antidiabetic effect, and inhibitory to the development of AIDS. Therefore, it is not a surprise that as food plants were developed in the cultivators, mushrooms were among those selected. Out of 38,000 varieties of mushrooms, 2000 of them are considered as edible, and 200 are found to have medicinal values (Panneerselvam et al. 2009).

Some tasty edible mushrooms primarily *Agaricus bisporus*, *Lentinus edodes*, *Volvariella volvacea*, *Pleurotus* sp., and *Auricularia* sp. are worth mentioning, and they are being cultivated throughout the world. The milky mushroom (*Calocybe* sp.) is being cultivated commercially. Of them, *P. sajor-caju*, *P. citrinopileatus*, *P. florida*, *P. platypus*, *P. djamor*, and *P. eous* are more suited or cultivated in South Indian climatic condition (Panneerselvam et al. 2009).

11.1.2 History: Scope of Edible Mushroom Cultivation

Mushrooms have been used as a delicacy for more than 2000 years (Bhavani Devi 1982). The ancient Indians, Iranians, and Chinese used certain mushroom in their ritualistic performances. In ancient Greek and Roman literatures, there are references about mushrooms. As early as 300 BC., Theophrastus recorded mushroom as valuable food (Panneerselvam et al. 2009). During seventeenth century (Louis XV 1638–1715), mushrooms were brought under cultivation as a crop grown in caves near Paris. Around 1800 the French were growing mushrooms underground in quarries, and mushroom growers in England developed brick spawn. In 1893 the French produced pure culture spawn. In 1953 it was initiated in Northern Taiwan. In India H.W. Newton was the first to grow mushroom during 1886. In 1917, Flack described the cultivation of *Pleurotus ostreatus* on tree stumps and logs (Panneerselvam et al. 2009).

Lambert (1938) discovered that productive spawn could also be prepared from single spore culture. Thomas et al. (1943) introduced *V. diplasia* in Tamil Nadu and did pioneering work on its cultivation. Tamil Nadu Agricultural University has taken up the mushroom research, and as a result, different varieties of mushroom such as *P. sajor-caju* (Rangaswamy et al. 1975), *P. djamor* (MDU1) (Geetha and Sivaprakasam 1993), *P. eous* (APK1) (Muthusamy et al. 1997), *Calocybe indica* (APK2) (Muthusamy and Krishnamoorthy 1996).

The production of mushroom in the world has been increasing steadily day by day. At present the mushroom is under consumed in our country on a worldwide basis. Being a rich source of good quality protein, mushroom is a part of vegetarian food in our country. For the successful and profitable cultivation of mushroom, logical and systematic approaches involving observation of experiments and scientific evolution are desirable for upgradation of quality and yield for various phases of mushroom cultivation. Such observations frequently suggest the growers for changing cultivation methods (Panneerselvam et al. 2009).

Quantitative measurements and objective evolutions are thus essential aspects for such experiments and observations. In this way, development of mushroom industry also requires gradual changes with the time. To promote the mushroom research, National Research Centre for Mushroom was established by ICAR in Chambaghat, Solan, where works are progressing on various aspects of mushrooms (Panneerselvam et al. 2009).

Important historical developments in India were as follows:

1896: Chemical analysis of local mushrooms by B.C. Roy
1918: Mushrooms from Calcutta recorded by Lt. Col. Kirtikar
1921: Two *Agaricus* cultured by S.R. Bose on a sterilized dung medium
1939: Experimental cultivation of paddy straw mushroom by DOA, Chennai
1943: Details of cultivation
1947: R.P. Asthana obtained higher yields of *Pleurotus flabellatus*
1961: First serious attempt on cultivation of *Agaricus bisporus* (a scheme was started by HP Govt. and ICAR at Solan)
1965: Construction on growing facilities and research on spawn production, synthetic compost, etc.
1971: ICAR Coordinated Scheme started
1974: W. A. Hayes, FAO expert guided on improved methods of compost preparation, pasteurization, casing, and environmental management for yield enhancement, etc.

11.1.3 History of Mushroom Cultivation

History goes back the ages of "VEDA" wherein the mention was made in the classical religious scriptures like "Rig Veda" and "Atharva Veda" about use of juice from fly agaric mushroom (*Amanita muscaria*) as an intoxicating drink, named as "Soma." However, the systematic research on mushroom domestication was not aimed for quite a long period (Prakasam 2012).

Newton (1886) exhibited some edible mushrooms in the annual flower show of Horticultural (1896–1897). Sir David Prain (1908) made thorough search for edible mushrooms from various parts of India. In 1918, Kirtikar of Imperial Mycological Society recorded the occurrence of mushrooms from Calcutta. Bose (1921–1926) successfully raised *Agaricus* on sterilized dung. Bose and Bose (1940) discussed some methods for growing mushrooms on horse manure. In 1940, Su and Seth described the procedure for spawn and cultivation of *Volvariella* (Prakasam 2012).

However, the first successful cultivation techniques for paddy straw mushroom, *Volvariella volvacea*, were demonstrated by Thomas in the year 1943 in AgriculturalCollege, Coimbatore. This is lead to the spread of cultivation of this mushroom in all parts of India. Asthana (1947) investigated on the supplementation of paddy straw mushroom beds with horse gram powder to boost the yield. But it is yet to be adopted on a commercial scale. Being atropical mushroom, it is highly suitable for northern Plains as well as coastal and plateaus of South India (Prakasam 2012).

However, a very little research support has been extended of this mushroom in India because of its very low yield potential coupled with highly perishable nature of the fruiting bodies. In 1961, the first cultivation on *Agaricus bisporus* at Solan in Himachal Pradesh started in collaboration with Government of Himachal Pradesh. Researchers conducted during 1960s and 1970s were more or less adaptive in nature and production technology developed lacked refinement (Prakasam 2012).

11.2 External Parameters

Mushrooms grow in nature almost in every country under varied agroclimatic and ecological niches (Nelson 2006). Mushrooms in general have a wider adaptability, and they are generally found to grow in all types of soils, forests and woods, open fields, mountains and hills, deserts, dead wood logs, tree stumps, or with some decomposed organic residues. They appear in all seasons of the year, found abundantly in natural state mainly during rainy season (Nelson 2006).

11.3 Importance of Mushrooms

Greeks believed from the very beginning that mushroom provides strength for warriors in battle. Some praised mushrooms as a delicacy and the Romans considered it as food of the Gods. Early man did not consume edible mushrooms for its nutritional composition because of the taste and flavor. Only after thorough analysis, it is known to all of us that mushrooms are a good source of delicious food with high nutritional attributes (Panneerselvam et al. 2009). Mushrooms can produce the highest quantity of protein per unit area and time from agro-wastes. Other major advantages are as follows (Panneerselvam et al. 2009):

1. In the process, environmental pollution may be reduced because disposal of these agricultural and industrial wastes and by-products may become less of a problem. Examples of such materials are straw, corncobs, sawdust, biogases, wood pulp, cotton waste, oil palm waste, banana levels, poultry wastes, coconut husks, and tree bark leaves. Mushroom grows independent of sunlight without fertile land. They do not compete with field or fruit crops and provide an additional avenue for increasing food supply.
2. Mushrooms have a huge export potential. There is a world market for 30 lake tones by 2005.
3. They offer vast rural employment potential. Mushroom cultivations involve various technologies. In the instances where limited capital is available, methods that require simple equipment can be used.
4. For outdoor cultivation, mushrooms may be grown on beds constructed from layers of various types of substrate or in tiers on wooden frames. Mushroom growing does not require large space as in case of other crops. They may be located on any unused land or available space between trees or beside buildings or houses, so that they do not compete with other crops for space.
5. They may be adopted as small-scale industry and may earn foreign exchange by commercial cultivation.

11.4 Different Types of Mushrooms

11.4.1 Agaricus *sp.*

Class: *Basidiomycetes*
Subclass: *Hymenomycetes*
Order: *Agaricales*
Family: *Agaricaceae*
Genus: *Agaricus*

The Latin name of the button mushroom is *Agaricus bisporus*. Different types of spawns are virgin spawn, flake spawn, brick spawn, and grain spawn. If the factors on which its production is dependent are provided by the controlling measure, its production is possible anywhere at any season. The related factors are the temperature, moisture, ventilation, and seed (spawn). On spawning hyphae begin to grow. When they form a network is called mycelium. This is the first phase of growth. In the second phase of growth, fruit bodies or pinheads grow from the mycelium. White button mushroom can be grown in the hills at an altitude of 2000 m for raising five crops in a year. From spawning to harvesting, the room temperature should not rise above 20 °C.

The work schedule for growing button mushroom (Dey 2013a, b):

1. Compost making by adopting correct methods
2. Collecting quality spawn and spawning
3. Casting the beds on spawning
4. Harvesting (Picking fruit bodies)

If the compost, casing material and the room for keeping trays are made germ-free, the incidence of diseases will not take place. The excess humidity, temperature, and poor ventilation are likely to cause the disease infection. The clean and tidy environment helps keep disease at day. The economics of inputs of the commercial culture of button mushroom is profitable. Even one can make profits by growing in the cottage scale. Moreover the profits come within a very short time (Dey 2013a, b).

*11.4.2 Oyster Mushroom (***Pleurotus** *sp.)*

Class:	*Basidiomycetes*
Subclass:	*Hymenomycetes*
Order:	*Poriales*
Family:	*Lentinaceae*
Genus:	*Pleurotus*

*11.4.3 Paddy Straw Mushroom (***Volvariella** *sp.)*

Class:	*Basidiomycetes*
Subclass:	*Agaricomycetes*
Order:	*Agaricales*
Family:	*Pluteaceae*
Genus:	*Volvariella*

Paddy straw is exclusively needed for growing *Volvariella* species. They cannot be grown in the straws of the other cereals. In the coastal region the continuous heavy rains decompose the paddy straw in the farmyard. *Volvariella volvacea* is found to grow on decomposed straw when the day temperature varies from 30 °C to 35 °C. Even in the concave joints, a paddy straw thatch where the straw decomposes fast *Volvariella* grows there. Nowadays it is grown with the help of the grain spawn and paddy straw commercially. It is known that *Volvariella* was grown in China in the eighteenth century. In this country the scientific culture of *Volvariella* was first experimentally started at Coimbatore. It requires high temperature more than any other cultivated mushroom. The congenial temperature for its ideal growth is being 30–35 °C (Dey 2013a, b).

11.5 Nutritive Value

In a country like India where the human population is steadily rising, it is inevitable to meet the increasing demand for high protein foods. It is also true that there is abundance availability of agricultural waste products and these agricultural wastes at present have limited or no use at all. Mushroom cultivation is an effective and economic way of upgrading agro-plant wastes into high value protein.

11.5.1 Vitamins

It is important to realize that mushrooms contain vitamins such as B1, B2, B3, B6, B12, D1, D2, and H and pantothenic acids which are not occurring in green plants. However the presence and quantity of vitamins on mushroom vary with regard species (Table 11.6a and b).Vitamins are more in *Calocybe indica* and they have rare type of vitamins also. According to Anderson and Fellers (1942), *A. bisporus* does not contain vitamins A, D, or E. They found 8.6 mg ascorbic acid, 5.82 mg nicotinic and 2.38 mg pantothenic acid, 0.12 mg thiamin, 0.52 mg riboflavin, and 0.018 biotin per 1000 mg fresh weight. Mushroom is reported to be an excellent source of riboflavin and nicotinic acid (niacin) and a good source of pantothenic acid (Tables 11.1 and 11.2).

Table 11.1 Vitamin content (Mg/100 g) of fresh mushrooms (Panneerselvam et al. 2009)

S. No	Mushroom	Thiamine	Riboflavin	Niacin
1.	*Pleurotus* sp.	0.5	0.510.9	–
2.	*Volvariella* sp.	0.14	0.61	2.40

Table 11.2 Vitamin content (Mg/100 g dry weight) of some of the edible mushrooms

S. No	Mushroom	Thiamine	Riboflavin	Niacin	Ascorbic acid
1.	*Agaricus bisporus*	1.1	5.0	55.7	81.9
2.	*Lentinus ostreatus*	7.8	4.9	54.9	–
3.	*Pleurotus* sp.	4.8	4.7	108.7	–
4.	*Volvariella* sp.	1.2	3.3	91.9	20.2

Table 11.3 Composition of cultivated mushroom and some common vegetables per 100 g of article (Wooster 1954)

S. No	Name	Calories	Moisture	Fat	Carbohydrates	Protein%
1.	Beetroot	42	87.6	0.1	9.6	12.9
2.	Lime beans	128	66.5	0.8	23.5	22.2
3.	Mushroom	16	91.1	0.3	4.4	26.9
4.	Potato	83	73.8	0.1	19.1	7.6

Table 11.4 Proximate analysis of edible mushrooms (Percent Fresh eight basis)

S. No	Name	Moisture	Ash	Protein	Fat	Crude fibers
1.	*Agaricus bisporus*	89.5	1.25	3.94	0.19	1.09
2.	*Lepiota* sp.	91.0	1.09	3.3	0.18	0.86
3.	*Pleurotus* sp.	90.0	0.97	2.78	0.65	1.08
4.	*P. ostreatus*	92.5	–	2.15	–	–
5.	*Termitomyces* sp.	91.3	0.81	4.1	0.22	1.13
6.	*Volvariella diplasia*	90.4	1.10	3.90	0.25	1.57
7.	*Volvariella volvacea*	88.4	1.46	4.98	0.74	1.38

11.5.2 Proteins

A sufficient calorie intake does not guarantee a good standard of nutrition. Food containing minerals, vitamins, and enough of the right kind of protein is necessary in addition to that furnishing energy. Protein deficiency is not only a future problem but also an existing reality. Generally protein is synthesized by green plants. But the presently available protein is not sufficient for human beings (Panneerselvam et al. 2009).

Mushroom was vegetables. To maintain nutrition, human being requires 100–200 gm of mushrooms (dry weight). The mushroom protein is similar to that of muscle protein. Hence, studies suggest that mushrooms are well suited to supplement diets which lack protein since they have rightly been called "vegetable meat" (Tables 11.3 and 11.4). Vast literature is available particularly on the protein content of mushroom. Generally mushrooms can be compared more favorably with other crops in terms of yield per unit area. Cereals can give an annual yield of 3000–6000 kg/ha, but mushrooms may give up to 2 million kg/ha. The product of an acre

Table 11.5 Amino acid composition of *A. bisporus* (per 100 g dry matter) (Hayes and Haddad 1976)

1.	Alanine	2.40
2.	Arginine	1.90
3.	Aspartic acid	3.14
4.	Cystine	0.18
5.	Glutamic acid	7.06
6.	Glycine	1.20
7.	Histidine	0.64
8.	Isoleucine	1.28
9.	Leucine	2.16
10.	Lysine	2.16
11.	Methionine	0.39
12.	Phenylalanine	1.55
13.	Proline	2.50
14.	Serine	1.89
15.	Threonine	1.89
16.	Tryptophan	3.94
17.	Tyrosine	0.78
18.	Valine	1.63

of land can be transformed into 10 tons as much fungal protein as meat protein. The following data show yields of dry protein per unit area utilized for forming beef and *A. bisporus* (Cooke 1977).

Approximate annual yield dry protein (kg/1 ha) (Panneerselvam et al. 2009)

11.5.3 Amino Acids

Edible mushroom contains maximum number of all the essential amino acids (Hayes and Haddad 1976). Tryptophan and lysine are present in high concentration when compared to cysteine and methionine. These amino acids are generally absent in vegetable problems. Mushroom protein is considered as significant in the vegetarian diet. The amino acid generally enhances the memory power, and it is called as fundamental unit of proteins, and the proteins are generally designated as body builders. A list of amino acid and their percentage composition is shown in Table 11.5.

11.5.4 Mineral Elements

Ash analysis given by Anderson and Fellers (1942) shows that *A. bisporus* contain maximum percentage of phosphorus (Table 11.6).

Table 11.6 Mineral content (Mg/100 g Dry Weight) of some other edible mushrooms (Chang and Hayes 1978)

Mushroom	Ca	P	Fe	Na	K
(On dry weight basis)					
Agaricus bisporus	23	1429	0.2	Ndb	4762
Lentinus edodes	33	1348	15.2	837	3793
P. ostreatus	98	476	8.5	61	Nd
V. volvacea	71	677	17.1	374	3455
(On wet weight basis)					
Pleurotus sp.	3.3	134.8	1.5	83.7	379.3
Volvariella sp.	5.60	100	1.70	80	320.0

Table 11.7 The fiber content of various mushrooms

S. No	Mushrooms	Fiber content (%)
1.	*Agaricus* sp.	8.0
2.	*Pleurotus* sp.	13.3
3.	*Calocybe* sp.	41.0
4.	*Volvariella* sp.	11.1
5.	*Lentinus edodes*	7.3
6.	*Auricularia* sp.	4.7

11.5.5 Fiber Content

Generally edible mushroom contains maximum amount of fiber. Their low sugar, low caloric, and high fibrous nature is in addition with various vitamins an added advantage. Due to this nature, it is believed to cure heart diseases, ulcers, diabetes, etc. (Table 11.7).

11.5.6 Energy Value of Mushrooms

Mushroom is the best source of energy. According to Hayes Haddad (1976), one pound (454 g) of fresh edible mushroom gives 120 kcal of energy. Since, ancient time mushroom have featured as choice dishes on many a regal table. In Italy Caesar relished the flavorful meadow mushroom. The Chinese of the orient added the delicate fungi to scent their sauces and soups. Mushroom possesses an appetizing property, and it is considered as potential protein source to bridge the protein gap. In India vegetarian peoples are more, popularization of mushroom as a vegetable protein sources among the vegetations is an essential need (Panneerselvam et al. 2009).

Table 11.8 China's production of edible mushrooms (Wu et al. 2013).

S. No	Years	Production in china 1000 ton	Share of world of products %	Values (in billion yuan)
1.	1978	60	6	–
2.	1986	586	27	–
3.	1994	2641	54	–
4.	2000	6636	64	23
5.	2001	7818	66	32
6.	2002	8764	71	41
7.	2003	10.387	73	48
8.	2004	11.600	68	46
9.	2005	13.340	70	48
10.	2006	14.000	70	64
11.	2007	16.820	75	80
12.	2008	17.300	80	82
13.	2009	22.203	80	110
14.	2010	22.012	80	141
15.	2011	25.717	80	149

11.6 Production of Edible Mushroom

Photochemicals are essentially useful mushrooms to the dietary supplements; medicinal properties and their cultivation to the sustainable agriculture were discussed below (Hobbs 1995). Mushroom is rich in protein and nutritional properties. It is a good food material for diabetic patients (Table 11.8).

11.7 Conclusion

Nowadays, the interest of people in utilization of mushrooms and their cultivation is greatly and steadily increasing due to the awareness about the usefulness of mushrooms and their nutrition. Since earlier times, mushrooms have been treated as special kind of food. Today we have the mushrooms for diet every day and contain vitamins and other important nutrients. Thus its immediate prospects contributions should be properly recognized. The fungi are the living organisms belonging to an altogether or separate class of organisms. Unlike plants, they are devoid of chlorophyll and, therefore, have to depend on outside source of food for their nutritional requirement.

As soon as the mankind started to gather their food from the nature and their surrounding environment, and subsequently grew them, they knew about mushrooms. Early records say that ancient Indians, Greeks and Romans, and others knew about mushroom and these were associated with the offerings to ancient gods and goddesses and also with royal classes. Their indoor cultivation first started in some European countries like France and Holland around the early part of the last country and on commercial scale at the turn of this century.

References

Anderson EE, Feller CR (1942) The food value of mushroom Agaricus Compestri. Pool Amer Soc Hort 41:3010–303

Asthana RP (1947) Mushroom cultivation in Central Provinces, Nag. Agric. Coll. Nagpur, 32:26–31

Alexopoulos CJ, Mims CW (1985) Introduction of Micologia Ediciones Omega. Barcelona 84(638):282–747

Bhavani Devi S (1982) Biology and cultivation of *Pleurotus florida*. Symp on plantation crops. C. F. T. R. I., Mysore

Bose SR, Bose AB (1940) An account of edible mushrooms of India. Science and Culture 6:141–149

Chang SI, Hayes WA (1978) The biology and Cultivation of Edible mushrooms. Academic Press, New York, p 819

Cooke RC (1977) Fungi man, and his environment Largman, London/New York, p 144

Dey SC (2013a) Mushroom growing series. Agros gardening series. Agrobios, Jodhpur, pp 27–43

Dey SC (2013b) Mushroom growing series. Agros gardening series. Agrobios, Jodhpur, pp 50–51

Geetha D, Sivaprakasam K (1993) Treating rice seeds with fungicides and antagonists to control seedborne diseases. Int Rice Res Notes (Philippines) 18:30–31

Hayes WA, Haddad N (1976) The food value of the cultivated mushrooms and its importance in industry. Mushroom J 40:104–110

Hobbs C (1995) Medicinal mushrooms: an exploration of tradition, healing and culture. Botanica Press, Santa Cruz

Lambert EB (1938) Principles and problems of mushroom culture. Bot Rev 4:397–426

Muthusamy S, Krishnamoorthy AM, Marimuthu T, Narasimhan V (1997) Temperature, pH and light requirement for the growth of Calocebe indica. Indian Mushroom Conference. 2:21

Nelson N (2006, August 13) *Parasola plicatilis*. Retrieved 2008 October 13

Panneerselvam A, Nadimuthu N, Ambikapathy V, Murugan R, Gopinathan M (2009) Hand book on edible mushroom cultivation. KP Printers, Thanjavur, pp 71–99

Prakasam V (2012) Current scenario of mushroom research in India. Indian Phytopath 65(1):1–11

Punia BS (1992) Microbial colonization in the rumen and utilization of ligno-cellulosic feeds. Indian J Microbiol 32:235–247

Rangaswamy G, Kandaswamy TK, Ramaswamy TK (1975) Pleurotus sajor-caju (Fr.) Dinger. A protein rich nitrogen fixing mushroom fungus. Curr Sci 44:403–4

Stamets P, Chilton J (1983) Casing layer in: The mushroom cultivator. Chapter- VIII. Agaricon Press Olympia. WA, USA 127–138

Sivapraksam K, Ramaraj B (1986) Studies on some factors influencing the yield of Pleurotus mushroom. Indian Mushrooms. Proc. Nat. Symp. Mushrooms 127–132

Thomas KM, Ramakrishan TS, Narsimhalu IL (1943) Paddy straw mushroom. Madras Agric J 31:57–59

Wooster HA Jr (1954) Nutritional data, 2nd edn. H. J. Heirriz Co., Pitsburg

Wu SR, Zhao CY, Hou B, Tai LM, Gui MY (2013) Analysis on Chinese edible fungus production area layout of nearly five years. Edible Fungi China 1:51–53

Chapter 12
Contributions to a Sustainable Production of Food of Animal Origin

Gerhard Flachowsky, Dirk von Soosten, and Ulrich Meyer

Abstract Sustainability in human food chain characterizes the global balance/equilibrium between efficient use of limited natural resources (such as arable land, water, fuel etc.), emissions (e.g. carbon dioxide, methane, laughing gas etc.) and socio-economic and ethical aspects as base for the existence of future generations.

Sustainability in the production of food of animal origin or edible protein means an efficient production. Such calculations should not only include the food chain links "feed – animal – food of animal origin" but the whole food chain. A system has the highest efficiency or the largest sustainability if it is impossible to improve one parameter without deterioration of one or more other parameters.

After introduction, the authors define the term sustainability and deduce the objective of the review paper. Protein of animal origin is the main point of the paper, and it is in the focus of the following sections. Resource inputs in form of edible land, water, fuel, etc. and outputs in form of animal yields (e.g. milk, eggs, meat, fish, etc.) and emissions are described, and reduction potentials for emissions are measured. Some potentials to improve sustainability of production of food of animal origin, such as feeds, which do not compete with human nutrition, plant and animal breeding, potentials of other protein sources and alternatives of animal products in nutrition including reduction of feed/food losses are discussed in the paper.

More complex calculations under consideration of parameters of efficient use of limited resources and reduction of emissions seem to be helpful to find out a certain optimum in production of food of animal origin.

G. Flachowsky (✉) • D. von Soosten • U. Meyer
Institute of Animal Nutrition, Friedrich-Loeffler-Institute (FLI), Federal Research Institute of Animal Health, Bundesallee 50, 38116 Braunschweig, Germany
e-mail: Gerhard.flachowsky@t-online.de

A. Dhanarajan (ed.), *Sustainable Agriculture towards Food Security*,
https://doi.org/10.1007/978-981-10-6647-4_12

12.1 Introduction

The energy and protein conversion from feed into food of animal origin is low and may vary between 3% (energy – beef) and up to 40% (energy – dairy; protein – chicken for fattening; Smil 2000; Cassidy et al. 2013). In some countries (e.g. USA) between 67% (energy) and 80% (protein) of the crops are used as animal feed (Cassidy et al. 2013). The high need for feed connected with low efficiency of conversion and high emissions present the following question: "Is there any need for food of animal origin?"

As vegans demonstrate, there is no essential need for food of animal origin, if the human diets contain all or are supplemented with all essential nutrients. But on the other side, the consumption of meat, fish, milk, eggs and other protein sources of animal origin may contribute significantly to meet the human requirements of amino acids (e.g. WHO et al. 2007; Smith et al. 2013; Thompson and Amoroso 2014; Wu et al. 2014a) and some important trace nutrients (such as Ca, P, Zn, Fe, I, Se, vitamins A, D, E, B_{12}), especially for children and juveniles as well as for pregnant and lactating women (Wennemer et al. 2006). Human nutritionists have recommended that about one third of the daily protein requirements (0.66–1 g per kg body weight; e.g. Rand et al. 2003; Jackson 2007; WHO et al. 2007, Bauer et al. 2013) of adults should originate from protein of animal origin. Consequently, about 20 g of the daily intake of about 60 g protein should be of animal origin, which is lower than the present average consumption throughout the world. During the last few years, there was an average consumption of protein of animal origin (without fish) of about 24 g per capita and day, ranging between 1.7 (Burundi) and 69.0 g (USA). It is a challenge for the future to overcome these discrepancies. Meat, milk and eggs provide around 13% of the energy and 28% of protein consumed globally, with the higher share in the so-called developed countries (around 20 and 40%, respectively, FAO 2009).

Other reasons for consumption of food of animal origin are the high bioavailability of most nutrients and their considerable "enjoyment value". Such food is presently also considered as an indicator for the "standard of living" in many regions of the world. Further reasons for the higher demand of food of animal origin in some countries are the increased income of the population (Keyzer et al. 2005; Tilman et al. 2011; Kastner et al. 2012) and the imitation of the so-called "Western style of life" (nutrition). Many developing countries continue to consume more animal products than they produce. Therefore, they will continue to drive the world demand for all agricultural products, including food of animal origin (Guyomard et al. 2013). Wu et al. (2014b) estimate that with exponential growth of the global population and marked rises in meat consumption per capita, demands for animal-source protein are expected to increase by 72% between 2013 and 2050. Higher food amounts of animal origin require higher plant yields and/or more area for feed production (Wu et al. 2014b; Flachowsky and Meyer 2015a, b) and more animals and/or higher animal yields as well an increase in agricultural trade. Therefore, some authors propose a redefinition of agricultural yield and agriculture in general:

"from tons to people nourished per hectare" (Kastner et al. 2012; Cassidy et al. 2013; Flachowsky et al. 2017b) and ask for more sustainable animal agriculture (e.g. Kebreab 2013; FAO 2014). In addition, livestock production is more and more also seen in connection with climate change (e.g. Malik et al. 2015).

On the other hand, changing the eating patterns (Guyomard et al. 2012) and eating less or no livestock products, especially meat, are often seen as possible solutions to reduce the environmental impact of animal agriculture (Pimentel and Pimentel 2003; Baroni et al. 2007) and to reduce the per capita land requirements (e.g. Peters et al. 2007; Flachowsky et al. 2017b).

These developments raise concerns about the sustainability and environmental impact of animal agriculture. In summary, food security and optimal human nutrition under consideration of limited resources, increased emissions and expected climate change can be considered as large challenges for all those dealing with feed and food production and nutrition (NRC 2015).

12.2 Definition of Sustainability and Objective of the Paper

The large need of wood in Europe in the middle age resulted in over-exploitation of forests.

Hans Carl von Carlowitz, a German administrator and scientist, formulated on the base of its own experiences the first time in the "Sylvicultura Oeconomica" (von Carlowitz 1713) the so-called principle of sustainability in forestry: "The most important objective of science/management is the conservation and cultivation of forests that a sustainable utilization (not more harvest than growing up) can follow that the county (Saxonia, a German county) may existent in the future".

The discussion of the "Club of Roma" about "borders of growth" sustainability was the first time described as a "condition of global equilibrium" (Meadows et al. 1972). Later, Brundtland (1987) introduced as chair of the UN commission on "sustainability and sustainable development" the term sustainability in the political language. Environmental objectives are combined with socio-economic objectives to arrive stable societies and a balance between economy, ecology and social aspects as shown in Fig. 12.1.

Later, but based on the developments mentioned above, philosophers and natural scientists of various disciplines studied and analysed more and more present global developments. The balance between planet (global resources and emissions), people (social aspects of population all over the world) and profit (economic aspects, moneymaking) in the so-called 3**P** concept (IUCN 2005; Boonen et al. 2012; Aiking 2014) is an important prerequisite for a sustainable life and development on the earth. Some authors are afraid that the balance between the 3P would be more and more disturbed and an ethical dimension should be introduced as the fourth dimension (IUCN 2005). Profit should not and cannot be the only one objective of production. Ethical aspects of food production should be also considered as parameter of sustainability (Millar et al. 2009; Casabona et al. 2010; Potthast and Meisch 2012;

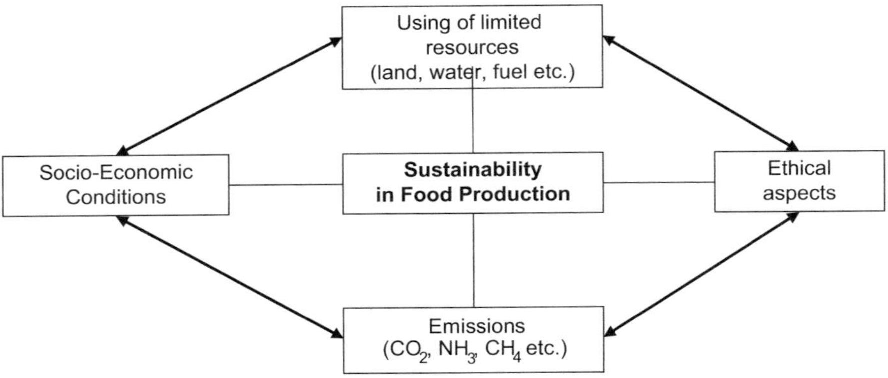

Fig. 12.1 Sustainability as balance between using of limited natural resources, emissions, socio-economic and ethical conditions to produce food of animal origin (Flachowsky and Hachenberg 2009)

Viljoen and Wiskerke 2012; Wals and Corcoran 2012; see Fig. 12.1). We have to find a balance between a careful, ethical responsible and sustainable use of limited resources (see above) on the one hand (Fedoroff et al. 2010; Giovannucci et al. 2012) and low emissions with local and global consequences for later generations (Foley et al. 2011) on the other hand.

Under consideration of the situation mentioned above, a sustainable agriculture including a sustainable production and also a distribution of food or protein of animal origin should be characterized by:

- Efficient use of limited resources and low emissions (planet, ecological aspects)
- Socio-economic and ethical responsible production (economy, people)
- Base for existence of future generations (ethical aspects)
- Fairness in food production and distribution

Ruttan concluded already in Ruttan 1999 that the transition to sustainable growth in agricultural production during the twenty-first century will take place within the context of a transition to a stable population and a possible transition to a stable level of material consumption. Global food sustainability is considered to be a great challenge for mankind (Lawrence et al. 2011).

Based on previous reports from our group (e.g. Flachowsky and Schulz 2011; Niemann et al. 2011; Flachowsky and Kamphues 2012; Flachowsky et al. 2013a, b; Flachowsky and Meyer 2015a, b, c; Flachowsky et al. 2017a, b), the objective of the paper is to analyse the present stage of production of edible protein of animal origin under aspects of sustainability.

12.3 Resource Need and Emission to Produce Protein of Animal Origin

In the future, there will be a strong competition for arable land (about 1.5 billion ha; FAO 2013) and further non-renewable resources such as fossil carbon sources, water (e.g. Molden et al. 2010; Schlink et al. 2010), some minerals (such as phosphorus; Hall and Hall 1984; Scholz and Wellmer 2013; see Table 12.1) as well as between feed, food, fuel, fibre, flower and fun (6 Fs concept; Aerts 2012) and areas for settlements and natural protected areas.

In this connection, more attention should be paid to the need of limited natural resources per amount of animal product, expressed as footprints per product such as "water footprints" (see Mekonnen and Hoekstra 2010), "mineral (especially phosphorus, P) footprints" and "land (arable or total land) footprints" (see De Vries and de Boer 2010; Nijdam et al. 2012; Flachowsky et al. 2017a). These footprints are given in kg, L, or tons per unit product and characterize the efficiency of various production processes.

On the other side, special attention has also been paid to the outputs from agriculture (e.g. FAO 2006; Vermeulen et al. 2012) including livestock keeping especially the so-called greenhouse gas (GHG)-relevant emissions such as CO_2, CH_4, N_2O and further gases (see IPCC 2006). All the climate-relevant emissions are summarized to the so-called carbon footprints (CF). They have also been modified or called ecological footprint (EF), eco-balances (EB), life cycle assessments (LCA) or life cycle impact assessment (LCIA). In all cases the term means a summarized parameter for all gaseous emissions with greenhouse gas potential to sensitize producers and consumers (e.g. Upham et al. 2010; Young et al. 2010) for an efficient use of fossil carbon sources and to reduce GHG emissions per product (see also De Alvarenga et al. 2012). CF or LCA are used as a tool for estimating environmental effects caused by products or processes. Furthermore, CF may also contribute to assessing the resource and feed efficiency between various regions and production systems (see FAO 2015; Flachowsky et al. 2017b). All these activities are summarized to contribute to a more sustainable production of food of animal origin (e.g. Makkar and Ankers 2014; Flachowsky and Meyer 2015a, b).

Sustainable production of food of animal origin means to contribute to an efficient use of limited natural resources and to minimize emissions during food production along the whole food chain (Fig. 12.2). In conclusion, plant and animal

Table 12.1 Limited resources and emissions to produce food of animal origin

Limited resources	Emissions
Land (especially arable land)	Carbon dioxide (CO_2)
Water	Nitrogen compounds (NH_3, N_2O etc.)
Fuel/energy	Methane (CH_4)
Some minerals (e.g. P, …)	

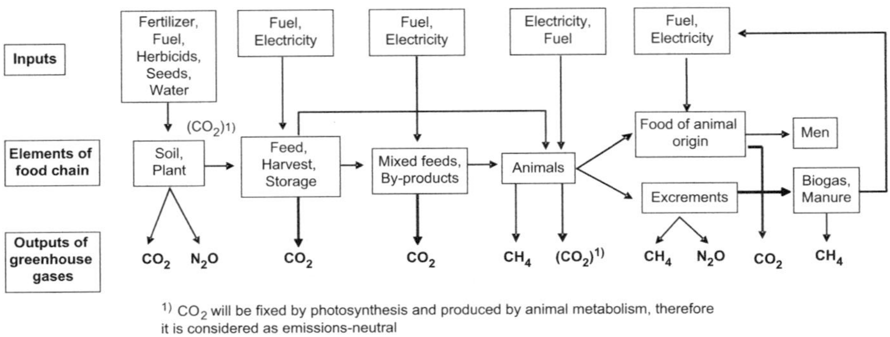

Fig. 12.2 Substantial elements of the food/supply chain to produce food of animal origin as well as selected inputs of resources and outputs of greenhouse gases (base for system boundaries)

breeding for lower inputs of limited resources, a more efficient conversion of limited resources in feed and food and a reduction of emissions along the human food chain could be considered as a real challenge for all these, working along the food chain (Flachowsky and Meyer 2015c).

12.4 Protein Content of Various Animal Products

The production of protein of animal origin is one of the most important goals of animal husbandry (De Vries and de Boer 2010; Lesschen et al. 2011; Flachowsky and Kamphues 2012; Nijdam et al. 2012; Flachowsky et al. 2017b). On the other hand, the efficiency and the emissions of food of animal origin can be also compared on the base of edible protein. The N or protein (N × 6.25) content of various food of animal origin may vary from values used for calculations in Table 12.3. These data do not substantially disagree with values from human food tables (e.g. Souci et al. 2008).

Quantification of protein yield varies depending on some influencing factors. For example, milk and eggs are clearly defined as food of animal origin, and the yield can be measured (and expressed as kg or L per animal or per day), and therefore, it is relatively easy to use the yield of lactating and laying animals for further calculations. The edible fraction is nearly adequate to the yield. Only minor fractions are not be consumed by humans (e.g. colostrum, milk samples at the beginning of milking, egg membranes and shells).

It is much more difficult to quantify and characterize the yield from the animal body after slaughtering and processing. The following endpoints can be measured in the case of animals for meat production (see Flachowsky and Kamphues 2012):

- Weight gain of the animal (per day or per growing period) during the whole life span
- Weight gain of animal without content of the gastrointestinal tract

Table 12.2 Advantages and disadvantages of various outputs/endpoints of animal yields (by Flachowsky and Kamphues 2012)

Animal yields	Advantages	Disadvantages
Milk, eggs	Easily measurable, almost complete edible	Variation in protein, fat and energy yield, analyses may be useful
Body weight gain	Easily measurable	High portion of nonedible fractions in the gains
Carcass weight	Easily measurable	Still contains fractions, which are not edible (e.g. bones)
Meat, edible fraction	Completely edible	Categorization and separation not easy
Edible protein	Most important objective of animal production; comparison of various ways and sources to produce protein of animal origin	Categorization of various fractions as edible and difficulties to measure; additional analytical work; variation in N/protein content

– Empty body weight (or carcass weight; meat and bones; warm or cold)
– Meat (empty body weight minus bones)
– Edible fraction (meat plus edible organs and tissues)
– Edible protein (edible fractions of the carcass multiplied with their specific protein content)

All endpoints are characterized by some advantages and disadvantages. From nutritional and scientific points of view, edible protein seems to be the most favourable measurement, but in the case of meat production, its measurement is not easy and requires some analytical work (Table 12.2).

For practical reasons carcass weight or weight gain (warm or cold) would be the most important endpoint to measure the yield of slaughtered animals. This weight is measurable in the slaughtering houses (Peters et al. 2010) and can be used for further calculations.

Mostly the term “meat” is used, but what is exactly meant (meat with or without bones) is sometimes not clearly described. Peters et al. (2010) introduced the term “hot standard carcass weight” (HSCW) as the weight at the exit gate of the meat processing plant. FAO (2013) defines meat from animals as fresh, chilled or frozen meat with bones. FAO (2013) data on meat are given in terms of dressed carcass weight excluding offals and slaughter fats. The HSCW varies between 50 and 62% of the live weight of cattle before slaughter, but it may vary between 50% in the case of sheep and up to 80% for turkeys (e.g. Williams et al. 2006; Peters et al. 2010). Nijdam et al. (2012) used the following killing out factors (carcass weight in % of live weight): 53% for beef, 75% for pork, 46% for mutton, 70% for poultry and 40% for fish. The edible meat yield (retail meat of carcass) is given by the same authors with 70% for beef, 75% for pork and mutton, 80% for poultry and 100% for fish.

Large differences exist between countries and also between population groups within one country, when it comes to the definition of “edible”. Therefore, it is difficult to compare results by various authors and to find out the actual protein yields. If other percentages of edible fractions are used for calculations (see Flachowsky

Table 12.3 Protein content[a] of some edible animal products/food by various authors (in g per kg edible product)

Product/food; Authors	Milk (Cows)	Beef	Pork	Poultry	Eggs
Flachowsky (2002)	34	190	150	200	120
GfE (1995, 1999, 2001, 2008)	34	170–210	157 (129–178)	n.d.	121 (110–124)
Souci et al. (2008)	33.3 (30.8–37.0)	220[b] (206–227)	220[b] (195–240)	199	125
De Vries and de Boer (2010)	30	190	190	190	130
Mekonnen and Hoekstra (2010)	33	138	105	127	111
Andersen (2011)	34	206–212	183–216	182–242	125
Lesschen et al. (2011)	34.4	206	156	206	119
Nijdam et al. (2012)	35	200	200	200	130
USDA (2016)	34	173	139	186	126

n.d. no data
[a]N-content × 6.25
[b]Muscles only

and Kamphues 2012), the authors should clearly describe the used values to understand and interpret the results.

Another important factor for a reliable calculation of protein yields of food of animal origin is the protein content of edible fractions as shown in Table 12.3.

The results of specific studies (see Table 12.3) as well as values from food tables (e.g. Lawrie and Ledward 2006, Souci et al. 2008, USDA 2016) agree more or less within the ranges of the protein content of milk (between 30.8 and 37.0 g/kg), beef (170–227 g/kg), pork (129–220 g/kg), poultry meat (182–242 g/kg) and eggs (110–130 g/kg). Except for milk, Mekonnen and Hoekstra (2010) used for their calculations the lowest values for all other foods of animal origin, compared with other authors (see Table 12.3). The protein yield in the edible fractions of milk, meat and eggs was calculated based on these data.

12.5 Resource Inputs

12.5.1 Land

Land, especially arable land, is one of the most important limiting factors (Bruinsma 2009). Only a small portion of the global surface (about 13.4 billion ha) is available as arable land (more than 1.5 billion ha, about 12% of the world's land area; FAO 2013). This area could be extended to a certain level (about 120 Mio. ha; FAO 2013), but some other areas cannot be used because of limited water resources, forests, urban settlements, protected for environmental reasons, deserts, mountains etc.

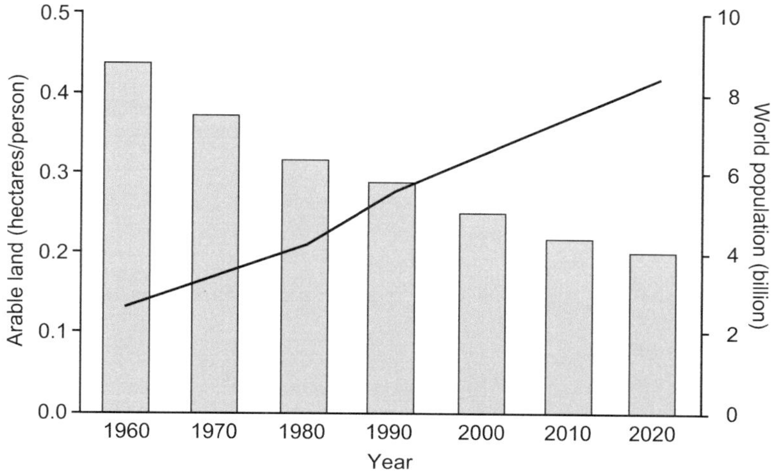

Fig. 12.3 Development of world population and arable land per inhabitant between 1960 and 2020 (By FAO 2013)

In consequence of the limited area of arable land and the increase of population, the area of arable land per inhabitant decreased from about 0.45 ha (1960) to about 0.25 ha (2010) and will further decrease to below 0.20 ha per inhabitant after 2020 (see Fig. 12.3).

This situation and the increasing area for biofuel production (Bessou et al. 2011), organic farming, settlements, natural protected areas and further purposes have consequences for feed and food production. Because of the high need of limited resources, attention has been spent to the area need for animal production. But only some authors considered the land use and also described as land or area footprint for food of animal origin (e.g. Peters et al. 2007; de Vries and de Boer 2010; Nijdam et al. 2012; Flachowsky et al. 2017b). They proposed to distinguish in areas (mostly arable land) which can be used also for other purposes (6 F concept) as for feed production and in typical feed areas (grassland or perennial crops). There is no reason to believe that the public interest in the use of limited resources and the high emissions (CF) of food of animal origin will diminish in the near future because of the worldwide increase in land need and emissions.

Arable land use per unit product or protein of animal origin depends mainly on animal species and category and plant and animal yields, using of grassland as well as portion of coproducts in the rations.

Except Roughages (grassland) and grains, coproducts from agriculture, such as cereal straw and sugar beet leaves, from food production (e.g. cereal coproducts, oilseed coproducts, etc.) and biofuel industry (e.g. DDGS, rapeseed cake; Ros et al. 2010; Wilkinson 2011; Makkar 2012; Knaus 2013; Ertl et al. 2015), are used in animal nutrition and make the calculation of arable land need for food of animal origin more complicated (see Bockisch et al. 2000; Flachowsky et al. 2017b). Some authors estimated human-edible (protein) fractions of feeds and calculated with

Table 12.4 Crude protein content of some feeds (Jeroch et al. 1993) and their human-edible fraction (hef; in %) by Wilkinson (2011) and for three different estimation scenarios (Ertl et al. 2015)

Feedstuff	Crude Protein (g/kg DM)	hef-fractions (in %) by Wilkinson (2011)	hef-fractions (in %) by Ertl et al. (2015)		
			Low	Medium	High
Barley	125	80	40	65	80
Maize	106	80	70	80	90
Wheat	138	80	60	80	100
Soybeans	404	80	50	92	93
Rapeseed meal	406	20	30	59	87
Soybeans meal	513	80	50	71	92
Wheat bran	160	20	0	10	20
Maize silage	86	0	19	29	45
Others[a]		0	0	0	0

[a]Other coproducts (e.g. sugar beet pulp; brewers grains; dried distiller's grains with solubles, etc.) and roughages (grass; silages, hay, etc.)

these values (see Table 12.4) in rations of animals. Coproducts in animal feeding may decrease the pressure on the global grain demand.

From our view, it is incorrect to give the hef-fraction for forages and silages with 0 (see Table 12.4; Wilkinson 2011), because such arable land can also be used for cultivation of cereals or other edible cultures as considered by Ertl et al. (2015); see Table 12.4.

The results presented here show that a clear description of study conditions or the bases of calculations is a very important prerequisite to compare results concerning land use.

More complex calculations under consideration of parameters of efficient use of limited resources and reduction of emissions seem to be helpful to find out a certain optimum in production of food of animal origin. The following parameters should be considered in future calculations:

- The use of arable land (competition between various users)
- Efficient use of water for feed and animal production
- Minimization of the use of fuel and other limited natural resources in the food chain
- Utilization of permanent grassland and coproducts from agriculture and industry (see Table 12.4)
- Reduction of greenhouse gas emissions per product or per kg edible protein and along the whole food chain
- Conservation of the biodiversity
- Plant and animal breeding as the starting points of human food chain
- Comparison of production of food of animal origin with other protein sources including vegetarian foods (e.g. milk, meat based on soybeans)
- Calculation of land use per inhabitant under consideration of eating patterns of population
- Reduction of food wastage

Producing of food of animal origin is a very complex process (see food chain; Fig. 12.2), and selective consideration, i.e. focussing on single factors, does not provide an assessment that reflects the complexity of the subjects.

A cooperation of animal scientists (nutritionists, breeders, animal keepers/farmers, veterinarians, etc.) with scientists working in the fields of plant and feed science, ecology and economy seems to be necessary to solve the problems and to develop better and loadable land footprints.

In summary, more (feed) for more (people) with less (resources and emissions) is one of the most important challenge for all those involved in feed/food science and production.

12.5.2 Water

Water is one of the most limiting factors for feed and food production in many regions (Pimentel et al. 2004; Molden 2007; Gordon et al. 2010). Furthermore, water for drinking should be considered as one of the most important feed and nutrient for animals. Mostly, animal nutritionists do not spend adequate attention to this feed.

An adequate water supply for plants and animals is a very important prerequisite for healthy plants and animals and stable and high yields. There exist various calculations/estimations for water needs for adequate plants growth. The group by Hoekstra (e.g. Mekonnen and Hoekstra 2010; Hoekstra et al. 2011) is distinguished in green (naturally infiltrated into the soil), blue (water in rivers and aquifers) and grey (water required to assimilate the load of pollutants) water and calculated water footprints (WF) for various animal feeds (Table 12.5).

The so-called water productivity is given in kg product per m^3 water. For cereals, about 1 m^3 water is required per kg product. For more general calculations, Mekonnen and Hoekstra (2010) give also summarized values for concentrate and

Table 12.5 Water footprint (sum of green, blue and grey water) of some selected plant products (Mekonnen and Hoekstra 2010; Mekonnen and Hoekstra 2012)

Plant product	WF (m^3/kg)	WF (m^3/MJ)
Sugar crops	0.20	2.9
Vegetables (for food)	0.32	5.6
Potatoes (starchy roots)	0.39.	2.0
Cereals	1.64	2.1
Wheat	1.83	
Barley	1.35	
Maize	1.22	
Oil crops	2.36	3.4
Soybeans	2.14	
Pulses	4.06	5.0

Table 12.6 Average water footprint (green, blue, grey and total WF; m^3/t) for concentrates and roughages (Mekonnen and Hoekstra 2010)

Type of feed	Green WF	Blue WF	Grey WF	Total WF
Concentrates	849	78	122	1048
Roughages	199	1.8	2	203

roughages (Table 12.6). Such values are used for further calculation in animal feeding and calculation of WF. According to Hoekstra (2016), the WF was developed and applied within the water resources research and should be successfully used in the future.

Compared with the water need for feed production (s. Tables 12.5 and 12.6) only small amounts of total water for animals are required for animal management and drinking. The WFs for animal products base on the water amount for feed production (about 98% of total water; Mekonnen and Hoekstra 2010; Gerbens-Leenes et al. 2013), water for drinking by animals (about 1%: Meyer et al. 2004, 2006) and management water (<1%; Schlink et al. 2010; Krauß et al. 2016).

12.5.3 Further Inputs

Fuel is used in various forms (e.g. diesel, coal, gas, electricity) in many fields of agriculture (see Bockisch et al. 2000; Barnett and Russell 2010; Frorip et al. 2012). Because of many influencing factors, details should not be considered in the following calculations (e.g. Fritsche et al. 1997; Bockisch et al. 2000; Kool et al. 2012).

Jokiniemi et al. (2012) concluded that the highest energy consumptions in plant production originate from agrochemicals, such as fertilizers, lime and pesticides.

Another point is the limited availability of some plant nutrients. This limitation is mainly true for phosphorous (Hall and Hall 1984; Scholz and Wellmer 2013). Therefore, animal excreta should be efficiently used in order to save inorganic resources (e.g. using of animal manure or green manure, plant species, etc.; e.g. Thorup Kristensen et al. 2003; Talgre et al. 2009). Nitrogen is available in large amounts in the air, but the potential as plant nutrient from the air is presently only used by legumes. A large potential for plant production and energy saving is seen by microbial N-fixation from the air (Nocera 2017).

12.6 Outputs

Production of food of animal origin includes various processes with greenhouse gas emissions such as soil cultivation, fertilization and manuring of soil, harvesting, storing and processing of feed as well as animal feeding and keeping. Figure 12.2 shows greenhouse gas emissions and processes in a managed ecosystem according

Table 12.7 Influence of animal species, categories and performances on yield of edible protein (By Flachowsky and Kamphues 2012)

Protein source (body weight)	Performance per day	Dry matter intake (kg per day)	Roughage to concentrate ratio (on DM base, %)	Edible fraction (% of product or body mass)	Protein in edible fraction (g per kg fresh matter)	Edible protein (g per day)	Edible protein (g per kg body weight and day)
Dairy cow (650 kg)	10 kg milk	12	90/10	95	34	323	0.5
	20 kg milk	16	75/25			646	1.0
	40 kg milk	25	50/50			1292	2.0
Dairy goat (60 kg)	2 kg milk	2	80/20	95	36	68	1.1
	5 kg milk	2.5	50/50			170	2.8
Beef cattle (350 kg)	500g[a]	6.5	95/5	50	190	48	0.14
	1000g[a]	7.0	85/15			95	0.27
	1500g[a]	7.5	70/30			143	0.41
Growing/fattening pig (80 kg)	500g[a]	1.8	20/80	60	150	45	0.56
	700g[a]	2	10/90			63	0.8
	1000g[a]	2.2	0/100			81	1.0
Broiler (1.5 kg)	40g[a]	0.07	10/90	60	200	4.8	3.2
	60g[a]	0.08	0/100			7.2	4.8
Laying hen (1.8 kg)	50%[b]	0.10	20/80	95	120	3.4	1.9
	70%[b]	0.11	10/90			4.8	2.7
	90%[b]	0.12	0/100			6.2	3.4

[a]Daily weight gain
[b]Laying performance

to IPCC (2006) along the food chain. Before calculation of CF per unit edible protein, the edible protein yield in dependence on animal species/categories and animal yields are shown (Table 12.7).

12.6.1 Edible Protein

The production of protein of animal origin is one of the most important goals of animal husbandry. Under consideration of various influencing factors such as animal yields, feeding, edible fractions and protein content in the edible fractions, the yield of edible protein per day and per kg body weight of animals is given in Table 12.7. There are large differences in animal protein yield per animal per day or per kg body weight and day depending on animal species and categories as well as their performances and the fractions considered as edible. Table 12.7 shows the highest protein yields per kg body weight and day for high-yielding lactating ruminants per day and for growing broilers as well for laying and lactating animals and the lowest values for growing/fattening ruminants.

12.6.2 Carbon Dioxide (CO_2)

The direct carbon dioxide emission from the animals can be considered as emission neutral. CO_2 will be fixed by photosynthesis of plants and excreted by the animals as result of animal metabolism. But nevertheless, the CO_2 emission must be seen along the whole food chain and based on burning of fossil carbon during feed production and land-use changes (LUC; Kim et al. 2009; Hergoualch and Verchot 2011; Caffrey and Veal 2013; MacLeod et al. 2013). In general, non-carbon dioxide GHG such as methane (CH_4) and nitrous oxide (N_2O) come directly from animals or from animal manure practices.

12.6.3 Methane (CH_4)

Methane is emitted under anaerobic conditions from the enteric fermentation in the digestive tract of animals, mainly in the rumen, but also during the manure management. Excess of hydrogen during anaerobic fermentation in the rumen is catalyzed by various reduction processes. The last step is catalyzed with methyl-coenzyme M reductase of reduction of CO_2 to CH_4 by hydrogenotrophic methanogenic archaea. Details about the enteric methane production are described in many papers (e.g. Flachowsky and Brade 2007; Tamminga et al. 2007; Bannink et al. 2008; Joany 2008; Beauchemin et al. 2009), and prediction equations are given (e.g. IPCC 2006; Ellis et al. 2010; Hristov et al. 2013; Ricci et al. 2013). Methane contributes not only to the greenhouse effect; between 2 and 12% of the ingested cross energy of ruminants can be lost by methane (Johnson and Johnson 1995) depending on diet composition and other influencing factors. Apart from the environmental effect, this energy could be potentially used by the animals for growth and lactation as shown in a model calculation in Table 12.8.

The methane emissions from the manure management are generally not directly associated with animals, but the losses can be considerably high (Hristov et al. 2013; Montes et al. 2013), especially if the excreta are stored under anaerobic conditions.

Table 12.8 Model calculation to show the influence of methane reduction on the energy available for dairy cows and milk yields

Methane reduction				
(g per kg DMI)	30	25	20	15
(g per cow per day)	600	500	400	300
Energy intake (MJ NEL per day)	130	135	140	145
Milk yield (kg per day)	28.0	29.5	31.0	32.5
Methane emission (g per kg milk)	21.4	17.0	12.9	9.2
Carbon footprint (g CO_{2eq} per kg milk)	735	585	440	315

Conditions for calculation: DMI, 20 kg per cow per day; body weight, 650 kg per cow, 7 MJ NEL per kg DM with 20 g CH_4 emission; Niemann et al. (2011)

12.6.4 Nitrous Oxide (Laughing Gas, N_2O)

Animals do not excrete nitrous oxide directly, but it can be formed in manure depending on the storage conditions and following land application (e.g. Flachowsky and Lebzien 2006; Hristov et al. 2013; Montes et al. 2013). N_2O is mainly produced in soils by microbial nitrification (the oxidation of ammonium NH_4^+ to nitrate NO_3^-) and denitrification (reduction of NO_3^- to N_2; Stevens et al. 1997). These microbial processes depend on temperature, moisture content and oxidation status of the environment. High N-fertilization and soil compaction are important factors that increase N_2O emissions. Since 1750, the tropospheric concentration of N_2O increased from 270 to 320 ppb (IPCC 2006). More details about N_2O production and emission from the soil are described by many authors and should not be considered in further details in the present paper (van Groningen et al. 2005; Lampe et al. 2006; Bessou et al. 2010; Schmeer et al. 2014).

12.6.5 Calculations of Carbon Footprints (CF)

Beginning with one and two studies per year in 1998–2000, about 20 studies per year were published in the last years (Avadi and Fréon 2013). The studies dealt with calculations of CF for nearly all types of food of animal origin (see summaries by Williams et al. 2006; Leip et al. 2010; Lesschen et al. 2011; Gerber et al. 2013; MacLeod et al. 2013; Opio et al. 2013; FAO 2015; Flachowsky 2015).

Results of CF calculation for food of animal origin depend on many influencing factors such as animal species and categories, animal yields and endpoints of animal production. Carbon footprints are defined as the total amount of GHG emissions (under consideration of their GHG potential) associated with a product along its supply (human food) chain "Plant production incl. harvesting, storing and treatment – Feed preparation – Feeding of food producing animals – Preparation of food (milk, meat, eggs etc.) – Distribution, market – Households".

Sometimes CF includes also emissions from consumption, end-of-life recovery and disposal. Usually, CFs are expressed in kg or t of carbon dioxide equivalents ($CO_{2\text{-eq}}$) per unit product (Opio et al. 2013). Studies and publications about CF have increased dramatically during the last few years, demonstrating that the interest for more resources of efficient and cleaner production enhanced. Agriculture and especially animal husbandry are considered as important GHG sources because of the high greenhouse potential of their emissions (e.g. $CO_2 \times 1$; $CH_4 \times 23$ and $N_2O \times 296$; (IPCC 2006)). Carbon footprints consider the GHG potential of climate-relevant gases and are given in $CO_{2\text{-eq}}$ per g or kg per product. Various authors calculated such CF for agriculture in general but also for separate segments.

The public interest in CF is discussed in the context of global warming and possible climate changes (IPCC 2006, 2014). Consequences of land-use change (LUC; e.g. change of forest into cropland or pasture) for CF calculations should also be considered, but in some cases, the values are not known or not considered in calculations (e.g. import of feeds).

A number of factors (e.g. plant yield, animal species and performances, type of production) cannot be ignored when taking into account the greenhouse gas potential of the various gases (see above) to derive CF and for the comparison of values along the food chain. The origin of the most important GHG, such as carbon dioxide, methane and nitrous oxide, was shown by the FAO (2015) recently and is discussed for food of animal origin in the next paragraphs.

12.6.6 *CF for Edible Protein by Land Animals*

Feeding and animal yields may influence CF of food of animal origin. In the case of ruminants, higher amounts of concentrate are required with higher animal yields. The proportion of coproducts (e.g. Bockisch et al. 2000; Makkar 2012) used in animal nutrition has not only nutritional implications, but it also affects the results of calculations on land use (Flachowsky et al. 2017b). Higher portions of edible fractions or higher protein content may increase the protein yield and reduce the CF per product. At high levels of performance, there are remarkable differences in CO_2 emissions due to a human consumption of 1 g protein from food of animal origin (eggs and meat from poultry < pork < milk < beef, see Table 12.9).

On the other hand, the efficiency and the emissions of food of animal origin can be also compared on the base of edible protein.

The highest CF per unit edible protein can be calculated for low-yielding beef cattle, followed by low-yielding lactating ruminants and pigs with low daily weight gain (see Table 12.9). Edible protein by broilers and laying hens is produced with the lowest CF per kg edible protein (Table 12.10).

12.6.7 *CF of Further Protein Sources*

Aquaculture is a strong upcoming way to produce food protein of animal origin. Recently some authors tried to determine CF for various forms of aquaculture. Mungkung et al. (2013) carried out a case study of combined aquaculture systems for carp and tilapia. The studied system included fingerling production in hatcheries, fish rearing in cages and transport of feed as well as that of harvested fish to markets.

Avadi and Fréon (2013) reviewed 16 LCA studies applied to fisheries and considered in the comparison the following aspects: scope and system boundaries, functional unit allocation strategies for coproducts, conventional and fishery-specific impact categories, fuel use, impact assessment methods, level of detail of inventories, normalization of results and sensitive analysis. Fishery-specific impact categories and fuel use in fishing operation were identified as the main contributors to environmental impact. Nijdam et al. (2012) analysed 18 and 11 studies for seafood from fisheries and agriculture, respectively. The authors summarized CF

Table 12.9 Influence of animal species, categories and performances on emissions

Protein source (body weight)	Performance per day	N-excretion (% of intake)	Methane emission (g per day)[c]	Emissions in kg per kg protein			
				P	N	CH_4[c]	$CO_{2\text{-eq}}$
Dairy cow (650 kg)	10 kg milk	75	310	0.10	0.65	1.0	30
	20 kg milk	70	380	0.06	0.44	0.6	16
	40 kg milk	65	520	0.04	0.24	0.4	12
Dairy goat (60 kg)	2 kg milk	75	50	0.08	0.5	0.8	20
	5 kg milk	65	60	0.04	0.2	0.4	10
Beef cattle (350 kg)	500 g[a]	90	170	0.30	2.3	3.5	110
	1000 g[a]	84	175	0.18	1.3	1.7	55
	1500 g[a]	80	180	0.14	1.0	1.2	35
Growing/fattening pig (80 kg)	500 g[a]	85	5	0.20	1.0	0.12	16
	700 g[a]	80	5	0.12	0.7	0.08	12
	900 g[a]	75	5	0.09	0.55	0.05	10
Broilers (1.5 kg)	40 g[a]	70	Traces	0.04	0.35	0.01	4
	60 g[a]	60		0.03	0.25	0.01	3
Laying hen (1.8 kg)	50%[b]	80	Traces	0.12	0.6	0.03	7
	70%[b]	65		0.07	0.4	0.02	5
	90%[b]	55		0.05	0.3	0.02	3

Per kg edible protein, own calculations on the base of data from Tables 12.11 and 12.12, see Flachowsky and Kamphues (2012)

[a]Daily weight gain

[b]Laying performance

[c]CH_4 emission depending on composition of diet

Table 12.10 Carbon footprints of protein of food of animal origin according to several LCA studies summarized by Nijdam et al. (2012)

Protein source (studies)	kg $CO_{2\text{-eq}}$ per kg product	kg $CO_{2\text{-eq}}$ per kg protein
Cow milk (*n* = 14)	1–2	28–43
Beef, intensive system (*n* = 11)	9–42	45–210
Meadow, suckler herds (*n* = 8)	23–52	114–250
Extensive pastoral systems (*n* = 4)	12–129	58–643
Mutton and lamb (*n* = 5)	10–150	51–750
Pork (*n* = 11)	4–11	20–55
Poultry meat (*n* = 5)	2–6	10–30
Eggs (*n* = 5)	2–6	15–42
Seafood from fisheries (*n* = 18)	1–86	4–540
Seafood from agriculture (*n* = 11)	3–15	4–75

between 1–86 for seafood from fisheries and 3–15 kg $CO_{2\text{-eq}}$ for seafood from agriculture, respectively. Nijdam et al. (2012) used for calculation of CF for seafood from fisheries 160–200 g and for seafood from agriculture 170–200 g protein per kg food. The results indicate that large differences exist between the studies and the

Table 12.11 Examples for ranges of proximate body composition of different insect species by various authors (Crude nutrients by Weende analysis; in % of dry matter)

Authors	No. of species or species (e.g. Makkar et al. 2014)	Crude protein ($N \times 6.25$)	Crude fat (ether extract)	Carbohydrates (NFE, fibre, NDF)	Crude ash
Bukkens (1997)	50	7.5–79.6	2.2–61.1	0–11.4	1.1–15.8
Finke (2002)	75	22.5–80.0	2.2–48.0	5.1–34.8	1.0–28.9
Grabowski et al. (2008)	17 (only Ensivera and Caelifera)	40.0–86.6	4.2–44.0	3.7–30.9	2.4–11.1
Rumpold and Schlüter (2013)	234 (samples)	4.9–74.8	0.7–67.2	3.0–86.3	0.6–26.0
Sanchez-Muros et al. (2014)	72	9.5–70.1	1.5–56.1	1.8–77.7	0.6–26.0
Makkar et al. (2014)	Black soldier fly larvae (1–5)[a]	41.1–43.6	15.0–34.8	7.0	14.6–26.8
	Housefly maggot meal (19–29)	42.3–60.4	9.0–26.0	1.6–8.6	6.2–17.3
	Tenebrio molitor (2–10)	47.2–60.3	31.1–43.1	7.4–15.0	1.0–4.5
	Locust or grasshopper meal (7–9)	29.2–65.9	4.2–14.1	2.4–14.0	4.4–10.0
	House cricket (2–4)	55.0–67.2	9.8–22.4	15.7–22.1	3.6–9.1
	Silkworm pupae meal (6–11)	51.6–70.6	6.2–37.1	2.5–5.8	3.3–10.6

[a]Number of samples

products. The outcomes for milk, pork, poultry meat and eggs show much more homogeneity than those for beef, mutton, lamb and seafood. This is largely because of the very wide variety in production systems of the last food groups.

These authors as well Avadi and Fréon (2013) define the need for standardization of fisheries life cycle assessments (LCA) research for further studies on sustainability of seafood and fishery-based agri-food.

Apart from milk, meat, eggs and fish, other sources of protein of animal origin, such as wild animals and insects, are also consumed by humans. Nothing is known about CF of food from wild animals.

More than 1900 insect species in various development stages (van Huis 2013) are eaten by man worldwide. The most commonly eaten insect groups are in the Coleoptera (beetles), Lepidoptera (caterpillars of butterflies and moths), Hymenoptera (bees, wasps, ants), Orthoptera (grasshoppers, locusts, crickets, termites), Hemiptera (cicades, leaf and plant hoppers, true bugs, scale insects), Odonata (dragonflies) and Diptera (flies) (EFSA 2015).

Because of the large number of insects consumed, it seems to be very difficult to review representatively nutritional composition of insects. They are rich in protein (mainly 20–70%) and contain considerable amounts of fat (mainly 10–50% of dry matter; Table 12.11). More than two billion people include processed insects in their diets (van Huis et al. 2013).

Some authors analysed the feed conversion, the land use and the GHG emissions of insects (Oonincx et al. 2010; Oonincx and de Boer 2012) and compared their data with values of "traditional" food-producing animals. Most studies with insect were done under laboratory conditions for a short time. Lundy and Parrella (2015) suggest that the laboratory scale rearing of crickets (*Acheta domestica*) was influenced by the type of diet used to rear the insects. The authors concluded that crickets reared on poultry feed showed similar feed conversion and emissions as poultry. More studies under field conditions are necessary to allow a conclusive evaluation of the sustainability of insects as a protein-rich feed and food sources (van Huis et al. 2013; EFSA 2015).

12.7 Challenges and Alternatives to Produce Food of Animal Origin

12.7.1 Plant Breeding

Plant breeding can be considered as the starting point of the food chain (see The Royal Society 2008; Flachowsky et al. 2013b, NASEM 2016). Traditional breeding as well as "green" biotechnology or green chemistry (Guillou and Matheron 2014) may result in changing of composition and nutritive value of feed plants. High and stable yields, resistant against biotic and abiotic stressors and with a low content of undesirable substances, can be considered as the most important objectives of plant breeding. Lower fibre content and higher digestibility of plants may reduce methane emission from the rumen. Presently, special attention is given to the adaptation of plants to expected climate changes (e.g. Reynolds 2010; Newman et al. 2011) and to improve their yield and the nutritive value for global food security (Fischer et al. 2014).

Genetically modified plants may contribute to achieve these objectives (see Flachowsky 2013; NASEM 2016) but are under critically public discussion presently. In 2016, 185 mio. ha (more than 12% of arable land) of genetically modified plants were cultivated in the global view (ISAAA 2017). Nutritionists distinguish genetically modified (GM) plants into plants of the first and second generation. This designation is purely pragmatic or historical; it does not reflect any particular scientific principle or technological development.

The first generation of GM plants is generally considered to be crops carrying simple input traits such as increased resistance to pest or tolerance against herbicides. Other inputs, such as more efficient use of water and/or nutrients or an increased resistance against heat and drought are not expected to cause any substantial change in composition and nutritive value. The newly expressed proteins that confer these effects occur in GM plants in very low concentrations and do not change their composition or feeding value significantly when compared with isogenic lines.

Table 12.12 Objectives of inputs and outputs of future plant breeding (by NASEM 2016)

Plants with input traits	Plants with output traits
Biotic stress tolerance	*Higher nutrient content*
Microbiological resistance, resistance against insects	Amino acids, trace elements, vitamins, fatty acids, non-essential ingredients etc.
Abiotic stress tolerance	*Higher food and feed safety*
Drought resistance, increased water utilization efficiency, cold, heat and salt tolerance	Lower content in mycotoxins and further undesirable ingredients
Nutrient intake and utilization	*Higher nutritive value*
N, P, CO_2, trace nutrients	High feed intake and high digestibility
Postharvest behaviour	*Biofuel and industrial crude products*
Microbiological resistance, improved storage properties, increase silage quality	High yield, improved properties of biofuel, Using of by-products as valuable feeds

GM plants of the so-called second generation (or plants with output traits or substantial changes in composition) are being developed with specific benefits for the consumer or the animals. Such biofortified crops contain higher amounts of desirable nutrients/substances such as proteins/amino acids, specific fatty acids, minerals, vitamins, enzymes, antioxidative substances, etc. or lower contents of undesirable substances, such as fibre/lignin, phytates, glucosinolates, mycotoxins, etc. Tillie et al. (2013) give a review about new events of GM crops in the pipeline as feed for animal nutrition. Adequate feeding studies for nutritional assessment of such feed of the second generation of GM crops are required (NASEM 2016).

Attention should be also spent to changes in plant/feed composition in consequence of traditional plant breeding.

Andersen et al. (2015) analysed new breeding techniques for organic farming and came to the conclusion that the most efficient methods are based on modern biotechnology techniques, which have yet to be embraced by the organic farm movement. The questions arise of whether the adoption of biotechnological methods is feasible not only from the perspective of sustainability but also from conceptual, socio-economic, ethical and regulatory perspectives.

In conclusion, from the view of animal nutrition, reduction of the content of undesirable substances from feed plants via plant breeding seems to be more important than biofortification. Objectives for plant breeding in the future are summarized in Table 12.12.

Further challenges of plant breeding may be summarized as follows:

- Plants must be made more resistant against biotic and abiotic stressors.
- More low input varieties of plants are needed (SCAR 2008; The Royal Society 2009; NASEM 2016).
- Losses during and post-harvest of feeds must be avoided.
- A more efficient conversion of feed into animal-derived food should be achieved.

More studies and analyses should be the base for consequences for a more efficient use of limited resources and lower emissions per product.

12.7.2 Animal Breeding

The global number of livestock animals used in agricultural production has been estimated to be about 1.8 billion large ruminants, including cattle, buffaloes and camels, 2.4 billion small ruminants (sheep and goats), nearly 1 billion pigs and ~20 billion of poultry (FAO 2013).

The breeding of domestic animals has a long-standing and successful history, starting with domestication several thousand years ago, by which men kept animals in his proximity and used products thereof (Larson et al. 2007). Using the technical options that were available in each time period, humans have propagated those populations that seemed useful for their respective needs and purposes. Selection mostly occurred according to the phenotype and/or because of specific traits. A scientifically based animal breeding only exists for ~50 years, mainly on the basis of population genetics and statistics. A vast number of phenotypically different breeds with desirable traits resulted from this process in an evolutionarily short time period. A good example is cattle, for which nowadays >800 cattle breeds can be found worldwide (Beja-Pereira et al. 2006). These breeds are not just variations of the same archetype but differ in qualitative and quantitative characteristics, including specific disease resistance, climate adaptation, lactation performance, fertility, meat quality and nutrient requirements. Modern animal breeding programs are based on biotechnological procedures, of which artificial insemination (AI) is the most prominent one. These breeding strategies are predominantly based on population genetics, marker-assisted selection or nowadays genomic selection. AI and ET (embryo transfer technology) technologies led to significant increases in the performances in domestic animals and are the basis for the regular supply with high-quality animal-derived food at acceptable prices.

Recent breakthroughs in reproductive technologies, such as somatic cloning and in vitro embryo production and their merger with molecular genetic tools, already developed in the mouse, will further advance progress in this field. The resulting targeted and even conditional transgenesis will be critical for the safe practical application. The efficient use of domestic animals is urgently needed in light of the worldwide shortage of arable land and the ever-increasing human population.

Strategies to produce animals with better utilization of feed into animal-derived food while concomitantly decreasing emission per product include (see Niemann et al. 2011):

- Higher feed intake of animals to improve the ratio between energy/nutrient requirements for maintenance and animal yields.
- Higher digestibility of feed to make energy/nutrients more available from the same feed amount.
- Reduction of energy losses in the digestive tract.
- Higher absorption of the digested nutrients.
- Lower energy/nutrient requirements for maintenance of the animals (GfE 2001; NRC 2001).

- Lower energy need for protein synthesis in the body or increase of anabolic processes and lower catabolic processes in the animal.
- Lower fat content in animal bodies or lower excretion of fat in milk and eggs or lower excretion of lactose in milk (lower energy content in products).
- Improved animal health, specifically, animals with higher resistance against biotic and/or abiotic stressors and lower losses during production may also contribute to a more efficient conversion of feed.

In summary, the above discussion shows that reducing specific nutrients in food of animal origin may contribute to lower energy requirements of animals and to a more efficient feed conversion. In the future, existing and emerging breeding technologies could be instrumental to produce livestock animals with much greater feed efficiency.

12.7.3 Using of Feeds Which Do Not Compete with Human Nutrition

Some feeds such as grassland, co- or by-products from agriculture and food and biofuel industry do not compete with human nutrition. Therefore, they do not compete with arable land and further limited resources and have a large potential for animal nutrition.

Ruminants are generally independent on grains/concentrates because their microbial population in the rumen is capable of digesting plant fibre. They are able to produce edible protein of animal origin (milk and meat) from permanent meadows and pastures. Pastoral systems may contribute to meeting human need for food of animal origin (Gill 2013; Taube et al. 2013). Kratli et al. (2013) pointed out that pastoralists are more efficient at producing food per unit area of dryland than other forms of agricultural land use under such conditions. Pastoralist systems are also efficient users of resources such as manure (Powell et al. 2013). Because of the predominantly high fibre content of roughage from grassland (see above) and the low animal yields, the methane emissions and the CFs per kg edible protein may be higher under such extensive conditions (FAO 2010; Gill et al. 2010), and GHG mitigation measurements are very important (e.g. Gerber et al. 2013; Hristov et al. 2013; Opio et al. 2013).

Apart from roughages and concentrates, coproducts from agriculture, such as cereal straw (e.g., Sundstol and Owen 1984; Flachowsky 1987), but also from food production (e.g., Kling and Wöhlbier 1983; Jeroch et al. 1993) and the biofuel-industry (Makkar 2012; Capper et al. 2013) are commonly used in animal feeding. Coproducts are by-products of main processes such as grain production (e.g. straw, stalks, husks), processing of raw products in the food industry (e.g. extracted oil meals from the oil industry, bran from the cereal processing, beet pulp or bagasse from the sugar industry, animal coproducts from milk, fish or meat processing) or from the biofuel industry (e.g. DDGS, rapeseed cake/meal or rapeseed-extracted oil meal as well as cakes and meals from other oil seeds). According to the FAO (2010), between 10 and 50% of the estimated concentrate feed comes from coproducts in various global regions (Gerber et al. 2011). In some countries, up to 100% of concentrate may base on coproducts.

Coproducts are inedible to humans and would otherwise mainly be wasted. They are used in various amounts and proportions in animal diets. Cereal straws and other coproducts rich in plant cell walls are mostly characterized by a low digestibility and are thus poor in energy and protein delivery to the animal. They are fed to ruminants with low animal yields or to meet their maintenance requirements. For high-yielding ruminants, they can only be considered as a source of fibre. Normally, they are not used in the feeding of non-ruminants.

Coproducts from food and fuel industry contain two to three times as many of the nutrients, which are not removed by processing (e.g. protein in the case of DDGS). They can be used as valuable sources of protein, minerals and other nutrients depending on the source material and the chemical or physical processing without any land footprint (see Flachowsky et al. 2017b). In the future more grains will be used for food and fuel, and more coproducts will be available for animal nutrition. More details about the nutritive value and the utilization of coproducts from biofuel industry in animal nutrition were recently compiled by Makkar (2012).

12.7.4 Further Alternatives to Save Protein Sources

To overcome food deficiency and to produce food in a sustainable way, many alternatives/challenges/developments are presently possible, such as feed without competition to human nutrition (e.g. grassland, coproducts of agriculture, food and biofuel industry; see above), reduction of losses in agriculture, food processing and trade and household, or are under discussion such as plant and animal breeding (e.g. Flachowsky and Meyer 2015a, b, c; Niemann et al. 2013), insects as food (van Huis et al. 2014) or other alternatives to produce or to save food such as:

- Cultured "lab-grown" meat or "cell-cultured", "synthetic" or "clean meat" (e.g. Dodson et al. 1997; Benjaminson et al. 2002; Post 2013, 2014)
- "Simulated" food on the base of plant products (cereal, legumes, plant leaves, etc.; e.g. McPherson 1986; Aiking 2011; Day 2013)
- Single-cell proteins, such as microbes, algae, yeasts, etc. (e.g. Bekatorou et al. 2006; Tredici 2010; Kovac et al. 2013; Enzing et al. 2016)
- Changing of eating pattern (e.g. Pimentel and Pimentel 2003; Baroni et al. 2007; Guyomard et al. 2012)
- Reduction of food losses and waste (FAO 2011)
- Feed and food losses should be also considered with a large influence on sustainability of food production

12.8 Conclusions

Sustainability in the production of food of animal origin or edible protein means an efficient use of limited resources as feed and low emissions per product of animal origin/edible protein. Such calculations should not only include the food chain links "feed – animal – food of animal origin" but the whole food chain (see Fig. 12.2). A system has the highest efficiency or the largest sustainability, if it is impossible to improve one parameter without deterioration of one or more other parameters.

Sustainability in production of food of animal origin required a complex consideration of the whole situation. Mono-causal assessments may improve some segments of food chain but do not essentially improve the whole system. Improvement in farm animal productivity or feed efficiency may also decrease the emissions of gases and the contamination of soils and water per unit food of animal origin.

References

Aerts S (2012) Agriculture's 6 F's and the need for more intensive agriculture. In: Potthast T, Meisch S (eds) Climate change and sustainable development. Wageningen Acad Publ, Wageningen, pp 192–195

Aiking H (2011) Future protein supply. Trends Food Sci Tech 22:112–120

Aiking H (2014) Protein production: planet, profit, plus people. Am J Clin Nutr 100(Suppl. 1):4835–4895

Andersen G (2011) The little Souci/Fachmann/Kraut: Food table for the practise, 5th edn. Deutsche Forschungsanstalt für Lebensmittelchemie/Wissenschaftliche Verlagsgesellschaft mbH, Freising/Stuttgart, 483 p

Andersen MM, Landes X, Xiang W, Anyshchenko A, Falhof J, Osterberg JT, Olsen LI, Edenbrandt AK, Vedel SE, Thorson BJ, Sandoe P, Gamborg C, Kappel K, Palmgren MG (2015) Feasibility of new breeding techniques for organic farming. Trends Plant Sci 20:426–434

Avadi A, Fréon P (2013) Life cycle assessment of fisheries: a review for fisheries scientists and managers. Fish Res 143:21–38

Bannink A, France J, Lopez S, Gerrits WJJ, Kebreab E, Tamminga S, Dijkstra J (2008) Modelling the implication of feeding strategy on rumen fermentation and functioning of the rumen wall. Anim Feed Sci Technol 143:3–26

Barnett J, Russell J (2010) Energy use on dairy farms. Environmental issues at dairy farm level. Bull Int Dairy Fed 443:23–32

Baroni L, Cenci L, Tettamanti M, Berati M (2007) Evaluating the environmental impact of various dietary patterns combined with different food production systems. Eur J Clin Nutr 61:279–286

Bauer J, Biolo G, Cederholm T, Cesari M, Cruz-Jentoft AJ, Morley JE, Phillips S, Sieber C, Stehle P, Teta D (2013) Evidence-based recommendations for optimal dietary protein intake in older people: a position paper from the Prot-Age Study Group. J Am Med Dir Assoc 14:542–559

Beauchemin KA, McAllister TA, McGinn SM (2009) Dietary mitigation of enteric methane from cattle. CAB Rev Perspect Agric Vet Sci Nutr Nat Resour 4(035):1–18

Bekatorou A, Psarianos C, Koutinas AA (2006) Production of food grade yeasts. Food Technol Biotechnol 44:407–415

Benjaminson MA, Gilchriest JA, Lorenz M (2002) In vitro edible muscle protein system. MPPS Stage 1 (fish). Acta Astronaut 51:879–889

Bessou C, Mary B, Leonhard J, Roussel M, Grehan E, Gabrielle B (2010) Modelling soil compaction impacts on nitrous oxide emissions in arable fields. Eur J Soil Sci 61:348–363

Bockisch F-J, Ahlgrimm H-J, Böhme H, Bramm A, Dämmgen U, Flachowsky G (2000) Bewertung von Verfahren der ökologischen und konventionellen landwirtschaftlichen Produktion im Hinblick auf Energieeinsatz und bestimmte Schadgasemissionen. Landbauforschung Völkenrode, Sonderheft 211, 206 p

Boonen R, Aerts S, De Tavernier L (2012) Which sustainability suits you? In: Potthast T, Meisch S (eds) Climate change and sustainable development. Wageningen Acad. Publ, Wageningen, pp 43–48

Bruinsma J (2009) The resource outlook to 2050: By how much do land, water and crop yields need to increase by 2050? FAO Expert meeting on how to feed the world in 2050? June 2009. FAO, Rome, p 33

Brundtland GH (1987) Report of the World Commission on Environment and Development: Our Common Future. UN Documents: Gathering a Body of Global Agreements; Annex to document A/42/427

Bukkens SGF (1997) The nutritional value of edible insects. Ecol Food Nutr 36:287–319

Caffrey KR and Veal MW (2013) Conducting an agricultural life cycle assessment: challenges and perspectives. Sci World J 2013, 472431, 13 p

Capper JL, Berger L, Brashears MM (2013) Animal feed vs. human food: challenges and opportunities in sustaining animal agriculture toward 20150. Council. Agric Sci Technol 53:1–16

Cassidy ES, West PC, Gerber JS, Foley JA (2013) Redefining agricultural yields: from tonnes to people nourished per hectare. Environ Res Lett 8:034015. https://doi.org/10.1088/1748-9326/8/3/034015

Casabona CMR, Epifanio LES, Cirion AE (2010) Global food security: ethical and legal challenges. Wageningen Academic Publishers, Wageningen, 532 p

Day L (2013) Proteins from land plants – potential resources for human nutrition and food security. Trends Food Sci Tech 32:25–42

De Alvarenga RAF, Silva Junior VP d, Soares SR, de Alvarenga RAF, da Silva Junior VP (2012) Comparison of the ecological footprint and a life cycle impact assessment method for a case study on Brazilian broiler feed production. J Clean Prod 28:25–32

De Vries M, de Boer IJM (2010) Comparing environmental impacts for livestock products: a review of life cycle assessments. Livest Sci 128:1–11

Dodson MV, Vierck JL, Hossner KL, Byrne K, McNamara JP (1997) The development and utility of a defined muscle and fat co-culture system. Tissue Cell 29:517–524

EFSA (European Food Safety Aurhority) (2015) Risk profile related to production and consumption of insects as food and feed. EFSA J 13 (10):4257, 60 p

Ellis JL, Bannink A, France J, Kebreab J, Dijkstra J (2010) Evaluation of enteric methane prediction equation for dairy cows used in the whole farm model. Glob Chang Biol 16:3246–3256

Enzing C, Ploeg M, Barbosan M, Sijtsma L, Vigani M, Parisi C, Rodriguez Cerezo E (2016) Microalgae-based products for the food and feed sector: an outlook for Europe, Scientific and Technical Research Papers, JRC Scientific and Polica Reports. EU Science Hub, Brussels, 75 p

Ertl P, Klocker H, Hörtenhuber S, Knaus W, Zollitsch W (2015) The net contribution of dairy production to human food supply: the case of Austrian dairy farms. Agric Syst 137:119–125

FAO (Food and Agriculture Organisation of the United Nations) (2006) Livestock's long shadow. Environmental issues and options. FAO, Rome, 406 p

FAO (2009) How to feed the world in 2050. FAO, Rome, 120 p

FAO (2011) Global food losses and food waste. FAO, Rome, 30 p

FAO (2013) FAO statistical yearbook, 2013; World food and agriculture. FAO, Rome, 289 p

FAO (2015) Estimating Greenhouse gas emissions in agriculture. A manual to address data requirements for developing countries. FAO, Rome, 180 p

Fedoroff NV, Battisti DS, Beachy RN, Cooper PJM, Fischhoff DA, Hodges PC, Knauf VC, Lobell D, Mazur BJ, Molden D, Reynolds MP, Ronald PC, Rosengrant MW, Sanches PA, Vonshak A, Zhu JK (2010) Radically rethinking agriculture for the 21st century. Science 327:833–834

Finke MD (2002) Complete nutrient composition of selected invertebrates commonly fed to insectivores. Zoo Biol 21:269–285

Fischer T, Byerlee D, Edmeades G (2014) Crop yields and global food security: will yield increase continue to feed the world? Australian Centre for International Agricultural Research, Canberra, 634 p

Flachowsky G (1987) Straw as feed (in German). Deutscher Landwirtschaftsverlag, Berlin, 255 p

Flachowsky G (2002) Efficiency of energy and nutrient use in the production of edible protein of animal origin. J Appl Anim Res 22:1–24

Flachowsky G (2013) Animal nutrition with transgenic plants. CAB International, Wallingford, 234 p

Flachowsky G (2015) Carbon footprints for food of animal origin. In: Malik PK et al (eds) Livestock production and climate change. CAB International, Wallingford, pp 125–145

Flachowsky G, Brade W (2007) Reduction potentials for methane emissions from ruminants (in German). Züchtungskunde 79:417–465

Flachowsky G, Hachenberg S (2009) CO_2-footprints for food of animal origin – present stage and open questions. J Consumer Prot Food Saf 4:190–198

Flachowsky G, Kamphues J (2012) Carbon footprints for food of animal origin: what are the most preferable criteria to measure animal yields? Animals 2:108–126

Flachowsky G, Lebzien P (2006) Possibilities for reduction of nitrogen (N) excretion from ruminants and the need for further research – a review. Landbauforschung Völkenrode 56:19–30

Flachowsky G, Meyer U (2015a) Sustainable production of protein of animal origin – the state of knowledge. Part 1: Resources and emissions as factors affecting sustainability. J Anim Feed Sci 24:273–282

Flachowsky G, Meyer U (2015b) Sustainable production of protein of animal origin – the state of knowledge. Part 2: Aquirements, objectives and ways of sustainability improvements. J Anim Feed Sci 24:283–294

Flachowsky G, Meyer U (2015c) Challenges for plant breeders from the view of animal nutrition. Agriculture 5:1252–1272

Flachowsky G, Schulz E (2011) Sustainability in animal feeding (in German). Annual book. In: Mühle and Mischfutter. Schäfer, Detmold, pp 189–209

Flachowsky G, Gruen M, Meyer U (2013a) Feed efficient ruminant production: opportunities and challenges. J Anim Feed Sci 22:177–187

Flachowsky G, Meyer U, Gruen M (2013b) Plant and animal breeding as starting points for sustainable agriculture. In: Lichtfouse E (ed) Sustainable agriculture reviews, vol 12. Springer, Dordrecht, pp 201–224

Flachowsky G, Meyer U, Behl RK (2017a) In: Behl RK et al (eds) Objectives for plant breeding from the view of animal nutrition in Biotechnology and Bioscience Perspectives. Agrobiost Int. Publishers, Jodhpur, pp 17–29

Flachowsky, G., Meyer, U., Südekum, K.-H. (2017b) Land use for edible protein of animal origin – a review. Animals 7, 25. https://doi.org/10.3390/ani7030025, 19 p

Foley JA, Ramankutty N, Brauman KA, Cassidy ES, Gerber JS, Johnston M, Mouller ND, O'Conell C, Ray DK, West PC, Balzer C, Bennett EM, Carpenter SR, Hill J, Monfreda C, Polasky S, Rockström J, Sheehan J, Siebert S, Tilman D, Zaks DPM (2011) Solutions for a cultivated planet. Nature 478:337–342

Fritsche UR, Leuchtner J, Matthes FC, Rausch L, Simon HH (1997) Gesamt-Emissions-Model Integrierter Systeme (GEMIS). Öko-Institut, Wiesbaden

Frorip J, Kokin E, Praks J, Poikalainen V, Ruus A, Veermäe I, Lepasalu L, Schäfer W, Mikkola H, Ahokas J (2012) Energy consumption in animal production – case farm study. Agron Res Biosyst Eng 10(Special Issue 1):39–48

Gerbens-Leenes PW, Mekonnen MM, Hoekstra AY (2013) The water footprint of poultry, pork and beef: a comparative study in different countries and production systems. Water Res Ind 1–2:25–36

Gerber PJ, Steinfeld H, Henderson B, Mollet A, Opio C, Dijkman F, Falcucci A, Tempio G (2013) Tackling climate change through livestock – a global assessment of emissions and mitigation opportunities. Food and Agric Org of the UN (FAO), Rome, 116 p

GfE (Gesellschaft für Ernährungsphysiologie) (1995) Recommendations for the supply of energy and nutrients to beef cattle. (in German). DLG-Verlag, Frankfurt am Main, p 85
GfE (1999) Recommendations for the supply of energy and nutrients to laying hens and chicken for fattening (broilers; in German). DLG-Verlag, Frankfurt am Main, 185 p
GfE (2001) Recommendations for the supply of energy and nutrients to dairy cows and heifers (in German). DLG-Verlag, Frankfurt am Main, 136 p
GfE (2008) Recommendations for the supply of energy and nutrients to pigs. DLG-Verlag, Frankfurt am Main, 245 p
Gill M, Smith P, Wilkinson JM (2010) Mitigating climate change: the role of domestic livestock. Animal 4:323–333
Giovannucci D, Scherr S, Nierenberg D, Hebebrand C, Shapiro J, Milder J, Wheeler K (2012) Food and Agriculture: the future of sustainability. A strategic input to the sustainable development in the 21st Century (SD21) project. United Nations Department of Economic and Social Affairs, Division for Sustainable Development, New York
Gordon LJ, Finalayson CM, Falkenmark M (2010) Managing water in agriculture for food production and other ecosystem services. Agric Water Manag 97:512–519
Grabowski NT, Nowak B, Klein K (2008) Proximate chemical composition of long-horned and short horned grasshoppers (Acheta domesticus, Schistocerca gregaria and Phymateus saxosus) available commercially in Germany. Arch Leb 59:204–208
Guillou M, Matheron G (2014) The world's challenge – Feeding 9 billion people. Springer Netherlands; Ed. QuE; Springer Berlin; 226 p
Guyomard H, Darcy-Vrillon B, Esnouf C, Marin M, Goillou M (2012) Eating patterns and food systems: critical knowledge requirements for policy design and implementation. Agric Food Secur 1:1–13
Guyomard H, Manceron S, Peyraud JL (2013) Trade in feed grains, animals, and animal products: current trends, future prospects, and main issues. Animal. Frontiers 3:14–18
Hall DC, Hall JV (1984) Concepts and measures of natural-resource scarcity with a summary of recent trends. J Environ Econ Manage 11:363–379
Hergoualch K, Verchot LV (2011) Stocks and fluxes of carbon associated land use change in Southeast Asia tropical peatlands: a review. Global Biochem Cycl 25 GB2001, 1–13
Hoekstra AY (2016) A critique on the water-scarcity weighted water footprint in LCA. Ecol Indic 66:564–573
Hoekstra AY, Chapagain AK, Aldaya MM, Mekonnen MM (2011) The water footprint assessment manual: setting the global standard. Earthscan, London, 203 p
Hristov AN, Oh J, Lee C, Meinen R, Montes F, Ott T, Firkins J, Rotz A, Dell C, Adesogan A, Yang W, Tricario J, Kebreab E, Waghorn G, Dijkstra J, Oosting S (2013) Mitigation of greenhouse gas emissions in livestock production – a review of technical options for non-CO_2 emissions. In: Gerber PJ, Henderson B, Makkar HPS (eds) FAO animal Production and Health Paper No. 177. FAO, Rome, 206 pp
IPCC (2014) Climate change 2014: mitigation of climate change. Contribution of working group III to the fifth assessment report of the intergovernmental panel on climate change. Cambridge University Press, Cambridge/New York
IPCC (Intergovernmantal Panel on Climate Change) (2006) IPCC Guidelines for National Greenhouse Gas Inventories: Glossery. IPCC, Japan. Available at http://www.ipcc-nggip.iges.or.jp/public/gpglulucf/gpglulucf_files/Glossary_Acronyms_BasicInfo/Glossary.pdf
ISAAA (International Service for the Acquisition of Agri-Biotechnology Applicationns) (2017) Global Status of Commercialised Transgenic Crops: 2016. ISAAA, Ithaca
IUCN (World Conservation Union) (2005) The IUCN Programme 2005–2008. Many voices, one earth Bangkok, Thailand, 17–25 Nov. 2004. Available at: https://cmsdata.iucn.org/downloads/programme-english.pdf
Jackson AA (2007) Protein. In: Mann J, Truswell S (eds) Essentials of human nutrition, 3rd edn. Oxford University Press, Oxford, pp 53–72
Jeroch H, Flachowsky G, Weissbach F (1993) Feed Science (in German). Fischer Verlag, Jena und Stuttgart, 510 p

Jokiniemi T, Mikkola H, Rossner H, Talgre L, Lauringson E, Hovi M, Ahokas J (2012) Energy savings in plant production. Agron Res Biosyst Eng 10(Special Issue 1):85–96
Kastner T, Rivas MJI, Koch W, Nonhebel S (2012) Global changes in diets and the consequences for land requirements for food. Proc Natl Acad Sci 109:6868–6872
Kebreab E (2013) Sustainable animal agriculture. CAB International, Wallingford, 321 p
Keyzer MA, Merbis MD, Pavel IFPW, van Westenbeeck CFA (2005) Diets shifts towards meat and the effects on cereal use: can we feed the animals in 2030? Ecol Econ 55:187–202
Kim H, Kim S, Dale BE (2009) Biofuels, land use change, and greenhouse gas emissions: some unexplored variables. Environ Sci Technol 43:961–967
Knaus W (2013) Re-thinking dairy cow feeding in light of food security. AgroLife Scientific Journal 2(1):36–40
Kool A, Marinussen M, Blonk H (2012) LCI data for the calculation tool Feedprint for greenhouse gas emissions of feed production and utilization – GHG emissions of N, P and K fertilizer production, Blonk Consultants, Gouda, 15 p
Kovac DJ, Simeunovic JB, Babic OB, Misan AC, Milovanovic IL (2013) Algae in food and feed. Food and Feed Research 40:21–32
Krauß M, Drasting K, Prochnow A, Rose-Meierhöfer S, Kraatz S (2016) Drinking and cleaning water use in a dairy cow barn. Water 8:302. doi: 3390/w8070302
Lampe C, Dittert K, Sattelmacher B, Wachendorf M, Loges R, Taube F (2006) Sources and rates of nitrous oxide emissions from grassed grassland after application of 15^{N}-labelled mineral fertilizer and slurry. Soil Biol Biochem 38:2602–2613
Lawrence D, Beddington J, Godfray C, Crute J, Haddad L, Muir J, Pretty J, Robinson S, Toulmin C (2011) The challenge of global food sustainability. Food Policy 36(Suppl. 1):S1–S2
Lesschen JP, van den Berg M, Westhock HJ, Witzke HP, Oenema O (2011) Greenhouse gas emission profiles of European livestock sectors. Anim Feed Scie Technol 166-167:16–28
Lundy ME, Parrella MP (2015) Crickets are not a free lunch: protein capture from scalable organic side-streams via high-density populations of *Acheta domesticus*. PLoS One 10(4):e0118785. https://doi.org/10.1371/journal.pone.0118785
MacLeod M, Gerber P, Mottet A, Tempio G, Falcucci A, Opio C, Vellinga T, Henderson B, Steinfeld H (2013) Greenhouse gas emissions from pig and chicken supply chains – a global life cycle assessment. Food and Agriculture Organization of the United Nations (FAO), Rome, 171 p
Makkar HPS (2012) Biofuel co-products as livestock feed. Opportunities and challenges. Food and Agriculture Organization of the United Nations, Rome, p 533
Makkar HPS, Ankers P (2014) Towards sustainable animal diets: a survey based study. Anim Feed Sci Technol 198:309–322
Makkar HPS, Tran G, Heuze V, Ankers a P (2014) State-of-the-art on use of insects as animal feed. Anim Feed Sci Technol 197:1–33
Malik PK, Bhatta R, Takahashi J, Kohn RA, Prasad CS (2015) Livestock production and climate change. CAB International, Wallingford/Boston, 385 p
McPherson AT (1986) Food for man from the past to the future: toward "a revolutionary advance in technology" a complete synthesis of food. Awasthi Associates, Allahabad, 83 p
Meadows DL, Meadows DH, Zahn E (1972) Die Grenzen des Wachstums. Bericht des Club of Rome zur Lage der Menschheit. Dt. Verlags-Anstalt, New York
Mekonnen MM, Hoekstra AY (2010) The green, blue and grey water footprint of farm animals and farm animal products. Value of Water Research Report Series No. 48. UNESCO-IHE, Delft
Mekonnen MM, Hoekstra AY (2012) A global assessment of the water footprint of farm animal products. Ecosystems 15(3):401–415
Meyer U, Everinghoff M, Gaedeken D, Flachowsky G (2004) Investigations on the water intake of lactating dairy cows. Livestock Production Sci 90:117–121
Meyer U, Stahl W, Flachowsky G (2006) Investigations on the water intake of growing bulls. Livest Sci 103:186–191
Millar K, Hobson-West P, Nerlich B (2009) Ethical futures: bioscience and food horizons. Wageningen Academic Publishers, Wageningen, 448 p

Molden D (ed) (2007) Water for food, water for life: a comprehensive assessment of water management in agriculture. IWMI/Earthscan, London, 645 p
Molden DT, Oweis P, Steduto P, Bindraban MA, Hanjra RA, Kijne J (2010) Improving agricultural water productivity: between optimism and caution. Agr Water Manage 97:528–535
Montes F, Meinen R, Dell C, Rotz A, Hristov AN, Oh J, Waghorn C, Gerber PJ, Henderson B, Makkar HPS, Dijkstra J (2013) Mitigation of methane and nitrous oxide emissions from animal operations: II. A review of manure management mitigation options. J Anim Sci 91:5070–5094
Mungkung R, Aubin J, Prihadi TH, Slembrouck J, van der Werf HMG, Legendre M (2013) Life cycle assessment for environmentally sustainable aquaculture management: a case study of combined aquaculture systems for carp and tilapia. J Clean Prod 57:249–256
NASEM (Committee on Genetically Engineered Crops: Past Experience and Future Prospects, Board on Agriculture and Natural Resources; Division on Earth and Life Studies; National Academies of Sciences, Engineering, and Medicine, National Academies of Sciences, Engineering, and Medicine) (2016) Genetically engineered crops: experiences and prospects. The National Academies Press, Washington, DC, 584 p. 10.17226/23395
Newman JA, Anand M, Henry HAL, Hunt S, Gedalof Z (2011) Climate change biology. CABI, Wallingford, 289 p
Niemann H, Kuhla B, Flachowsky G (2011) Perspectives for feed-efficient animal production. J Anim Sci 89:4344–4363
Nijdam D, Rood T, Westhoek H (2012) The price of protein: review of land use and carbon footprints from life cycle assessments of animal food production and their substitutes. Food Policy 37:760–770
Nocera D (2017) Sustainable solar-to-fuels and solar-to-fertilizer production. Annual Meeting of the American Chemical Society, Division of Inorganic Chemistry, Session 538
NRC (National Research Council) (2015) Critical role of animal science research in food security and sustainability. The National Academies Press, Washington, DC, 460 p
Oonincx DGAB, De Boer LJM (2012) Environmental impact of the production of mealworms as a protein sources for humans – a life cycle assessment. PLoS One 7:e51145
Oonincx DGAB, van Itterbeeck J, Heetkamp MJW, Van den Brand H, Van Loon JJA, Van Huis A (2010) An exploration on greenhouse gas and ammonia production by insect species suitable for animal and human consumption. PLoS One 5:e14445
Opio C, Gerber P, Mottet A, Falcucci A, Tempio G, MacLeod M, Vellinga T, Henderson B, Steinfeld H (2013) Greenhouse gas emissions from ruminant supply chains – a global life cycle assessment. Food and Agric Org of the UN (FAO), Rome, 191 p
Peters CJ, Wilkins JL, Fick GW (2007) Testing a complete-diet model for estimation the land resource requirements of food consumption and agricultural carrying capacity: the New York state example. Renew Agric Food Syst 22:145–153
Peters GM, Rowley HV, Wiedemann S, Tucker R, Short MD, Schulz M (2010) Red meat production in australia: life cycle assessment and comparison with overseas studies. Environ Sci Technol 44:1327–1332
Pimentel D, Pimentel M (2003) Sustainable of meat-based and plant-based diets and the environment. Am J Clin Nutr 78:6605–6635
Pimentel D, Berger B, Filberto D, Newton M, Wolfe B, Karabinakis E, Clark S, Poon E, Abbett E, Nandagopal S (2004) Water resources: agricultural and environmental issues. Bioscience 54:909–918
Post MJ (2013) Medical technology to produce food. J Sci Food Agric. https://doi.org/10.1002/jsfa.6474. Retrieved 20 June 2016
Post MJ (2014) An alternative animal protein source: cultured beef. Ann N Y Acad Sci 1328:29–33
Potthast T, Meisch S (2012) Climate change and sustainable development. Ethical perspectives on land use and food production. Wageningen Academic Publ, Wageningen, 526 p
Reynolds MP (2010) Climate change and crop production. CAB International, Wallingford/Cambridge, 292 p

Ricci P, Rooke JA, Nevison I, Waterhouse A (2013) Methane emissions from beef and dairy cattle: quantifying the effect of physiological state and diet characteristics. J Animal Sci 91:5379–5389
Ros J, van der Born G, Notenboom J (2010) The contribution of by-products to the sustainability of biofuels. Netherlands Environmental Assessment Agency, Bilthoven, 8 p
Rumpold BA, Schlüter OK (2013) Nutritional composition and safety aspects of edible insects. Mol Nutr Food Res 57:802–823
Ruttan VW (1999) The transition to agricultural sustainability. Proc Natl Acad Sci 96:5960–5967
Sanchez-Muros M-J, Barroso FG, Manzano-Agugliaro F (2014) Insect meal as renewable source of food for animal feeding: a review. J Clean Prod 65:16–27
SCAR (Scientific Committee for Agricultural Research) (2008) New challenges for agricultural research. Climate change, rural development, agricultural knowledge systems. The 2nd SCAR Foresight Exercise, Brussels, 112 p
Schlink AC, Nguyen M-L, Viljoen GJ (2010) Water requirements for livestock production: a global perspective. Rev Sci Tech Off Int Epiz 29:603–619
Schmeer M, Loges R, Dittert K, Senbayram M, Horn R (2014) Legume-based forage production reduce nitrous oxide emissions. Soil & Tillage Res 143:17–25
Scholz RW, Wellmer F-H (2013) Approaching a dynamic view on the availability of mineral sources: what we may learn from the case of phosphorus. Glob Environ Chan 23:11–27
Smil V (2000) Feeding the world: a challenge for the twenty-first century. MIT Press, Cambridge, MA, p 360
Smith J, Sones K, Grace D, MacMillan S, Tarawali S, Herrero M (2013) Beyond milk, meat, and eggs: role of livestock in food and nutrition security. Anim Front 3:6–13
Souci SW, Fachmann W, Kraut H (2008) Food composition and nutrition tables. MedPharm Scientific Publishers; 7th rev. and compl. Ed., Taylor & Francis; A CRC Press Book, 1364 p
Stevens R, Lauglin R, Burns L, Arah J, Hood R (1997) Measuring the contributions of nitrification and denitrification to the flux of nitrous oxide from soil. Soil Biol Biochem 29:139–151
Sundstol F, Owen E (1984) Straw and other fibrous by-products as feed. Developments in animal and veterinary sciences, 14. Elsevier, Amsterdam, 604 p
Talgre L, Lauringson E, Rosstalu H, Astover A, Eremeev V, Selge A (2009) The effects of pure and undersowing green manure on yields of succeeding spring cereals. Acta Agric Scand Sect B Soil Plant Sci 59(1):70–76
The-Royal-Society (2009) Reaping the benefits: science and the sustainable intensification of global agriculture. Royal Society Policy document 11/09, issued Oct. 2009, RS 1608
Thompson B, Amoroso L (2014) Improving diets and nutrition – food-based approaches. CAB International, Wallingford/Boston, 403 p
Thorup Kristensen K, Magid J, Jensen LS (2003) Catch crops and green manures as biological tools in nitrogen management in temperate zones. Adv Agron 79:227–302
Tillie P, Dillen K, Rodriguez-Cerezo E (2013) The pipeline of GM crops for improved animal feed: challenges for commercial use. In: Flachowsky G (ed) Animal nutrition with transgenic plants, CABI biotechnology series. CABI, Wallingford/Boston, p 166
Tilman D, Balzer C, Hill J, Befort BL (2011) Global food demand and the sustainable intensification of agriculture. Proc Natl Acad Sci 108:20260–20264
Tredici MR (2010) Photobiology of microalgae mass cultures: understanding the tools for the next green revolution. BioFuels 1:143–162
Upham P, Dendler L, Bleda M (2010) Carbon labelling of grocery products: public perceptions and potential emissions reductions. J Clean Prod 19:348–355
USDA (United State Department of Agriculture); (2016) National Nutrient database for Standard Reference. Release 28. http://ndb.nal.usda.gov/ndb/foods
Van Groningen J, Velthof G, van der Bolt F, Vos A, Kuikman P (2005) Seasonal variation in N_2O emissions from urine patches: effects of urine concentration, soil compaction and dung. Plant Soil 273:15–27
Van Huis A (2013) Potential of insects as food and feed in assuring food security. Annu Rev Entomol 58:563–583

Van Huis A, van Itterbeek J, Klunder H, Mertens E, Halloran A, Muir G, Vantome P (2013) Edible insects – future prospects for food and feed security, FAO Forestry Paper 171. Food and Agriculture Organization of the United Nation, Rome, 187 p

Van Huis A, Van Gurp H, Dicke M (2014) The insect cookbook – food for a sustainable planet. Columbia University Press, New York, 216 p

Vermeulen SJ, Campbell BM, Ingram JSI (2012) Climate change and food systems. Annu Rev Environ Resour 37:195–222

Viljoen A, Wiskerke JSC (2012) Sustainable food planning: evolving theory and practice. Wageningen Academic Publ, Wageningen, 600 p

Von Carlowitz HC (1713) Sylviculture oeconomica, Anweisung zur wilden Baum-Zucht. Braun Verlag, Leipzig, pp 105–106

Wals AEJ, Corcoran PB (2012) Learning for sustainability in times of accelerating change. Wageningen Academic Publishers, Wageningen, 550 p

Wennemer H, Flachowsky G, Hoffmann V (2006) Protein, population, politics – how protein can be supplied sustainable in the 21st century. Plexus Verlag, Mittenberg/Frankfurt/Main, 160 pp

WHO, FAO, UNU (2007) Protein and amino acid requirements in human nutrition. Report of a Joint WHO/FAO/UNU Expert Consultation, World Health Organization Technical Report Series, vol 935. WHO, Geneva, pp 1–265

Wilkinson JM (2011) Re-defining efficacy of feed use by livestock. Animal 5:1014–1022

Williams AG, Audsley E, Sanders DL (2006) Determining the environmental burdens and resource use in the production of agricultural and horticultural commodities. Main report. Defra Research Project IS0205. Cranfield University and Defra, Bedford, 97 p. Available on www.silsoe.cranfied.ac.uk, and www.defra.gov.uk

Wu G, Bazer FW, Cross HR (2014a) Land-based production of animal protein: impacts, efficiency, and sustainability. Ann N Y Acad Sci 1328:18–28

Wu G, Fanzo J, Miller DD, Pingali P, Post M, Steiner JL, Thalacker-Mercer AE (2014b) Production and supply of high-quality food protein for human consumption: sustainability, challenges and innovations. Ann N Y Acad Sci 1321:1–19

Young W, Hwang K, McDonald S, Oates CJ (2010) Sustainable consumption: green consumer behaviour when purchasing products. Sustain Dev 1:20–31

Part V
Biotechnology

Chapter 13
Role of Plant Tissue Culture for Improving the Food Security in India: A Review Update

Chinnasamy Ragavendran and Devarajan Natarajan

Abstract The availability of food is an essential situation for food protection. India is more or less self-sufficient in cereals but insufficient in pulses and oilseeds. Due to the changes in consumption patterns, demand for fruits, vegetables, dairy, meat, poultry and fisheries has been increasing. Hence, a need to raise crop diversification and improve allied activities. It may be noted that the slowdown in agriculture growth could be attributed to structural factors on the supply side, such as public investment, credit, technology, land and water management, etc., rather than to globalization and trade reforms. In this situation, plant tissue culture offers remarkable opportunities in in vitro propagations, plant quality improvement and production of plants with desirable agronomical quality and quantity. It's now possible to develop virus-free plant regeneration, herbicide resistance, salinity tolerance, disease resistance, incorporation of high protein content and genetically engineered plants for desirable traits. Micropropagated plant cells and tissues have been widely used for the production of secondary metabolites, which are the rich source of many pharmaceutical and industrial products. Crop plants play a major role in the human nutrition and health by providing carbohydrates, proteins, fats, minerals, antioxidants, vitamins, phytoconstituents and dietary fibres. Plants can be used to produce pharmaceutically important proteins for immunization, enzyme therapy or pharmaceutical products. Mainly transgenic plants are used as bio-factories for producing pharmaceutical and industrial chemicals and raw materials. This paper mainly deals with a comprehensive review on the role of plant tissue culture in crop improvements and food security in Indian context.

C. Ragavendran • D. Natarajan (✉)
Natural Drug Research Laboratory, Department of Biotechnology, School of Biosciences, Periyar University, Salem 636 011, Tamil Nadu, India
e-mail: natarajpu@gmail.com

A. Dhanarajan (ed.), *Sustainable Agriculture towards Food Security*,
https://doi.org/10.1007/978-981-10-6647-4_13

13.1 Introduction

13.1.1 Food Security in India

Food security is the capability to assure, on a long-term basis, that the system provides the total population's access to a timely, consistent and nutritionally adequate supply of foods. Food security can be visualized in four main stages: (1) an adequate quantity of cereals available to all to ensure survival; (2) sufficient amount of availability of cereals and pulses; (3) food security to include cereals, pulses, milk and milk products; and (4) food security to include vegetables, fruits, milk and milk products (fish, egg and meat in case of non-vegetarians).

13.1.2 Availability of Food Grains

In 1950–1951, the annual net imports of cereals amounted to 4.1 million tonnes (Source: Directorate of Economics and Statistics, MoA, GoI). After 1995–1996 India became an exporter of cereals. In the last 50 years, there has been an increase in the per capita availability of cereals to the extent of 9%. However, the country has failed to increase the production of pulses consistent with the needs of the growing populations. This is important for many vegetarians in the country which depend on pulses for their protein requirements. Tenth plan data indicates that consumption of milk and meat products as well as vegetables and fruits has increased as a natural outcome of economic development (Chowdhry 2014).

13.1.3 Sources of Food Grains Across the Nation

Rice is mainly grown in Assam, West Bengal, Bihar, Eastern Uttar Pradesh, Kashmir Valley, Eastern Madhya Pradesh, Andhra Pradesh, Orissa, Tamil Nadu, Kerala, Karnataka and coastal areas of Maharashtra. Rice productions touched the figure 863.5 lakh tonnes in 2003–2004 (Hanafi et al. 2009).Wheat-growing areas include Uttar Pradesh, Punjab, Haryana, parts of Rajasthan and Bihar. Wheat production for 2003–2004 was estimated at 727.4 lakh tonnes (Nelson et al. 2010).

Millets include jowar, bajra and ragi. Bajra is a crop of dry and warm regions of Rajasthan. Ragi is a rain-fed crop grown in drier parts of Karnataka and Tamilnadu. Maize is mainly produced in Karnataka, Uttar Pradesh, Bihar, Andhra Pradesh and Madhya Pradesh. Pulses are grown both as rabi and kharif crops. The rabi (winter season) crops are masoor and peas. The kharif crops (sown around April and harvested in September–October) include arhar, urad and moong (Source for current level: NFHS 1998–1999). The major grain-producing areas are Madhya Pradesh, Uttar Pradesh and Rajasthan (Sources: Directorate of Economics and Statistics, Ministry of Agriculture). Food grain production touched 229.9 million tonnes in

2008–2009. Moreover, the production of oilseeds, sugarcane, fruits, vegetables and milk has also increased significantly (Chowdhry 2014).

13.1.4 The Food Security Concept

Food security is defined by FAO as the access by all the people at all times to the food needed for a healthy and active life (WHO Repository). The concept means the accomplishment of the food self-sufficiency and guarantees that this condition will be sustained in the future. Food security implies attainment of a productive growth, compatible with the economic status of the producers and the preservation of the environment (World Food Summit 1996). The factors that determine the degree of food security, in any region, country or zone in particular, are food availability, production stability and access to food by all members of the community (FAO 1998).

Food availability is a factor directly associated with the productivity of raw food supplies, as grains, roots and tubers, oil and molasses. Variations in yield due to limitations of the crop genetic potential for a given environment and losses due to biotic, abiotic and postharvest factors greatly affect these issues in common. The lack of access to food is often exacerbated by the fact that producers try to keep higher prices by destroying large amounts of stocked foods or livestock. In addition, few countries that generate food surplus deny their access to other countries as a measure of political pressure. The causes and consequences of food insecurity and poverty are inextricably linked (FAO; Garett 1997).

Indeed, by 2020, the projection suggests that food supply will be inadequate to meet the demand of a growing population, with a stagnant agricultural system unable to keep pace with both diversified and increased demand, with a very real scenario of starvation as a potential consequence (Rama Mohan Rao et al. 2003). The Indian government has not adequately addressed issues of hunger and food security. 'Despite persisting food insecurity, efforts by the Indian government to eliminate poverty and hunger are still lacking. Political and social mobilization to make food security a resonant demand that cannot be ignored is therefore essential' (Jayati Ghosh 2010).

13.1.5 Biotechnology and Agriculture

Plant biotechnology offers several possibilities for increasing productivity, diversification and productions while developing a more sustainable agriculture (Izquierdo et al. 1995). This technology includes biopesticide production, plant tissue culture techniques and the use of superior molecular biology techniques for plant transformation, genomic analysis coupled with breeding and plant disease diagnoses. It can be an efficient instrument to mitigate the consequences of climatic change and can offer cropping alternatives in lands degraded by erosion and desertification or by careless agricultural use (FAO 1999).

13.1.6 Opportunities and Scenario for Plant Biotechnology Applications

The first experiment to culture plant cells under in vitro conditions was conducted many years ago (Haberlandt 1902). However in sorghum, the earliest work on in vitro culture was reported by Strogonov et al. (1968) who found callus induction from aseptically germinated sorghum seedlings. Masteller and Holden (1970) have reported that the callus growth may be the growth of aberrant meristematic tissues and not undifferentiated cells. They also showed that callus growth generally forms at the basal node of the sorghum seedlings in response to 2,4-D auxin analogues and the growth regulator of choice.

Gamborg et al. (1977) have experimented the morphogenesis and new plant productions from callus cultures of immature healthy embryo of sorghum. The genetic improvement of the major cereals such as wheat (*Triticum aestivum*), rice (*Oryza sativa*), maize (*Zea mays*), barley (*Hordeum vulgare*), sorghum (*Sorghum bicolor*), millet (*Pennisetum* sp.), oat (*Avena sativa*) and rye (*Secale cereale*) has been particularly important for plant breeders for decades, since these crops supply more than half of the food consumed by mankind being the major sources of plant proteins and carbohydrates. They are also the basis for production of animal feed oil, starch, flour, sugar, alcoholic beverages, renewable energy, etc. (FAO 2007).

13.1.7 Applications of Molecular Markers in Plant Tissue Culture Research

- Fingerprinting of selected genetic stocks
- Assessment of genetic diversity
- Increasing the efficiency of selection for difficult traits
- Makes environment neutral selection possible
- Selection for desirable genotypes
- Manipulation of qualitative test loci that conditions complex economic traits
- Correctly mapping or placing various interacting genes that condition complex agronomic traits which are in turn useful for effective manipulation of imported genes

Plant tissue culture includes different methods/techniques like embryo culture, anther culture, endosperm culture, protoplast culture, somatic hybridization, synthetic seed production, in vitro secondary metabolite production, micropropagation, cryopreservation, etc. (Basavaraju 2005).

13.2 Methods of Tissue Culture

13.2.1 Micropropagation of Plants

The multiplication of plant materials through tissue culture referred to as micropropagation has many advantages over conventional methods of propagation such as rapid and large-scale multiplication of productive plants under in vitro conditions, irrespective of season, conservation of space and time as well as production of virus-free plants or disease-free plants. The technique essentially involves removal of a bit of tissues or cells from the leaf, stem, root, etc. Disease free healthy mother plant and grow each of them into whole plants into appropriate artificial nutrient medium in a culture vessel under in vitro-controlled conditions of light, temperature and humidity. Murashige (1974) outlined three major stages involved in micropropagation via artificial nutrient preparation, establishment, proliferation, multiplication, rooting and hardening. In vitro-grown plants are transferred to the greenhouse for acclimatization and put forward to cultivation. For example, disease-resistant crop plants have been produced through in vitro culturing in potato (*Solanum tuberosum*) against *Phytopthora infestans* (late blight of potato), in (*Zea mays*) maize against *Helminthosporium maydis* and in tobacco (*Nicotiana tabacum*) against *Pseudomonas tabaci*. At the Biotechnology Centre, IARI, plants resistant to toxin produced by *Alternaria brassicae*, a blight-causing organism, have been isolated by Kahia (1999).

13.2.2 Embryo Culture

Embryo culture is a type of plant tissue culture technique used to grow embryos from seeds and ovules in an artificial or natural nutrient medium. In embryo culture, the plant develops directly from the embryo or indirectly through the formation of callus and subsequently formation of shoots and roots. The technique has been developed to break seed dormancy and production of haploid plants (Holeman 2009; Burun and Poyrazoglu 2002). Wild species of crop plants possess several useful genes for disease, insect/pest resistance, male sterility, quality and stress tolerance. Embryo culture involving the growth of isolated immature embryos on a nutrient medium produces successful interspecific and intergeneric hybrids (cotton, barley, tomato, rice, maize, wheat, legumes, barley, rye, etc.) with recombined desirable genes for earliness and resistance to fungi, bacteria, nematode, pests and diseases in crops like tomato, maize, rice, brassica, etc. (Chawla 2009).

13.2.3 Anther Culture

Anther culture has become the most popular method for production of homozygous lines for rice cultivars worldwide. Haploids can be produced by culturing anthers or haploid plant explants. Induction of haploidy was first reported in *Datura innoxia* by Guha and Maheshwari (1964). Research on pea (*Pisum sativum* L.) haploidy began in the 1960–1980s. Calli, roots, shoots and embryos were produced in anther culture (Gupta et al. 1972; Gupta 1975). More recently haploid plant recovery from culture of isolated anthers and microspores is attempted. Calli, embryo-like structures, regenerated shoots and plants were produced in anther culture (Sidhu and Davies 2005).

13.2.4 Protoplast Culture and Somatic Hybridization

Protoplasts can be obtained from many plants, including mostly crop species (Gamborg et al. 1981; Lal and Lal 1990; Feher and Dudits 1994). Protoplast fusion is an alternative to conventional cross-hybridization method for plant improvement. In this system, fertilization is bypassed, and new characters are introduced in plants by artificial fusion of two plant cells in the form of protoplast fusion. By this somatic hybridization method, more than 400 plant species of 146 genera and 50 families of crop plants have been obtained, and these include cereals, legumes, vegetables, fruits, medicinal plants and other important species (Basavaraju 2011). Somatic hybridization is also used for gene transfer for resistance to important diseases, to improve quality and higher yield of many crop plants (Singh et al. 2015).

The families *Solanaceae* and *Brassicaceae* that contain the most commonly used species for somatic hybridization have been achieved at the Biotechnology Centre, IARI, for both interspecific and intergeneric hybrids. Many disease-resistance genes for potato leaf roll virus, leaf blight, verticillium wilt, etc. have been transferred to *Solanum tuberosum* from their species; normal crossings would not be possible due to a taxonomic barrier. Likewise many disease-resistance genes are transferred from wild species to cultivated varieties of many crop plants of cereals and vegetables. Cytoplasmic male sterility has been successfully transferred in various crop plants like *Oryza* sp., *Lycopersicum* sp., *Brassica* sp., *Nicotiana*, etc. Resistance to herbicides and antibiotics has been introduced to some species of *Brassica* (Chowdhury et al. 1997).

13.2.5 Synthetic Seed Production

A synthetic or artificial seed production has been defined as a somatic embryo encapsulated inside a coating, and it's considered to be analogous to a zygotic seed (Redenbaugh 1993). There are several types of synthetic seed: somatic embryos encapsulated in a water gel, dried and coated somatic embryos, dried and uncoated somatic embryos, somatic embryos suspended in a fluid carrier and shoot bud encapsulated in a water gel (Sanada et al. 1993). Many applications for synthetic seeds include the maintenance of male sterile lines, the maintenance of parental lines for hybrid crop production and the preservation and multiplication of specific elite genotypes of woody plants that have long juvenile developmental phases (Villalobos and Engelmann 1995; Deunff 1993).

13.2.6 Secondary Metabolites

Plants produce two types of metabolites, i.e. primary and secondary metabolites. Primary metabolites are important for the growth and development of the plants. Secondary metabolites are considered as end products, not involved in metabolic activity of growth and mostly accumulated plant cells in smaller quantities.

In vitro-regenerated plant cells and tissues have been extensively used for the production of secondary metabolites. Depending on the objectives, biotechnological methods are used for understanding metabolic pathways and improvement of plants for the production of secondary metabolites (Ferrante and Simpson 2001). Plants are a rich source of various pharmaceutical and industrial products. Nearly 30% of the drugs produced are of plant origin. Tissue culture of medicinal plants provides a continuous and reliable source of natural products round the year without the destruction of the entire plants.

Nowadays, one of the most promising methods for producing proteins, antibodies and vaccines is the use of transgenic plants (James 2008). Transgenic plants represent an economical alternative to fermentation-based production systems. Plant-based vaccines or antibodies (plantibodies) are especially striking, as plants are free of human diseases, thus reducing screening costs for viruses and bacterial toxins (James 2008).

13.2.7 Particle Bombardment Transformation

Particle bombardment technology was conceived by John Sanford et al. (1987) and allows the delivery of naked DNA into intact plant cells. It is one of the extensively adopted direct gene transfer methods by plant biotechnologists. Crops which typically have been difficult to transform using conventional *Agrobacterium*-mediated and protoplast transformation techniques may be better suited for transformation via particle bombardment or shotgun method (Finer 1990). This technique was reported to be a method of choice for engineering ergonomically important crops. Worldwide important crops, including legumes (soybean, common bean and peanut), cereals (sorghum, maize, rice, wheat, barley, oat and sugarcane) and tree species (poplar, white spruce and papaya), have been genetically engineered using particle bombardment (Christou 1993) transformation technique.

13.3 Tissue Culture of Various Indian Food Grains

13.3.1 Tissue Culture of Wheat

Wheat (*Triticum aestivum* L.) belonging to family Poaceae is growing in many regions of the world. In India it is one of the most important cereal crops and a major staple food grown all over the country. Its average yield is very low and uncertain due to biotic and abiotic stresses (Rashid et al. 2012). The population growth rate is increasing day by day, so there exists a gap between wheat yield and its demand throughout the world (Bhalla 2006). Wheat crop are regenerated plants from tissue culture, which are mainly derived from calli of immature embryos or inflorescences. It has also been reported that wheat plants were regenerated from leaf basal segments of seedlings germinated for 3–5 days (Chugh and Khurana 2003; Kopertekh and Stribnaya 2003). The inclusion of the auxin (NAA at 1.0 mg/l) in half-strength MS medium was found to be efficiently inducing roots in in vitro-produced wheat plantlets (Fennell et al. 1996; Eapen and Rao 1982).

13.3.2 Tissue Culture of Oryza sativa

Rice has been cultivated for more than 7000 years as a major crop and currently supports more than 50% of the world's population (Izawa and Shimamoto 1996). *Oryza sativa* contains a high content of dietary carbohydrates and is a staple food for more than three billion of people, supplying 50–80% of their daily calorie need (Khush 2005). Hence, there is an urgent need to improve global rice productions to meet the demand of ever-increasing populations.

An efficient plant regeneration method is a prerequisite for genetic transformation of plants and crop improvement (Raemakers et al. 1997; Dabul, 2009). Various protocols have been developed to initiate callus from explants such as immature embryos (Seraj et al. 1997; Li et al. 2007), mature embryos (Seraj et al. 1997, Ramesh and Gupta, 2006, Wang et al. 1987, Azria and Bhalla 2000), root segment (Abe and Futsuhara 1984; Mandal et al. 2003), coleoptile (Oinam and Kothari 1995) and leaf bases (Ramesh et al. 2009). A major drawback in tissue culture is conservation of genetic constancy (Rout et al. 2006).

13.3.3 In Vitro Culture of Ragi

Ragi (*Eleusine coracana* [L.] Gaertn.) is an important cereal crop and is grown worldwide (more than 4 million ha), and it's a staple food for millions of people in less-developed countries of Asia and Africa (Sastri 1989). Finger millet is considered as a traditional food crop and has higher nutritional values and ability to grow under harsh, severe drought conditions for farmers in drought-prone regions, and its seed can be stored safely for many years without any harm by insects/pests (Latha et al. 2005).

It is a recommended highly nutritious food for children and diabetic patients; in addition, its grain powder is provided as a healthful food for infants as it is digested easily and is rich in calcium, proteins, phosphorus, iron and amino acids like tyrosine, cysteine, methionine and tryptophan (Ignacimuthu and Ceasar 2012).

The efficient plant regeneration technique is a requirement for genetic transformation of different variety crops including ragi (Sharma et al. 2011). Various regeneration and genetic transformation protocols are available for cereal crops; finger millet has received minor attention compared with rice, wheat, oats, maize and barley (Ceasar and Ignacimuthu 2009). The earlier regeneration studies of finger millet use various types of explants, viz. embryogenic seed (Kothari et al. 2004; Sharma et al. 2011; Babu et al. 2012), shoot tips (Eapen and George, 1990; Ceasar and Ignacimuthu, 2011), somatic tissues (Rangan 1976), leaf segments (George and Eapen, 1990), mature embryos (Kumar et al. 2001), immature embryos (Kumar et al. 2001; Yemets et al. 2003), immature inflorescences (George and Eapen 1990; Kumar et al. 2001), mesocotyl (Rangan 1976; Mohanty et al. 1985), root (Mohanty et al. 1985) and leaf segments (Rangan 1976), respectively.

13.3.4 In Vitro Propagation of Sorghum

In vitro propagation of important Indian cereal crop *Sorghum bicolor* (L.) Moench. is done by enhancing shoot proliferation in shoot tip segments, and it's the world's fifth major cereal crop and holds importance as a construction material, food and fodder source. The crop is well adapted to tropical and subtropical regions in the entire world with vast areas under its cultivation. The potential use of this plant as biofuel source has been identified by Baskaran and Jayabalan (2005). Sorghum species are sources of fibre, fuel and secondary metabolites and are largely used in alcohol industry (sweet sorghum) as it contains high amount of starch. In addition, it was used as flour and as preparation of porridge and unleavened bread. In sorghum, plant regeneration (via callus) has chosen different explants (Thomas et al. 1977; Gamborg et al. 1977; Ma et al. 1987).

In sorghum, immature inflorescences were used as budding explants for regeneration (Elkonin et al. 1996); somatic embryogenesis has been raised from shoot tip explants of sorghum (Seetharama et al. 2000), and few reports on in vitro multiplication of *S. bicolor* from shoot tip (Zhong et al. 1998).

13.4 Tissue Culture of Fruit Varieties

13.4.1 Micropropagation of Banana

Banana is one of the most significant major fruit crops grown in India. Botanically, banana is a monocotyledonous herbaceous plant belonging to family Musaceae (Purseglove 1976). Almost every part of the banana plant is used some way or the other and it is rightly called as 'poor man's apple'. It is most likely the cheapest fruit available throughout the year. Banana occupies an important position in the Indian economy, and it contains an excellent source of carbohydrates, proteins, vitamins and minerals. Bananas are grown in 128 countries with a total cultivated area of 4.92 million hectares and globally 97.38 million metric ton productions. India ranked first all over the world in banana production, which produces 27 million metric tonnes (FAO 2009).

Several pests and diseases threaten the production in the high quantities of pesticides with severe consequences for the environment. Because of high degree of sterility, polyploidy of edible varieties and classical breeding are difficult (Stover and Simmonds 1987). In order to enhance conventional breeding and to avoid constraints imposed by some pests and pathogens, transgenic and in vitro approaches are being considered (Tripathi 2003). Mass propagation of certain genotypes, somaclonal variation techniques, genetic engineering and other biotechnological applications can be utilized for banana improvement which is based on reliable plant regeneration methods (Iqbal et al. 2013).

Banana micropropagation using shoot tip culture has long been established (Vuylsteke 1998). Previously, many researchers have reported the regeneration of *Musa* sp. via in vitro propagation (Jarret, 1986; Diniz et al. 1999; Krishnamoorthy et al. 2001; Kagera et al. 2004; Roels et al. 2005). In this view of increasing human populations and associated problem of food security, the value and cost-effective importance of bananas as a food and fruit crop cannot be overemphasized. The use of plant tissue culture has significantly enhanced the quality of planting resources and enlarged the productivity of bananas (Jain 2004).

13.4.2 Micropropagation of Guava

Guava (*Psidium guajava* L.), a few times called an *apple of tropics*, is a very precious tropical and subtropical fruit representing a staple food in various countries. It contains a rich natural source of vitamin C as well as a good source of minerals, calcium, phosphorous, iron and pectin (Singh 2005) and also contains a lot of high-grade antioxidants such as lycopene, carotenoids and polyphenols (Jiménez-Escrig et al. 2001).

Being a higher nutritionally valuable fruit that can be used for both table and processing purposes, guava has many advantages as a crop species (Chandra et al. 2010). In recent years, because of its high nutritional value and uses in several processed products such as juice, preserves and dairy or bakery items, guava is a favourite of billions of people in the tropical and subtropical countries (Yadava 1996).

In fruit trees including guava, which is studied worldwide for breeding, genetic engineering, in vitro propagation purposes and synthetic seed or artificial seed technology can be beneficial in exchange of sterile and elite materials. In vitro plant regeneration of important *Psidium guajava* fruit species is via organogenesis (Amin and Jaiswal 1987; Yasseen et al. 1995; Singh et al. 2002) and somatic embryogenesis (Chandra et al. 2004; Rai et al. 2007); encapsulation of somatic embryos have also been reported by Akhtar and Jaiswal (1994) and Biswas et al. (2007). However, crop improvement of guava through traditional breeding approaches is a complicated and slow process because of the high heterozygosity and polyploidy, and it takes long generation times for commercial varieties of guava.

13.4.3 In Vitro Propagation of Mango

The mango (*Mangifera indica* L., Anacardiaceae) is one of the favourable fruit crops in tropical and subtropical regions of the world, particularly in Asia, because of its broad adaptability, high nutritive value, richness in variety, delicious taste, tremendous flavour, attractive appearance and commercial utility in India as well as in various regions of the world. Its popularity and importance can effortlessly be realized by the fact that it is often referred to as 'king of fruits' in the tropical country (Singh 1996). It occupies 3,794,741.00 ha area and contributes 27,181,020.00 tonnes of fruits globally (Anonymous 2004) and has very strong cultural links and economic importance with several civilizations particularly in Asia.

Mango crop improvement using biotechnological approach such as plant transformation, in vitro mutagenesis followed by selection, improvement of enhanced somaclonal variations from cell, protoplast culture, etc. requires the development of well-organized regeneration methods. Mango has been a hard-to-deal crop compared with other horticultural crops. Various attempts have been made for the in vitro regeneration of mango using leaves (Singh et al. 1991; Raghuvanshi and Srivastava 1995) and shoot explants (Thomas and Ravindra 1997; Sharma and Singh 2002).

13.4.4 Micropropagation of Apple

Malus domestica Borkh (apple) is a woody plant belonging to *Rosaceae* family (Brown 1992), and it's a primary fruit species present throughout the world, with an annual production of 64.3 million tonnes (Dobránszki and da Silva 2010). Apple is a very nutritious, aromatic, and delicious fruit and is rich in vitamins A, B and C, and it contains 11% of sugar besides essential minerals in appreciable amounts. It has an appealing colour, stimulates appetite and is generally refreshing (Annonymus 2001).

Apples also purify the blood by transforming toxic chemicals found in the blood into less harmful compounds and excreted via stools and urine, minimize appetite, lower blood cholesterol and remove different free radicals (Sabir and Shah 2004). Plant tissue culture or micropropagation is defined as the culture of different somatic cells, tissues or organs of plants under controlled aseptic conditions producing a huge number of progeny plants, which are genetically identical to the mother plant, in a relatively short time compared to conventional propagation.

In vitro cloning based on different plant parts used like buds, meristems, tissues and cells is capable of regenerating into whole plants under adequate in vitro conditions. A special characteristic of plant cells having an ability to produce a whole plant is called totipotency (Reinert and Backs 1968; Verdeil et al. 2007; George 2008). Lane and McDougald (1982) and Kovalchuk et al. (2009) studied the shoot multiplication of four apple cultivars ('M.27', 'M.9', 'M.26' and 'Macspur') and found that cultivars differed in their response to the concentration of a cytokinin, 6-benzylaminopurine (BA) in the culturing medium. Similar observations were made by Bahmani et al. (2009) who found that shoot regeneration and proliferation of 'MM.106' rootstock improved on the medium supplemented with 90 mM sorbitol.

13.4.5 In Vitro Propagation of Papaya

The papaya (*Carica papaya* L.) belongs to family Caricaceae. The edible fruits are available in *Carica* genus (Muthukrishnan and Irulappan 1990). There is a tremendous scope of developing fruit industry in India. The ripe fruit contains low calories and is rich in vitamin A and vitamin C (Farzana et al. 2008).

Papain is a proteolytic enzyme present in the latex of green fruits, and it has several uses in beverages, food and pharmaceutical industries like chill-proofing of beer, tenderizing of meat and drug preparation for alleviating digestive ailments (Nakasone and Paull 1998). Conventional vegetative propagation methods in papaya such as grafting (Allan et al. 2010) and rooted cuttings (Rajan and Markose 2007) exist, but they are often not carried out on a large scale. Hence, the alternative method is micropropagation techniques for mass multiplication of specific elite plant materials. Development of an efficient in vitro regeneration protocol would be a remarkable progress for mass propagation and uniform plants for both commercial and research purposes. Various attempts have been made to in vitro propagation by shoot tip or axillary bud (Teixeira da Silva et al. (2007); Kabir et al. (2007); Panjaitan et al. (2007)), callus regeneration, somatic embryogenesis (Litz and Conover 1982) and shoot proliferation (Litz 1984).

13.4.6 In Vitro Propagation of Watermelon

Watermelon is an economically important crop belonging to family Cucurbitaceae, and its fruits are rich in vitamins A, C and B_6 and mineral nutrients like potassium, iron and calcium (Anonymous 1992). It is widely grown in the tropics and subtropics, including many parts of Southeast Asia, Africa, the Caribbean and Southern United States. Among the watermelon producers of the world, India is ranked at the sixth spot with an annual production of 400,000 metric tonnes (FAO 2013). Watermelon breeders mainly focused on developing cultivars with improved resistance to abiotic stress and improved nutritional quality by the use of biotechnology methods (Compton et al. 2004).

Transgenic watermelons with improved/increased resistance to biotic and abiotic stress have been predominantly raised through adventitious shoot regeneration (Huang et al. 2011; Lin et al. 2012). Plant tissue culture can be used to produce disease-resistant (Xiao et al. 1999), fertile, non-chimeric tetraploid plants for use in triploid hybrid seed production. Triploid plants produced seedless fruits by currently available triploid hybrids. Adventitious shoot regeneration is reported in a broad range of diploid and tetraploid cultivars (Dong and Jia 1991; Srivastava et al. 1989).

In vitro regeneration of watermelon, different explants are used including shoot apex (Compton and Gray 1993), immature embryo (Ahad et al. 1994), cotyledons (Compton 1997), hypocotyl (Srivastava et al. 1989) and leaves (Sultana et al. 2004). In vitro regeneration of plants from cotyledons of mature seeds and young seedlings has received considerable attention in the recent years due to their easy accessibility and immediate response and high capacity for shoot organogenesis (Compton et al. 1996), somatic embryogenesis, protoplast culture (Tabei 1997) and transformation studies (Tabei 1997).

13.5 Micropropagation of Vegetables

13.5.1 In Vitro Propagation of Brinjal

Eggplant or brinjal (*Solanum melongena* L.) is an important solanaceous crop grown as vegetable. Brinjal is the most common and popular vegetable crop in India. Eggplant is more susceptible to many diseases and pests that cause severe crop losses, and these problems have been noticed using hybridizing eggplant with wild resistant *Solanum* species. Recently, tissue culture techniques have been generally used for the enhancement of various crops. In vitro shoot induction from callus culture can induce genetic and epigenetic changes in the newly regenerated plantlets, and these genetic changes have been coined as 'somaclonal variation' (Larkin and Scowcroft 1981), cell suspensions (Fassuliotis et al. 1981), anthers (Isouard et al. 1979; Tuberosa et al. 1987) and protoplasts (Saxena et al. 1981; Gleddie et al. 1986; Clark et al. 1988), hypocotyl (Kamat and Rao 1978), leaf (Gleddie et al.

1983; Mukherjee et al. 1991), root (Franklin and Sita 2003) and cotyledon (Saito and Nishimura 1994).

Genetic transformation of eggplant via *Agrobacterium* was first reported by Guri and Sink (1988b), and the organogenesis may be influenced by the antibiotic used to eliminate *A. tumefaciens.* For example, Augmentin can cause improved shoot proliferation by TDZ (Billings et al. 1997). Resistance to Colorado potato beetle (*Leptinotarsa decemlineata* Say) (CPB), a pest that has developed resistance to synthetic insecticides, became a severe problem in agriculture (Arpaia et al. 1997).

13.5.2 In Vitro Propagation of Tomato

Tomato (*Lycopersicon esculentum* Mill.) belongs to family Solanaceae. It's the most important vegetable crop and plays an essential role in maintaining human health and vigour and is helpful in healing wounds because of its antibiotic properties found in the ripe fruits. Tomato is used in preserved products such as sauce, chutney, soup, paste, etc. It is one of the most important protective foods as it possesses appreciable quantities of vitamin C, vitamin B, organic acid, essential amino acids, dietary fibres and β-carotene (Raziuddin et al. 2004), and sometimes it is referred to as the poor man's orange (Devi et al. 2008). Tomato production is mainly affected by different stresses such as diseases, high temperature, drought, salinity and its vulnerability to different insects and pests. Sherkar and Chavan (2014) reported that the major diseases of tomato are caused by fungi, bacteria, viruses and nematodes.

Micropropagation is an important alternative technique of biotechnology which can be used to improve productivity of crop via rapid accessibility of better planting materials (Bhatia et al. 2004). In vitro techniques are important tools for modern plant development program to develop suitable cultivars in a short time. The economic importance of this crop is responsiveness for further enhancement via genetic manipulation (Evans 1989; Taji et al. 2002). Earlier many researchers have reported that adventitious regeneration in tomato deals with induction of shoots on hypocotyls, apical meristem, stems, anthers and inflorescence explants (Compton and Veilleux 1991; Moghaleb et al. 1999; Raziuddin et al. 2004), leaf (McCormic et al. 1986; Gaffer et al. 1997) and cotyledon (vanRoekel et al. 1993), respectively.

13.5.3 Micropropagation of Potato

Potato (*Solanum tuberosum* L.) is an internationally important stable food crop plant among world populations. Techniques have been adopted for improving the yield of potato crops and in return preventing shortage of food. Potato is the fourth largely important crop after wheat, rice and maize and which biotechnology has been successfully implemented for seed productions (Jones 1973).

Earlier studies reported that the in vitro propagation of potatoes depends on the biological importance of cultivars and explant types (leaf, node, shoot tip, etc.). Potato is being seen as an essential food security crop and as a substitute for costly cereal imports. It is vulnerable to a number of biotic and abiotic stresses which limit production, particularly among small and marginal farmers with limited resources. Tissue culture technology has been the foundation of high-quality, healthy, disease-free planting material production at a mass scale, particularly in vegetative propagated crop. In vitro propagation is an alternative method to conventional propagation of potatoes using sprouts and nodal cutting which is more consistent for maintaining the genetic integrity of the multiplied clones (Chandra and Birhman 1994; Liljana et al. 2012).

13.5.4 Micropropagation of Carrot

Carrot (*Daucus carota* L.) is one of the essential major vegetable crops produced around the world, and it contains a high content of vitamin A and fibre for human nutrition (Horvitz et al. 2004). The phytochemicals in carrots such as β-carotene (provitamin A), lutein, lycopene and anthocyanins play an important nutritional role in human health (Seddon et al. 1994). Genetic transformation can be used as a corresponding technology to develop carrot quality and productivity (Jayaraj and Punja 2007).

In the last two decades, public concerns about the failure of various habitats and species have increased pressure for the development of new successful ways to reduce environmental damage and species extinction (Meyers et al. 2000). Traditional breeding methods have greatly contributed to the improvement of carrot desirable traits such as root shape, root colour, smooth skin, β-carotene levels and sugar content (Yau and Simon 2005). In situ conservation strategies have been applied with relative achievement to the management of plant genetic resources (Graudal et al. 2001). In addition, plant biotechnological protocols mostly become increasingly important for plant conservation of both cultivated crops and endangered species (Pence 1999). Micropropagation has been applied to produce a large number of crops and endangered plants (Chawla 2002).

Through these biotechnological methods, high rate of multiplications can be obtained, and the regenerated plants are genetically identical (George and Debergh 2008). In vitro culture of *D. carota* somatic embryogenesis was first achieved in carrot (Steward et al. 1958), and since then, several reports about in vitro propagation and somatic embryogenesis have been published for carrot (Makunga et al. 2005).

13.6 Tissue Culture of Nuts

13.6.1 In Vitro Propagation of Groundnut

Groundnut or peanut (*Arachis hypogaea* L.) is the principal economic crop of the world. The grain legumes are an important group crop with a major source of dietary protein and oil. India is a major groundnut-producing country, but the yielding of groundnut is especially low. The low yield is attributed to poor agricultural practices and biotic and abiotic stresses (Madhusudhana 2013). The requirement for oils and fats has increased at the rate of 6% per annum in the last 13 years (Veeramani and Subrahmaniyan 2011). Thus, crop improvement in terms of productions, desirable traits and resistance to biotic and abiotic stresses is a prerequisite for this crop. Recent biotechniques are alternative to conventional methods needed to boost the yield of the crops.

In vitro plant regeneration of wild species of *Arachis* has been achieved from different explant types including embryonic and mature tissues, through distinctive morphogenic pathways, aiming at conservation, multiplication and germplasm availability for improvement programs of the cultivated groundnut (Rey and Mroginski 2006). In vitro organogenesis is considered as well organized than somatic embryogenesis and callus culture for genetic transformation (Arockiasamy et al. 2002). Therefore, the development of in vitro organogenesis protocol would be important for improving groundnut production and quality through molecular breeding (Geng et al. 2011).

13.6.2 In Vitro Propagation of Cashew Nut

Cashew (*Anacardium occidentale* L.) is an important cash crop and grown in tropical countries. It is valued for its delicious and nutritive kernels that are the most economic part of crop and oil taken from the shell (CNSL). Additional low percentage of fruit set (3–4%) has been reported in cashew nuts, and conventional methods of propagations are not efficient enough to provide high-yielding plant materials (Sivantham et al. 1990).

To overcome these problems, plant tissue culture technique is being well studied for obtaining large number of plant materials within the quickest time. In vitro plant regeneration technique via organogenesis or embryogenesis provides rapid multiplication of superior varieties. Organogenesis has been used by earlier investigators (Falcone and Leva 1987; Lievens et al. 1989; Leva and Falcone 1990) to induce shoot elongation on microcuttings from seedlings and bud growth on cotyledons from mature seed and young leaves from seedlings. Das et al. (1996) and (D'Souza and D'Silva 1992) used nodal cuttings and shoot tip, cotyledonary nodes from in vitro-raised seedlings.

13.7 Selection for Abiotic Stress Tolerance

Developing abiotic stress-tolerant plants especially for salt and drought conditions by in vitro selection has been reported in a broad range of plant species including cereals, vegetables, fruits, crops and commercially important plant species (Rai et al. 2011). Unlike the conditions in the field or nursery, a better control of culture environment can be achieved through in vitro screening techniques. Salt and drought tolerance has been reported in many plants. The selection is applied to callus, cell suspension and protoplast cultures by the inclusion of growth inhibitory level of selection agents (sodium chloride, polyethylene glycol (PEG), sorbitol and mannitol) in culture medium (Table 13.1).

Selection for acid soil and aluminium tolerance can be made with aluminium chloride as the selection agent on the low-acid media as much as pH 4, and the method can be employed in isolating Al-tolerant plants. It is also possible to select cell lines resistant to proline analogue to develop mutants with increased free proline and abiotic tolerance to stresses such as salt, drought or cold (Widholm 1976). Both one-step and stepwise selection methods can be applied (Bressan et al. 1985; Nabors 1990). In a single-step selection, the callus or explant is exposed once or a few times to the inhibitory level of sodium chloride, and then resultant surviving tissues are isolated and plants are regenerated. Using such a criterion, salt-tolerant plantlets have been obtained in flax (Rowland et al. 1989), sugar beet (Freytag et al. 1990), brown mustard (*Brassica juncea*) (Kirti et al. 1991) and sorghum (Waskom et al. 1990).

The second method is the long-term stepwise selection in which cultures are allowed to grow over several subculture cycles; it contains high salt concentrations. Bressan et al. (1985) obtained salt adopted in tobacco cells, which were grown for at least 25 generations in 25 gl^{-1} sodium chloride. Ochatt and Power (1988) reported the long-term selection method in colt cherry cell lines that survived six transfers on the mannitol-containing medium and subjected to three cycles of direct recurrent selection, each consisting of 2–3-week subcultures on salt medium. NaCl-resistant cell lines were developed from *Nicotiana tabacum* L. Cell suspension culture is treated with mutagen ethyl methanesulfonate (EMS) and then grown in a medium containing 0.03 M NaCl and then on a medium as high as 0.09 M NaCl (Nabors et al. 1975). A third approach is the indirect way of selecting for a resistance to proline analogue or ABA insensitivity. Cultured cells of carrot (Ricardi et al. 1983), *Brassica napus* (Chandler and Thorpe 1986) and *Vigna radiata* (Kumar and Sharma 1989) exposed to proline analogues exhibited tolerance to salt stress. Stable NaCl-tolerant *Chrysanthemum* variants were developed through a whole plant or callus selection after in vitro mutagenesis using ethyl methanesulfonate (EMS) as the chemical mutagen (Hossain et al. 2006).

Embryogenic suspension cultures of sweet potato cv. 'Lizixiang' were exposed to 80 Gy gamma rays followed by in vitro selection with NaCl (He et al. 2009). A total of 276 plants regenerated from the irradiated 2783 cell aggregates by a two-step in vitro selection with 86, 171, 257 and 342 mM NaCl of the 18 plant lines

Table 13.1 Examples of in vitro treatment for abiotic stress tolerance in different crop varieties

Plant species	Selection agent and level used	Tolerance to selectable trait	References
Saccharum sp.	Mannitol (0.62, 0.84 and 1.08 MPa)	Drought	Errabii et al. (2006)
Oryza sativa	PEG (control and 100 gL^{-1})	Drought	Adkins et al. (1995)
Capsicum annuum L.	PEG (0, 5, 10, 15, 20, 25 or 30% PEG and gradually decreased to 0% by continuous sub-culturing)	Drought	Santos-Diaz and Ochoa-Alejo (1994)
Tagetes minuta	Mannitol (6–80 mM)	Drought	Mohamed et al. (2000)
Saccharum sp.	NaCl (42.8, 85.6, 128.3, 171.1, 213.9, 256.7, 299.5 or 342.2 mM)	Salt stress	Patade and Suprasanna (2008)
Oryza sativa L.	NaCl (1 and 2% for in vitro; 0.5% for natural conditions)	Salt stress	Vajrabhaya et al. (1989)
Brassica juncea	NaCl (in vitro proliferation 0, 1.0, 1.25, 1.50, 1.60, 1.80, 2.0% NaCl)	Salt stress	Kirti et al. (1991)
Vigna radiata	Mannitol (0, 180, 360, 449, 540, 629, 720 $molm^{-3}$ of mannitol)	Drought	Gulati and Jaiwal (1993)
Oryza sativa	NaCl (0, 0.5, 1.0, 1.5, 2.0% of NaCl)	Salt stress	Shankhdhar et al. (2000)
Oryza sativa	NaCl (EC at 6 and 12 dS/m by NaCl)	Salt stress	Lee et al. (2003)
Citrus limon	NaCl (0 and 170 mM NaCl)	Salt stress	Piqueras et al. (1996)
Citrus aurantium	NaCl (0, 100, 200 and 300 mM NaCl)	Salt stress	Koc et al. (2009)
Brassica napus	NaCl ((0, 0.1, 0.2, 0.3, 0.4,0.5, 0.6, 0.7% NaCl)	Salt stress	Rahman et al. (1995)
Glycine max	NaCl (0, 25, 50, 75, 100, 125, 150 mM NaCl)	Salt stress	Liu and van Staden (2000)
Strawberry	NaCl (200 mM NaCl)	Salt stress	Dziadczyk et al. (2003)
Vigna radiata	NaCl (0, 50, 100 and 150 mM NaCl)	Salt stress	Hassan et al. (2008)
Solanum tuberosum	NaCl (direct selection 60, 90, 120, 150, 300 or 450 mM)	Salt stress	Ochatt et al. (1999)
Saccharum sp.	NaCl (0 or 68 mM NaCl)	Salt stress	Gandonou et al. (2006)
Winter barley	Hydroxyproline (10–20 mM)	Frost	Tantau et al. (2004)
Oryza sativa	Al (0, 250, 500, 7500, 1000, 1250, 1500, 2000 μM of al)	Aluminium	Jan et al. (1997)
Oryza sativa	Al (0, 30 and 60 ppm of Al in the form of $Al_2(SO_4)_3 \cdot 18H_2O$)	Aluminium	Roy and Mandal (2005)

revealed significantly higher in vitro salt tolerance than control plants. Selection (in vitro) has been practised, and tolerant lines have been obtained in several crops including *Brassica*, bamboo, sunflower, strawberry, soybean, flax, rice, tomato, potato, sweet potato, sugarcane, wheat and rice (Rai et al. 2011), respectively.

13.8 Selection for Biotic Stress Tolerance

Plant diseases are caused by a variety of different pathogens, and selection systems to isolate tolerant lines have been considered using selection for resistance like culture filtrates, chemicals and toxins (Table 13.2). In a first report by Carlson (1973), chances of in vitro selection for disease resistance were explored in tobacco for selection against *Pseudomonas syringae* that causes wildfire disease. Few *Helminthosporium* toxins have successfully been used in crop plants. In maize, Gengenbach et al. (1977) applied *Helminthosporium maydis* T (HmT) toxin in the selection medium to select embryogenic callus and regenerated resistant plants that exhibited disease resistance. In sugarcane, Larkin and Scowcroft (1981) applied callus cultures of cultivar Q-101 to *Helminthosporium sacchari* toxin and recovered 480 regenerated plants, and several of these were resistant.

Chawla and Wenzel (1987) used 3000 callus cultures of barley and 2000 callus cultures of wheat for selection against *Helminthosporium sativum*, and in 6–17% of calli and regenerants, resistance was evident. The use of fusaric acid (FA) has been useful in the selection of *Fusarium*-resistant plants (Remotti and Loffler 1997).

Culture filtrates (CF) represent an easy and simple method of selection by incorporation into the culture media at appropriate concentrations. In several selection experiments, culture filtrates, purified and partially purified, have been successfully used. In vitro lines resistant to fungal, bacterial and viral pathogens have been isolated in several plant species (Table 13.2). In vitro selection of culture filtrates and phytotoxins is used for developing disease-resistant plants in various crops (Kumar et al. 2008; Rai et al. 2011).

13.9 Improving the Product Quality and Quantity

The plant tissue culture plays an important role in the improvement of sustainable agriculture though micropropagation, embryo culture, anther culture, protoplast culture and secondary metabolites and takes care of the health of mankind.

The fundamental idea behind the treatment of elicitors mainly focuses on the stress that induces upon administration in the cell cultures, which concurrently affects the yield and quality parameters of the secondary metabolites accumulated (Pullaiah and Rao 2009). Food crops play a very essential role in the human nutrition and health besides providing proteins, amino acids, minerals, micronutrients, vitamins, antioxidants, phytosterols and dietary fibres (Rui Hai Liu 2013).

Table 13.2 Example of in vitro selection for various disease resistances in important crops

Crops	Selective agents	Resistance	References
Annona comosus	Filtrate, FA	*Fusarium subglutinans*	Borras et al. (2001)
Arachis hypogaea	CF	*Cercosporidium personatum*	Venkatachalam and Jayabalan (1996)
Brassica napus	CF	*Phoma lingam*	Sacristan (1982)
Carthamus tinctorius	CF	*Alternaria carthami*	Kumar et al. (2008)
Citrus limon	CF	*Phoma tracheiphila*	Gentile et al. (1992)
Curcuma	CF	*Pythium graminicolum*	Gayatri et al. (2005)
Curcuma longa	CF	*Pythium graminicolum*	Gayatri et al. (2005)
Fragaria vesca	Partially purified toxins	*Phytopthora cactorum, Rhizoctonia fragariae, Botrytis cinerea*	Battistini and Rosati (1991)
Glycine max	CF	*Septoria glycines*	Song et al. (1994)
Gossypium hirsutum	CF	*Fusarium oxysporum, Alternaria macrospora*	Ganesan and Jayabalan (2005)
Hordeum vulgare	FA	*Fusarium* spp.	Chawla and Wenzel (1987)
Linum usitatissimum	CF	*Fusarium oxysporum*	Krause et al. (2003)
Lycopersicon esculentum	CF	*Pyrenochaeta lycopersici*	Fuime and Fuime (2003)
Medicago sativa	CF	*Fusarium oxysporum*	Hartman et al. (1984); McCoy (1988)
Musa spp.	CF	*Fusarium oxysporum*	Matsumoto et al. (1999)
Oryza sativa	CF	*Helminthosporium oryzae*	Vidhyasekaran et al. (1990)
Peach	Fractionated CF	*Xanthomonas campestris* pv. *pruni*	Hammerschlag (1988)
Solanum tuberosum	CF	*Phytophthora infestans*	Behnke (1980)
Triticum aestivum	DON	*Fusarium* sp., *Fusarium graminearum*	Maier and Oettler (1992); Yang et al. (1998)
Wheat	Syringomycin	*Pseudomonas syringae* pv. *syringae*	Pauly et al. (1987)
Zea mays	HmT toxin	*Helminthosporium maydis*	Gengenbach et al. (1977)

Biotechnological studies have shown that several agriculturally important traits can be used to improve different vegetable crops and for biotic and abiotic stresses and also for pharmaceutical important proteins for immunization, enzyme therapy or precursors for any pharmaceutical product (Basavaraju 2011).

Plant-derived antibodies have wide-ranging medicinal uses such as passive immunization and targeted drug delivery (Monika and Ananda Kumar 2005).

Compounds like phytoalexins (a category of secondary metabolites of plants) are the stimulators that induce the production of desired secondary metabolites in in vitro conditions (Keen et al. 1972). Depending on the usage of in vitro studies, elicitors are of two types, abiotic and biotic, the former being physical (UV, IR or gamma irradiations) or chemical (alkalinity, osmotic pressure, heavy metal ions, etc.) in nature and the latter being homogenates of fungal or bacterial cultures. Potential abiotic elicitors include diethyl amino ethyl dichloro phenyl ether catharanthin in the production of Ajmalicine (*Catharanthus roseus*), activated carbon for echinofuron (*Lithospermum erythrorhizon*), UV irradiation for flavonoid glycosides (*Haplopappus gracilis*) and ethephon for caffeine (*Coffea arabica*).

13.10 Conclusion

This chapter summarizes the micropropagation of food crops for fast in vitro propagation of better cultivars, for faster introduction of newer cultivars with specific desirable traits and for rapid multiplication of disease-free, healthy propagation materials and abiotic and biotic resistances. Recently, several reproducible methods have been developed for micropropagation of various domestic crops. *In vitro* shoot-developed plantlets have also served as valuable tools for basic and applied research, which is an attractive advantage of in vitro systems with well-controlled culture environment in the physiological condition of test plants. The role of plant tissue culture technique has a great profit making and high plant quality and has the ability to produce high-yield crop plants. It should be used on a large scale to develop new vegetables/crops to fulfil the requirement of growing world populations and ensure food security in the near future.

References

Abe T, Futsuhara Y (1984) Varietal difference of plant regeneration from root callus tissues in rice. Jpn J Breed 34:147–155

Adkins SW, Kunanuvatchaidach R, Godwin ID (1995) Somaclonal variation in rice - drought tolerance and other agronomic characters. Aust J Bot 43:201–209

Ahad A, Islam R, Hossain M, Khalekuzzaman M, Joarder O (1994) Plant regeneration from immature and mature embryo axes of watermelon [in Bangladesh]. Plant Tiss Cult 4:39–44

Akhtar N, Jaiswal VS (1994) Role of somatic embryogenesis and artificial seed production for micropropagation and improvement of guava (*Psidium guajava* L.). In: Proceedings of the symposium on role of plant tissue culture in biotechnology in India. Varanasi, India, (abst.), p 2

Allan P, Clark C, Laing M (2010) Grafting papayas (*Carica papaya* L.) Acta Hortic 851:254–258

Amin MN, Jaiswal VS (1987) Rapid clonal propagation of guava through *in vitro* shoot proliferation on nodal explants of mature trees. Plant Cell Tissue Org Cult 9:235–243

Annonymus (2001) Common marketing standard for apples and pears. Commission Regulation (EC). Off J L 215:3–16

Anonymous (1992) The wealth of India a dictionary of Indian raw materials and industrial products, vol 3. Publication and Information Directorate, CSIR, New Delhi, pp 606–609

Anonymous (2004) FAO Statistical database.. Available at: http://apps.fao.org/page/collections?subset=agriculture

Arockiasamy S, Prakash S, Ignacimuthu S (2002) Direct organogenesis from mature leaf and petiole explants of *Eryngium foetidum* L. Biol Plant 45:129–132

Arpaia S, Mennella G, Onofaro V, Perri E (1997) Production of transgenic eggplant (*Solanum melongena* L.) resistant to Colorado Potato Beetle (*Leptinotarsa decemlineata* Say). Theor Appl Genet 95:329–334

Azria D, Bhalla PL (2000) Plant regeneration from mature embryo-derived callus of Australian rice (*Oryza sativa* L.) varieties. Aust J Agr Res 51:305–312

Babu AG, Geetha KN, Manjunatha V, Shankar AG (2012) An efficient high throughput plant regeneration and transformation protocol for production of transgenics tolerant to salt in inger millet. Intl J For Crop Improv 3:16–20

Bahmani R, Karami O, Gholami M (2009) The effect of carbon source and concentration on *in vitro* shoot proliferation of MM.106 apple rootstock. Fruit Veg Cereal Sci Biotechnol 3:35–37

Basavaraju R (2005) Plant tissue culture in plant biotechnology. Proc of AP Academic of Sci 9:193–200

Basavaraju R (2011) Plant tissue culture-agriculture and health of man. Ind J Sci Technol 4:333–335

Baskaran P, Jayabalan N (2005) *In vitro* plant regeneration and mass propagation system for *Sorghum bicolor*-a valuable major cereal crop. J Agric Technol 1(2):345–363

Battistini C, Rosati P (1991) *In vitro* evaluation of somaclonal strawberry *(Fragaria ananassan* (Brighton) variant for susceptibility to *Phytopthora cactorum).* In: Dale A, Lubby WW (eds) The strawberry into the 21st century. Timber Press, Portland, pp 121–123

Behnke M (1980) General resistance to blight of *Solarium tuberosum* plants regenerated from callus resistant to culture filtrates of *Phytophthora infestans*. Theor Appl Genet 56:151–152

Bhalla PL (2006) Genetic engineering of wheat-current challenges and opportunities. Trends Biotechnol 24:305–311

Bhatia P, Ashwath N, Senaratna T, Midmore D (2004) Tissue culture studies of tomato (*Lycopersicon esculentum*). Plant Cell Tissue Organ Cult 78:1–21

Billings S, Jelenkovic G, Chin CK, Eberhadt J (1997) The effect of growth regulation and antibiotics on eggplant transformation. J Am Soc Hortic Sci 122:158–162

Biswas BK, Joshee N, Yadav A, Yadav AK (2007) Development and application of biotechnology in guava: a nutraceutical fruit. In: Desjardins Y (ed) ISHS Acta Horticulturae 744: I international symposium on human health effects of fruits and vegetables

Borras R, Santos AP, Matos RS, Cabral M, Arzola A (2001) First attempt to use a *Fusarium subglutinans* culture filtrate for the selection of pineapple cultivars resistant to fusarose disease. Plant Breed 120:435–438

Bressan RA, Singh NK, Handa AK, Kononowick A, Hasegawa PM (1985) Stable and unstable tolerance to NaCl in cultured tobacco cells. In: Freeling M (ed) Plant genetics: proceedings of the third annual ARCO plant cell research institute- UCLA symposium on plant biology, symposia on molecular and cellular biology (April 13–19,1985. ULCA), New Series, A.R. Liss, New York, pp 755–769

Burun B, Poyrazoglu EC (2002) Embryo culture in barley (*Hordeum vulgare* L). Turk J Biol 26:175–180

Carlson PS (1973) Methionine sulfoximine-resistant mutants of tobacco. Science 180:1366–1368

Ceasar SA, Ignacimuthu S (2009) Genetic engineering of millets: current status and future prospects. Biotechnol Lett 31:779–788

Ceasar SA, Ignacimuthu S (2011) *Agrobacterium*-mediated transformation of finger millet (*Eleusine coracana* (L.) Gaertn.) using shoot apex explants. Plant Cell Rep 30:1759–1770

Chandra R, Birhman RK (1994) *In vitro* micro propagation in relation to pedigree in potato. J Indian Potato Assoc 21:87

Chandra R, Bajpai A, Gupta S, Tiwari RK (2004) Embryogenesis and plant regeneration from mesocarp of *Psidium guajava* L. (guava). Indian J Biotechnol 3:246–248

Chandra R, Kamle M, Bajpai A, Muthukumar M, Kalim S (2010) *In vitro* selection: a candidate approach for disease resistance breeding in fruit crops. Asian J Plant Sci 9:437–446

Chandler SF, Thorpe TA (1986) Variation from plant tissue cultures; biotechnological application to improving salinity tolerance. Biotechnol Adv 4:117–135

Chawla HS (2002) Introduction to plant biotechnology, 2nd edn. Science, Enfield

Chawla HS (2009) Introduction to plant biotechnology. Oxford and IBH Publishing Company Pvt, New Delhi

Chawla HS, Wenzel G (1987) In vitro selection of barley and wheat for resistance against *Helminthosporium sativum*. Theor Appl Genet 74:841–845

Chowdhry NK (2014) Social sciences for class IX, S. Chand school books. https://books.google.co.in/books?isbn=8121927722

Chowdhury JB, Jain RK, Chowdhury VK (1997) Somatic hybridization as a tool in plant breeding. In: Kapoor RI, Saini MI (eds) Plant breeding and crop improvement. CBS Publishers & Distributors, New Delhi, pp 342–346

Christou P (1993) Particle gun mediated transformation. Curr Opin Biotechnol 4:135–141

Chugh A, Khurana P (2003) Regeneration via somatic embryogenesis from leaf basal segments and genetic transformation of bread and emmer wheat by particle bombardment. Plant Cell Tissue Organ Cult 74:151–161. https://doi.org/10.1023/A:1023945610740

Clark CB, Sihachakr D, Haicour R, Serraf E, Barrientos E, Herbreteau C, Ducreux G, Rossignol L, Souvannavong V (1988) Electrofusion for the production of somatic hybrid plants of *Solanum melongena* L. and *Solanum khasianum*. Plant Sci 57(3):215–224

Compton ME (1997) Influence of seedling pretreatment and explant type on watermelon shoot organogenesis. Hort Science 32(3):514–515

Compton ME, Gray D (1993) Shoot organogenesis and plant regeneration from cotyledons of diploid, triploid and tetraploid watermelon. J Am Soc Hortic Sci 118(1):151–157

Compton ME, Veillux RE (1991) Shoot root and flower morphogenesis on tomato inflorescence explants. Plant Cell Tissue Organ Cult 24:223–231

Compton ME, Gray DJ, Maynard DN (1996) Use of tetraploid somaclones in breeding seedless watermelons. In vitro 32:28

Compton ME, Gray DJ, Gaba VP (2004) Use of tissue culture and biotechnology for the genetic improvement of watermelon. Plant Cell Tissue Organ Cult 77:231–243

D'Souza L, D'Silva I (1992) *In vitro* propagation of *Anacardium occidentale* L. Plant Cell Tissue Organ Cult 29(1):1–6

Dabul ANG (2009) Screening of a broad range of rice (*Oryza sativa* L.) germplasm for *in vitro* rapid plant regeneration and development of an early prediction system. In vitro Cell Dev Biol Plant 45:414–420

Das S, Jha TB, Jha S (1996) In vitro propagation of cashewnut. Plant Cell Rep 15:615–619

Deunff YL (1993) Conclusions and future. In: Redenbaugh K (ed) SYll seeds: application of synthetic seeds to crop improvement. CRC Press, Boca Raton, pp 453–561

Devi M, Dhaliwal MS, Kaur A, Gosal SS (2008) Effect of growth regulators on in vitro morphogenetic response of tomato. Indian J Biotechnol 7:536–540

Diniz JDN, Goncalves AN, Hernandez FFF, Torres AC (1999) Macronutrient absorption by banana explants *in vitro*. Pesq Agrop Brasileira 34(7):1201–1209

Dobránszki J, Teixeira da Silva JA (2010) Micropropagation of apple – a review. Biotechnol Adv 28(4):462–488

Dong JZ, Jia SR (1991) High efficiency plant regeneration from cotyledons of watermelon (*Citrullus vulgaris* Schrad). Plant Cell Rep 9(10):559–562

Dziadczyk P, Bolibok H, Tyrka M, Hortynski JA (2003) *In vitro* selection of strawberry (*Fragaria ananassa* Duch.) clones tolerant to salt stress. Euphytica 132(1):49–55

Eapen S, George L (1990) Influence of phytohormones, carbohydrates, amino acids, growth supplements and antibiotics on somatic embryogenesis and plant differentiation in finger millet. Plant Cell Tissue Organ Cult 22:87–93

Eapen S, Rao PS (1982) Plant regeneration from callus cultures of durum and emmer wheat. Plant Cell Rep 1:215–218

Elkonin LA, Lopushanskaya RF, Pakhomova NV (1996) Embryogenic callus of *Sorghum* (*Sorghum bicolor* (L.) Moench) by amino acids. Maydica 40:153–157

Errabii T, Gandonou CB, Essalmani H, Abrini J, Idomar M, Senhaji NS (2006) Growth proline and ion accumulation in sugarcane callus cultures under drought-induced osmotic stress and its subsequent relief. Afr J Biotechnol 5:1488–1493

Evans DA (1989) Somaclonal variation--genetic basis and breeding applications. Trends Genet 5:46–50

Falcone AM, Leva AR (1987) Prove preliminary sulla morfogenesis in colture di anacardio. Rivista di Agricoltura Subtropicale e Tropicale 81:117–125

FAO (1998) Regional Development Partnerships Programme (RDPP)

FAO (1999) Committee on agriculture: biotechnology. 25–29 January

FAO (2013) Food and Agricultural Organization. http://faostat3faoorg. Cited 27 April 2015

Farzana ARF, Palkadapala PG, Meddegoda KM, Samarajeewa PK, Easwara JP (2008) Somatic embryogenesis in papaya (*Carica papaya* L.) cv. Rathna. J Natl Sci Found Sri Lanka 36(1):41–50

Fassuliotis G, Nelson BV, Bhatt DP (1981) Organogenesis in tissue culture of *Solanum melongena* cv. Florida Market. Plant Sci Lett 22:119–125

Feher A, Dudits D (1994) Plant protoplasts and cell fusion and direct DNA uptake: culture and regeneration systems. In: Vasil IK, Thorpe TA (eds) Plant cell and tissue culture. Kluwer Academic, Dordrecht, pp 71–118

Fennell S, Bohorova N, Van- Ginkel M, Crossa J, Hoisington D (1996) Plant regeneration from immature embryos of 48 elite CIMMYT bread Wheats. Theor Appl Genet 92:163–169

Ferrante E, Simpson D (2001) A review of the progression of transgenic plants used to produce plantibodies for human usage. Biol Biomed Sci 4:1–10

Finer JJ, McMullen (1990) Transformation of cotton (*Gossypium hirsutum* L.) via particle bombardment. Plant Cell Rep 8:586–589

Food and Agricultural Organization, Statistics (2009) www.fao.org

Franklin G, Lakshmi Sita G (2003) *Agrobacterium tumefaciens* mediated transformation of eggplant (*Solanum melongena* L.) using root explants. Plant Cell Rep 21:549–554

Freytag AM, Wrather JA, Eeichsen AW (1990) Salt tolerance sugar beet progeny from tissue cultures changed with multiple salts. Plant Cell Rep 8:647–650

Fuime F, Fuime G (2003) Use of culture filtrates of *Phyrenochaeta lycopersici* in tests for selecting tolerant varieties of tomato. J Plant Pathol 85:131–133

Gaffer J, Tiznado ME, Handa AK (1997) Characterization and functional expression of a ubiquitously expressed tomato pectin methylesterase. Plant Phys 114:1547–1556

Gamborg A, Shyluk JP, Brar DS, Constable F (1977) Morphogenesis and plant regeneration from callus of immature embryos of *Sorghum*. Plant Sci Lett 10:67–74

Gamborg OL, Shylak JP, Shahin EA (1981) Isolation, fusion and culture of plant protoplasts. In: Thorpe TA (ed) Plant tissue culture: methods and applications in agriculture. Academic, New York, pp 115–153

Ganesan M, Jayabalan N (2005) Carbon source dependent somatic embryogenesis and plant regeneration in cotton (*Gossypium hirsutum* L. cv. SVPR 2). Indian J Exp Biol 43:921–925

Gandonou CB, Errabii T, Abrini J, Idaomar M, Senhaji NS (2006) Selection of callus cultures of sugarcane (*Saccharum* sp.) tolerant to NaCl and their response to salt stress. Plant Cell Tissue Organ Cult 86:239–248

Garett J (1997) Challenges to the 2020 vision for Latin America: food and agriculture science 1970. Food, Agriculture and the Environment Discussion Paper 21. International Food Policy Research Institute, Washington, DC

Gayatri MC, Roopa Darshini V, Kavyasree R (2005) Selection of turmeric callus for tolerant to culture filtrate of *Pythium graminicolum* and regeneration of plants. Plant Cell Tissue Organ Cult 83:33–46

Geng L, Niu L, Shu C, Song F, Huang D, Zhang J (2011) High efficiency regeneration of peanut (*Arachis hypogaea* L.) plants from leaf discs. Afr J Biotechnol 10:12650–12652

Gengenbach BG, Green CE, Donovan CM (1977) Inheritance of selected pathotoxin resistance in maize plants regenerated from cell culture. Proc Natl Acad Sci U S A 74:5113–5117

Gentile A, Continella G, Tribulato E, Vardi A (1992) Differential responses of citrus calli and protoplasts to the culture filtrate and toxin of *Phoma tracheiphila*. Theor Appl Genet 83:759–764

George EF, Debergh PC (2008) Micropropagation: uses and methods. In: George EF, Hall MA, De Klerk GJ (eds) Plant propagation by tissue culture, 3rd edn. Springer, Dordrecht, pp 29–64

George L, Eapen S (1990) High frequency plant-regeneration through direct shoot development and somatic embryogenesis from immature inflorescence cultures of finger millet (*Eleusine coracana* Gaertn). Euphytica 48:269–274

Ghosh J (2010) The political economy of hunger in 21st century India. Econ Polit Wkly 45(44/45 (October 30–November 12, 2010):33–38

Gleddie S, Keller W, Setterfield G (1983) Somatic embryogenesis and plant regeneration from leaf explants and cell suspensions of *Solanum melongena* (eggplant). Can J Bot 61:656–666

Gleddie S, Keller W, Setterfield G (1986) Somatic embryogenesis and plant regeneration from cell suspension derived protoplasts of *Solanum melongena* (eggplant). Can J Bot 64:355–361

Guha S, Maheswari SC (1964) *In vitro* production of embryos from anthers of *Datura*. Nature 204:497

Gulati A, Jaiwal PK (1993) In vitro selection of salt-resistant callus line of *Vigna radiata* (L.) Wilczek. Res J Plant Environ 9:145–152

Gupta S (1975) Morphogenetic response of haploid callus tissue of *Pisum sativum* (var. B22). Indian Agriculturalist 19:11–21

Gupta S, Ghosal KK, Gadgil VN (1972) Haploid tissue culture of *Triticum aestivum* var. Sonalina and *Pisum sativum* var. B22. Indian Agriculturalist 16:277–278

Guri A, Sink KC (1988) *Agrobacterium* transformation of eggplant. J Plant Physiol 133:52–55

Haberlandt G (1902) Kulturversuche mit isolierten Pflanzenzellen; Sitzungsber. Akad Wiss Wien Math-Naturwiss Kl Abt J 111:69–92

Hammerschlag FA (1988) Selection of peach cells for insensitivity to culture filtrate of *Anthomonas campestris* pv. Pruni and regeneration of resistant plants. Theor Appl Genet 76:865–869

Hanafi MM, Hartinie A, Shukor J, Mahmud TMM (2009) Upland rice varieties in Malaysia: agronomic and soil physicchemical characteristics. Pertanika J Trop Agric Sci 32:225–246

Hartman CL, MeCoy TJ, Knous TR (1984) Selection of alfalfa (*Medicago sativa*) cell lines and regeneration of plants resistant to the toxin(s) produced by *Fusarium oxysporum* f. sp. Medicaginis. Plant Sci Lett 34:183–219

Hassan N, Serag M, El-Feky F, Nemat Alla M (2008) *In vitro* selection of mungbean and tomato for improving tolerance to NaCl. Ann Appl Biol 152:319–330

He S, Han Y, Wang Y, Zhai H, Liu Q (2009) In vitro selection and identification of sweet potato (Ipomoea batatas) plants tolerant to NaCl. Plant Cell Tissue Organ Cult 96:69–74

Holeman DJ (2009) Simple embryo culture for plant breeders: a manual of technique for the extracyion and *in vitro* germination of mature plant embryos with emphasis on the rose, 1st edn. Rose Hybridizers Association, 10

Horvitz MA, Simon PW, Tanumihardjo SA (2004) Lycopene and β-carotene are bioavailable from lycopene 'red' carrots in humans. Eur J Clin Nutr 58:803–811

Hossain Z, Mandal AKA, Datta SK, Biswas AK (2006) Development of NaCl-tolerant strain in *Chrysanthemum morifolium* Ramat through *in vitro* mutagenesis. Plant Biol 8:450–461

Huang YC, Chiang CH, Li CM, TA Y (2011) Transgenic watermelon lines expressing the nucleocapsid gene of Watermelon silver mottle virus and the role of thiamine in reducing hyperhydricity in regenerated shoots. Plant Cell Tissue Organ Cult 106:21–29

Ignacimuthu S, Ceasar SA (2012) Development of transgenic finger millet (*Eleusine coracana* (L.) Gaertn.) resistant to leaf blast disease. J Biosci 37:1–13
Iqbal MM, Muhammad A, Hussain I, Bilal H (2013) Optimization of *in vitro* micropropagation protocol for Banana (Musa Sapientum L.) under different hormonal concentrations and growth media. Int J Agric Innov Res 2(1):23–27
Isouard G, Raquin C, Demarly Y (1979) Obtention de plantes haploides et diploides par culture *in vitro* d'anthères d'aubergine (*Solanum melongena* L.) Comptes Rendus de l' Academie de Sciences de Paris 288:987–989
Izawa T, Shimamoto K (1996) Becoming a model plant: the importance of rice to plant science. Trends Plant Sci 1(3):95–99
Izquierdo J, Ciampi L, García E (1995) Biotechnología apropiable: racionalidad de su desarrollo y aplicación en América Latina y el Caribe. Red de Cooperación Technica and Biotecnologia Vegetal (REDBIO). Oficina Regional FAO para América Latina y el Caribe, Santiago, Chile
Jain SM (2004) Introduction: cellular biology and biotechnology including mutation techniques for creation of new useful banana genotypes. In: Jain SM, Swen-nen R (eds) Banana improvement: cellular molecular biology and induced mutations. Science Publishers Inc, Enfield, pp 15–16
James C (2008) Global status of commercialized biotech/GM crops. ISAAA Brief No. 39. Ithaca, p 243
Jan V, Macedo V, Kinet J, Bouharmont J (1997) Selection of al-resistant plants from a sensitive rice cultivar, using soma clonal variation, *in vitro* and hydroponic cultures. Euphytica 97(3):303–310
Jarret RL (1986) *In vitro* propagation and genetic conservation of bananas and plantains. In: IBPGR advisory committee on *in vitro* storage, report of the third meeting (appendix). IBPGR, Rome
Jayaraj J, Punja ZK (2007) Combined expression of chitinases and lipid transfer protein genes in transgenic carrot plants enhances resistance to foliar fungal pathogens. Plant Cell Rep 26:1539–1546
Jiménez-Escrig A, Rincón M, Pulido R (2001) Guava fruit (*Psidium guajava* L.) as a new source of antioxidant dietary fiber. J Agric Food Chem 49:5489–5493
Jones RL (1973) Gibberellins their physiological role. Agri Rev Pl Physiol 24:571–598
Kabir AH, Bari MA, Huda AKMN, Rezvy MA, Mahfouz I (2007) Effect of growth regulators and carbon sources on axillary shoots proliferation from shoot-tip explant and successful transplantation of papaya (*Carica papaya* L.) Biotechnology 6:268–272
Kagera AG, Kagera GR, Kegera CBM, Van den Houwe I, Swennen R (2004) Rapid mass propagation and diffusion of new banana verities among small. Scale farmers in north western Tazania. Afr Crop Sci J 12(1):7–17
Kahia WJ (1999) *In vitro* propagation of the new *Coffea arabica* cultivar- Ruiru 11. Ph D thesis, University of London
Kamat MG, Rao TS (1978) Vegetative multiplication of eggplants (*Solanum melongena*) using tissue culture techniques. Plant Sci Lett 13:57–65
Keen NT, Parirdge JE, Zake AL (1972) Pathogen produced elicitor of a chemical defense mechanism in soybeans monogenetically resistant to Phytophthora megasperma var. Sojae. Phytopathology 62:768–772
Khush GS (2005) What it will take to feed 5.0 billion rice consumers in 2030. Plant Mol Biol 59:1–6
Kirti PB, Hadi S, Kumar PA (1991) Production of sodium-chloride-tolerant *Brassica juncea* plants by *in vitro* selection at the somatic embryo level. Theor Appl Genet 83:233–237
Koc NK, Bas B, Koc M, Kusek M (2009) Investigations of *in vitro* selection for salt tolerant lines in sour orange (*Citrus aurantium* L.) Biotechnology 8:155–159
Kopertekh LG, Stribnaya LA (2003) Plant regeneration from wheat leaf explants. Russ J Plant Physiol 50:365–368. https://doi.org/10.1023/A:1023826304989
Kothari SL, Agarwal K, Kumar S (2004) Inorganic nutrient manipulation for highly improved in vitro plant regeneration in finger millet *Eleusine coracana* (L.) Gaertn. In vitro Cell Dev Biol Plant 40:515–519

Kovalchuk I, Lyudvikova Y, Volgina M, Reed BM (2009) Medium, container and genotype all influence in vitro cold storage of apple germplasm. Plant Cell Tissue Org Cult 96:127–136
Krause IR, Mankowska G, Lukaszewicz M, Szopa J (2003) Regeneration of flax (*Linum usitatissimum* L) plants from anther culture and somatic tissue with increased resistance to *Fusarium oxysporum*. Plant Cell Rep 22:110–116
Krishnamoorthy V, Kumar N, Soorianathasundaram K (2001) *In vitro* propagation of the popular desert cv. Poovan (AAB). First post graduate seminar. TNAU, Coimbatore, p 46
Kumar V, Sharma DR (1989) Isolation and characterization of sodium chloride-resistant callus culture of *Vigna radiata* (L.) Wilczek var. radiate. J Exp Bot 40(1):143–147
Kumar S, Agarwal K, Kothari SL (2001) *In vitro* induction and enlargement of apical domes and formation of multiple shoots in finger millet, *Eleusine coracana* (L.) Gaertn and crowfoot grass, *Eleusine indica* (L.) Gaertn. Curr Sci 81:1482–1485
Kumar JV, Ranjithakumari BD, Sujatha G, Castano E (2008) Production of plants resistant to *Alternaria carthami* via organogenesis and somatic embryogenesis of safflower cv.NARI-6 treated with fungal culture filtrates. Plant Cell Tissue Organ Cult 93:85–96
LaL R, LaL S (1990) Crop improvement using biotechnology. CRC Press, Boca Raton
Larkin PJ, Scowcroft WR (1981) Somaclonal variation- a novel source of variability from cell cultures for plant environment. Theor Appl Genet 60:197–214
Latha AM, Rao KV, Reddy VD (2005) Production of transgenic plants resistant to leaf blast disease in finger millet (*Eleusine coracana* (L.) Gaertn). Plant Sci 169:657–667
Lee SY, Lee JH, Kwon TO (2003) Selection of salt-tolerant doubled haploids in rice anther culture. Plant Cell Tissue Organ Cult 74:143–149
Leva AR, Falcone AM (1990) Propagation and organogenesis *in vitro* of *Anacardium occidentale* L. Acta Hortic 280:143–145
Li ZN, Fang F, Liu GF, Bao MZ (2007) Stable *Agrobacterium* mediated genetic transformation of London plane tree (*Platanus acerifolia* Willd.) Plant Cell Rep 26:641–650
Lievens C, Polser M, Boxus PH (1989) First results about micropropagation of *Anacardium occidentale* by tissue culture. Fruits 44(10):553–557
Liljana KG, Mitrev S, Fidanka T, Mite I (2012) Micropropagation of potato- *Solanum tuberosum* L. Electr J Biol 8(3):45–49
Lin CY, KuHM CYH, HoHY YTA, Jan FJ (2012) Development of transgenic watermelon resistant to cucumber mosaic virus and watermelon mosaic virus by using a single chimeric transgene construct. Transgenic Res 21:983–993
Litz RE (1984) Papaya. In: Handbook of plant cell culture. Macmillon Publishing Co, New York, pp 349–368
Litz RE, Conover RA (1982) *In vitro* somatic embryogenesis and plant regeneration from *Carica papaya* L. ovular callus. Plant Sci Lett 26(2–3):153–158
Liu RH (2013) Health-Promoting components of fruits and vegetables in the diet. Adv Nutr 4:384–392
Liu T, van Staden J (2000) Selection and characterization of sodium chloride-tolerant callus of *Glycine max* (L.) Merr cv.Acme. Plant Growth Regul 31:195–207
Ma H, Gu M, Liang GH (1987) Plant regeneration from cultured immature embryos of *Sorghum bicolor* (L.) Moench. Theor Appl Genet 73:389–394
Madhusudhana B (2013) A survey on area, production and productivity of groundnut crop in India. IOSR-JEF 1(3):01–07
Maier FJ, Oettler G (1992) Selection for the *Fusarium* toxin deoxynivalenol in callus cultures of triticale. In: Third European *Fusarium* seminar, IHARR, Adzikow, Poland, pp 43–49
Makunga NP, Jäger AK, Van Staden J (2005) An improved system for the *in vitro* regeneration of *Thapsia garganica* via direct organogenesis—influence of auxins and cytokinins. Plant Cell Tissue Organ Cult 82:271–280
Mandal AB, Maiti A, Biswas A (2003) Somatic embryogenesis in root derived callus of *indica* rice. Plant Tissue Cult 13(2):125–133

Masteller VM, Holden DJ (1970) The growth of and organ formation from callus tissue of *Sorghum*. Plant Physiol 107:679–685

Matsumoto K, Barbosa ML, Souza LAC, Teixeira JB (1999) *In vitro* selection for *Fusarium* wilt resistance to banana II. Resistance to culture filtrate of race I *Fusarium oxysporum* f. sp. Cubense. Fruits 54:151–157

McCormic S, Niedermeyer J, Fry B, Barnason A, Horch R, Farley R (1986) Leaf disk transformation of cultivated tomato (*L. esculentum*) using *Agrobacterium tumefaciens*. Plant Cell Rep 5:81–84

McCoy T (1988) Tissue culture selection for disease resistant plants. Iowa State J Res 62:503–521

Meyers N, Mittermeier A, Mittermeier CG, da Fonseca GAB (2000) Kent J biodiversity hotspots for conservation priorities. Nature 403:853–858

Moghaleb REA, Saneoka H, Fujita K (1999) Plant regeneration from hypocotyls and cotyledon explants of tomato (*Lycopersicon esculentum*). Soil Sci Plant Nutr 45:639–646

Mohamed MAH, Harris PJC, Henderson J (2000) *In vitro* selection and characterization of a drought tolerant clone of *Tagetes minuta*. Plant Sci 159:213–222

Mohanty BD, Gupta SD, Ghosh PD (1985) Callus initiation and plant regeneration in ragi (*Eleusine coracana* Gaertn.) Plant Cell Tissue Organ Cult 5:147–150

Monika D, Ananda Kumar P (2005) Biotechnology of vegetable crops: current trends. Proc AP Acad Sci 9:219–228

Mukherjee SK, Rathinasabapathi B, Gupta N (1991) Low sugar and osmotic requirements for shoot regeneration from leaf pieces of *Solanum melongena* L. Plant Cell Tissue Organ Cult 25:13–16

Murashige T (1974) Plant propagation through tissue culture. Annu Rev Plant Physiol 25:135–166

Muthukrishnan CR, Irrulappan I (1990) Papaya. In: Bose TK, Mitra SK (eds) Fruits: tropical and subtropical. Naya Prakash, Calcutta, pp 304–355

Nabors MW (1990) Environmental stress resistance. In: Dix PJ (ed) Plant cell line selection procedure and applications. VCH Weinheim, New York, pp 167–186

Nabors MW, Daniels A, Nadolny L, Brown C (1975) Sodium chloride tolerant lines of tobacco cells. Plant Sci Lett 4:155–159

Nakasone HY, Paull RE (1998) Papaya. In: Tropical fruits. CAB International, Wallingford, pp 239–269

Nelson GC, Rosegrant MW, Palazzo A, Gray I, Ingerstoll C, Robertson R (2010) Food security, farming, and climate change to 2050: scenarios, results, policy options. Research monograph. International Food Policy Research Institute, Washington, DC. doi:10:2499/9780896291867

Ochatt SJ, Power JB (1988) An alternative approach to plant regeneration from protoplasts of sour cherry (Prunus cerasus L.). Plant Sci 56:75–79

Ochatt SJ, Marconi PL, Radice S, Arnozis PA, Caso OH (1999) In vitro recurrent selection of potato: production and characterization of salt tolerant cell lines and plants. Plant Cell Tissue Organ Cult 55:1–8

Oinam GS, Kothari SL (1995) Totipotency of coleoptile tissue in *indica* rice (*Oryza sativa* L. cv. CH 1039). Plant Cell Rep 14:245–248

Panjaitan SB, Aziz MA, Rashid AA, Saleh NM (2007) *In vitro* plantlet regeneration from shoot tip of field-grown hermaphrodite papaya (*Carica papaya* L. cv. Eksotika). Int J Agric Biol 6:827–832

Patade VY, Suprasanna BVA (2008) Gamma irradiation of embryogenic callus cultures and in vitro selection for salt tolerance in sugarcane (*Saccharum officinarum* L.) Agric Sci (China) 7(9):101–105

Pauly MH, Shane WW, Gengenbach BG (1987) Selection for bacterial blight phytotoxin resistance in wheat tissue culture. Crop Sci 27:340–344

Pence VC (1999) The application of biotechnology for the conservation of endangered species. In: Benson EE (ed) Plant conservation biotechnology. Taylor and Francis, London, pp 227–241

Piqueras A, Hernandez JA, Olmos E, Hellin E, Sevilla F (1996) Changes in antioxidant enzymes and organic solutes associated with adaptation of citrus cells to salt stress. Plant Cell Tissue Organ Cult 45:53–60

Pullaiah T, Subba Rao MV (2009) Plant tissue culture theory and practice. Scientific Publishers (India), Jodhpur

Purseglove JWC (1976) Tropical crops: monocotyledons. Longmans, New York, pp 273–283

Raemakers CJJM, Sofiari E, Jacobsen E, Visser RGF (1997) Regeneration and transformation of cassava. Euphytica 96:153–161

Raghuvanshi SS, Srivastava A (1995) Plant regeneration of *Mangifera indica* using liquid shaker to reduce phenolic exudation. Plant Cell Tissue Organ Cult 41:83–85

Rahman MH, Krishnaraj S, Thorpe TA (1995) Selection for salt tolerance in vitro using microspore derived embryos of *Brassica napuscv*. Topas, and the characterization of putative tolerant plants. In vitro Cell Dev Biol Plant 31:116–121

Rai MK, Akhtar N, Jaiswal VS (2007) Somatic embryogenesis and plant regeneration in *Psidium guajava* L. cv. Banarasi local. Sci Hortic 113:129–133

Rai MK, Kalia RK, Singh R, Gangola MP, Dhawan AK (2011) Developing stress tolerant plants through in vitro selection – an overview of the recent progress. Environ Exp Bot 71(1):89–98

Rajan S, Markose BL (2007) Propagation of horticultural crops. Horticulture Ser 2:16

Rama Mohan Rao MS, Batchelor CH, James AJ, Nagaraja RJ, Seeley J, Butterworth A (2003) Andhra Pradesh rural livelihoods programme water audit report. APRLP, Rajendranagar, Hyderabad 500 030, India

Ramesh M, Gupta AK (2006) Genetic transformation of embryogenic callus cultures of *indica* rice using particle bombardment and selection of stably transformed hygromycin resistant plants. Asian J Microbiol Biotechnol Environ Sci 8(3):617–622

Rangan TS (1976) Growth and plantlet regeneration in tissue cultures of some Indian millets: *Paspalum scrobiculatum* L., *Eleusine coracana* Gaertn. and *Pennisetum typhoideum*. Pers Z Pflanzenphysiol 78:208–216

Rashid H, Khan MH, Chaudhry Z, Bano R, Raja NI (2012) An improved *Agrobacterium* mediated transformation system in wheat. Pak J Bot 44(1):297–300

Raziuddin SS, Shah HJ, Chaudhary T, Mohammad Ali S (2004) Hormonal effect on callus induction in tomato. J Agri 20:223–225

Redenbaugh K (ed) (1993) Synseeds: applications of synthetic seeds to crop improvement. CRC Press, Boca Raton

Reinert J, Backs D (1968) Control of totipotency in plant cells growing *in vitro*. Nature 220:1340–1341

Remotti PC, Loffler HJM, Vanvlotendoting L (1997) Selection of cell lines and regeneration of plants resistant to fusaric acid from *Gladiolus grandi-florus* cv. Peter pears. Euphytica 96:237–245

Rey HY, Mroginski LA (2006) Somatic embryogenesis and plant regeneration in diploid and triploid *Arachis pintoi*. Biol Plant 50:152–155

Ricardi GM, Cella RC, Ciferri O (1983) Resistance to azetidine-carboxylic acid and sodium chloride tolerance in carrot cell cultures and Spirulina platensis. Plant Cell Physiol 24:1073–1078

Roels S, EScalona M, Cejas I, Noceda C, Rodriguez R, Canal MJ, Sandoval D (2005) Optimization of plantain (Musa AAB) micropropagation by temporary immersion system. Plant Cell Tissue Organ Cult 82:57–66

Rout GR, Mohapatra A, Jain SM (2006) Tissue culture of ornamental pot plant: a critical review on present scenario and future prospects. Biotechnol Adv 24(6):531–560

Rowland GG, Mchughen A, McOnie C (1989) Field performance at saline-affected sites of a somaclonal variant of mcgregor flax selected for salt *tolerance in vitro*. Can J Plant Sci 69:49–60

Roy B, Mandal AB (2005) Towards development of Al-toxicity tolerant lines in indica rice by exploiting somaclonal variation. Euphytica 145:221–227

Sabir SM, Shah AS (2004) Effect of chemical treatment, wax coating, oil dipping and storage behavior of apple (*Malus domestica* Borkh). Pak J Nutr 3(2):122–127

Sacristan MD (1982) Resistance response to Phoma lingam of plant regenerated from selected cell and embryogenic cultures of haploid *Brassica napus*. Theor Appl Genet 61:193–200
Saito T, Nishimura S (1994) Improved culture conditions for somatic embryogenesis using an aseptic ventilative filter in eggplant (*Sollanum melongena* L.) Plant Sci 102:205–211
Sanada M, SakamotoY HM, Mashiko T, Okamoto A, Ohnishi N (1993) Celery and lettuce. In: Redenbaugh K (ed) Synseeds: applications of synthetic seeds to crop improvement. CRC Press, Boca Raton, pp 305–327
Santos-Diaz MS, Ochoa-Alejo N (1994) PEG-tolerant cell clone of chili pepper: growth, osmotic potentials and solute accumulation. Plant Cell Tissue Organ Cult 37:1–8
Saxena PK, Gill R, Rashid A, Maheshwari SC (1981) Plantlet formation from isolate protoplasts of *Solanum melongena* L. Protoplasma 106:355–359
Seddon JM, Ajani UA, Sperduto RD, Hiller R, Blair N, Burton TC, Farber MD, Gragoudas ES, Haller J, Miller DT, Yannuzzi LA, Willet W (1994) Dietary carotenoids, vitamins A, C and E and advanced age-related mascular degeneration. J Am Med Assoc 272:1413–1420
Seetharama N, Sairam RV, Rani TS (2000) Regeneration of sorghum from shoot tip cultures and field performance of the progeny. Plant Cell Tissue Organ Cult 61:169–173
Seraj ZI, Islam Z, Faruque MO, Devi T, Ahmed S (1997) Identification of the regeneration potential of embryo derived calluses from various indica rice varieties. Plant Cell Tissue Organ Cult 48:9–13
Shankhdhar D, Shankhdhar SC, Mani SC, Pant RC (2000) *In vitro* selection for salt tolerance in rice. Biol Plant 43:477–480
Sharma RR, Singh SK (2002) Etiolation reduces phenolic content and polyphenol oxidase activity at the pre-culture and *in vitro* exudation of phenols from mango explants. Trop Agric 79(2):94–99
Sharma M, Kothari-Chajer A, Jagga-Chugh S, Kothari SL (2011) Factors influencing *Agrobacterium tumefaciens*-mediated genetic transformation of *Eleusine coracana* (L.) Gaertn. Plant Cell Tissue Organ Cult 105:93–104
Sherkar HD, Chavan AM (2014) Studies on callus induction and shoot regeneration in tomato. Sci Res Rep 4(1):89–93
Singh RN (1996) Mango. ICAR, New Delhi
Singh G (2005) Strategies for improved production in guava. In: Kishun R, Mishra AK, Singh G, Chandra R (eds) Proceeding of 1st international guava symposium. CISH, Lucknow, India, pp 26–39
Singh SK, Sharma HC, Singh SP, Kishore PBK (1991) Callus induction and root initiation from explants of zygotic embryos in mango hybrids. Plant tissue culture and biotechnology: emerging trends. Proceeding of Symposium, pp 129–33
Singh SK, Meghwal PR, Sharma HC, Singh SP (2002) Direct shoot organogenesis on explants from germinated seedlings of *Psidium guajava* L. cv. Allahabad Safeda. Sci Hortic 95:213–221
Singh AK, Mishra NK, Kumar S, Pandey S (2015) Applications of somatic hybridization for the improvement of horticultural crops. Hortic Biotechnol Res 1:39–45
Sivantham M, Pugalendhi L, Jeeva S, Somasundaram D (1990) The Cashew 4:17
Song HS, Lim SM, Widholm JM (1994) Selection and regeneration of soy-beans resistant to pathotoxic culture filters of *Septoria glycines*. Phytopathology 84:948–951
Source for current level: NFHS (1998–1999), p 361
Source: Directorate of Economics and Statistics. MoA, GoI
Sources: Directorate of Economics and Statistics. Ministry of Agriculture http://mospi.nic.in/Mospi_New/upload/SYB2015/index1.html
Srivastava D, Andrianov V, Piruzian E (1989) Tissue culture and plant regeneration of watermelon (*Citrullus vulgaris* Schrad. cv. Melitopolski). Plant Cell Rep 8(5):300–302
Steward FC, Mapes MO, Mears K (1958) Growth and organized development of cultured cells. II. Organization in culture grown freely suspended cells. Am J Bot 45:705–708
Stover RH, Simmonds NW (1987) Bananas. Longman Scientific and Technical, Essex

Strogonov BP, Komizerko EI, Butenko RG (1968) Culturing of isolated glasswort, *Sorghum*, sweet clover and cabbage tissue for comparative study of their salt resistance. Soviet. Plant Physiol 15:173–177

Sultana R, Bari M, Rahman M, Rahman M, Siddique N, Khatun N (2004) *In vitro* rapid regeneration of plantlets from leaf explant of water melon (*Citrullus lanatus* Thumb.) Biotechnology 3(2):131–135

Tabei Y (1997) Study on breeding of cucurbitaceae using biotechnology. Bull Natl Inst Agrobiol Res 11(1):107

Taji A, Kumar PP, Lakshmanan P (2002) *In vitro* plant breeding. Food Products Press, New York, p 167

Tantau H, Balko C, Brettschneider B, Melz G, Dorffling K (2004) Improved frost tolerance and winter survival in winter barley (*Hordeum vulgare* L.) by in vitro selection of proline over accumulating lines. Euphytica 139:19–32

Teixeira da Silva JA, Rashid Z, Tan ND, Sivakumar D, Gera A, Teixeira SM Jr, Tennant PF (2007) Papaya (*Carica papaya* L.) Biol Biotechnol 1(1):47–73

Thomas P, Ravindra MB (1997) Shoot tip culture of mango, influence of media, genotype, explant factors, season, and decontamination treatments on phenolic exudation, explant survival and axenic culture establishment. J Hortic Sci 72:713–722

Thomas E, King PJ, Portykus I (1977) Shoot and embryo-like structure formation from cultured tissues of *Sorghum bicolor*. Naturwissenschaften 64:587

Tripathi L (2003) Genetic engineering for improvement of *Musa* production in Africa. Afr J Biotechnol 2(12):503–508

Tuberosa R, Sanghineti MC, Conti S (1987) Anther culture of eggplant *Solanum melongena* L. lines and hybrids. Genét Agrár 41(3):267–274

Vajrabhaya M, Thanapaisai T, Vajrabhaya T (1989) Development of salt tolerant lines of KMDL and LPT rice cultivars through tissue culture. Plant Cell Rep 8:411–414

Veeramani P, Subrahmaniyan K (2011) Nutrient management for sustainable groundnut productivity in India – a review. Int J Eng Sci Technol 3:8138–8158

Venkatachalam P, Jayabalan N (1996) *In vitro* screening of groundnut (*Arachis hypogaea* L.) cell lines and regeneration of plants resistant to pathotoxic culture filtrate of *Cercosporidium personatum*. Plant Tissue Tiss Cult 6:73–82

Verdeil JL, Alemanno L, Niemenak N, Tranbarger TJ (2007) Pluripotent versus totipotent plant stem cells: dependence versus autonomy? Trends Plant Sci 12:245–252

Vidhyasekaran P, Ling DH, Borromeo ES, Zapata FJ, Mew TW (1990) Selection of brown spot-resistant rice plants from *Helminthosporium oryzae* toxin-resistant calluses. Ann Appl Biol 117:515–523

Vuylsteke DR (1998) Shoot-tip culture for the propagation, conservation, and distribution of Musa Germplasm. International Institute of Tropical Agriculture, Ibadan, p 82

Wang MS, Zapaa FJ, Castro DCD (1987) Plant regeneration through somatic embryogenesis from mature seed and young inflorescence of wild rice (*Oryza perennis*). Plant Cell Rep 6:294–296

Waskom RM, Miller DR, Hanning GE, Duncan RR, Voigt RL, Nabors MW (1990) Field evaluation of tissue culture derived sorghum for increased tolerance to acid soils and drought stress. Can J Plant Sci 70:997–1004

Widholm JM (1976) Selection and characterization of cultured carrot and tobacco cells resistant to lysine, methionine and proline analogs. Can J Bot 54(13):1523–1529

World Food Summit 1996

Xiao GH, DX W, Liu JX, Zheng SO, Xiao LY (1999) Resistance to wilt disease of watermelon germplasm resources by exogenous DNA introduction. J Human Agric Univ 25:453–457

Yadava UL (1996) Guava production in Georgia under cold-protection structure. In: Janick J (ed) Progress in new crops. ASHS Press, Arlington, pp 451–457

Yasseen M, Barringer SA, Schnell RJ, Splittoesser WE (1995) *In vitro* shoot proliferation of guava (*Psidium guajava* L.) from germinated seedlings. Plant Cell Rep 14:525–528

Yang Z, Yang X, Huang D (1998) Studies on somaclonal variants for resistance to scab in bread wheat (*Triticum aestivum* L.) through in vitro selection for tolerance to deoxynivalenol. Euphytica 101:213–219

Yau YY, Simon PW (2005) Molecular tagging and selection for sugar type in carrot roots using co-dominant, PCR-based markers. Mol Breed 16:1–10

Zhong H, Wang W, Sticklen M (1998) *In vitro* regeneration of *Sorghum bicolor* (L.) Moench: efficient plant regeneration from shoot apices. J Plant Physiol 153:719–726

Chapter 14
Phytochemical Screening of Transgenic and Non-transgenic Leguminous Plant Species

Amal Thomas Cheeran, Dhandapani Gurusamy, and Krishnan Vasanth

Abstract Tissue culture technology through organogenesis by *in vitro* plant regeneration has been applied and successfully developed in several economically important legume and nonlegume plant species. This method has not provided robust regeneration protocol for application and limits the utilization of genetic manipulation to its full potential. Therefore, *in planta* methods are quite easy to perform and do not require much proficiency and labour and are also less cost-effective. Here, we have explained three economically important legumes such as horse gram, black gram and cowpea using *in planta* transformation methods. In the present study, methods of phytochemical screening of transgenic and non-transgenic leguminous plant species have been analysed, and observed results were reported. The results showed that metabolic changes were observed due to stress of *Agrobacterium tumefaciens*. Phytochemical analysis revealed that the total phenolic content was gradually increased in all the transgenic plants as compared to the control plant. The flavonoid and anthocyanin contents increased in black gram, whereas primary metabolic compounds such as protein, amino acids and chlorophyll contents were increased in the cowpea and decreased in the horse gram and black gram, respectively. But the amounts of carbohydrate and starch contents were decreased among all three transgenic plants. This simple procedure can be used for the genetic manipulation of genes to enhance the production of important leguminous species.

A.T. Cheeran • K. Vasanth (✉)
Molecular Biology Laboratory, Department of Botany, Bharathiar University,
Coimbatore 641 046, Tamil Nadu, India
e-mail: vasanthlabbu@gmail.com

D. Gurusamy
Department of Entomology, University of Kentucky, Lexington 40546-0091, KY, USA

A. Dhanarajan (ed.), *Sustainable Agriculture towards Food Security*,
https://doi.org/10.1007/978-981-10-6647-4_14

14.1 Introduction

About 10,000 years ago, a remarkable change has occurred in the way people procured the food from the natural sources. They started cultivation and domestication of plants and animals, respectively. Seed formation is a unique feature of higher plants especially in Spermatophyta (gymnosperms and angiosperms). The seeds are outcome product of sexual reproduction and forerunner of the next generation. Majority of the higher plants propagate by their seeds. The higher energy content and compactness of seed are the main factors which influenced the human population and shift the humans from hunter-gatherer society to agrarian society. This agricultural revolution has started mainly due to the impressive nature of seeds because they are more convenient to handle and easy for cultivation, based on their requirements and needs. The seeds are not only foundation of an agricultural revolution but also heralded a fundamental change in human society, material's wealth, social organization and cultural achievements. The farmers started to modify the natural ecosystem suited to the human need and interfered with the normal flow of energy in the biosphere (Chrispeels and Sadava 1994). People have cultivated around 400 species, and only 15 species are more often used for their regular dietary purpose. These 15 plants come under the four major categories on the basis of the utilitarian background cereals, legumes (pulses), stem/root crops (sugarcane/potato, tapioca) and fruit-yielding plants. The cereals (wheat, rice, corn, rye, oats, barley, sorghum and millets) and legumes (beans, peas, peanuts, soybean and forage species) are the two main/important groups of plants that provide maximum dietary product for the people throughout the world (Chrispeels and Sadava 1994).

Plants are the major potent sources of biochemical compounds and produce phytomedicine for human welfare. The phytomedicine derived from plant products is a part since eternity and phytochemical compounds are basically synthesized by primary or rather secondary metabolism of living organisms. There is an obscure function of chemically diverse secondary metabolic compounds, although the precise boundaries between the two groups in some instances are blurred. Primary metabolic compounds are essential for association with plant developmental process such as photosynthesis, respiration and developmental growth. These include phytosterol, nucleotide, amino acid and organic acid synthesis. Secondary metabolites gain great interest of study because of their derived products as dyes, fibres, glues, oils, waxes, flavouring agents, drugs and perfumes. They are viewed as potential sources of new natural drugs, antibiotics, insecticides and herbicides for application and utility for the human welfare.

Legumes are the one of the largest and third families (Leguminosae) of flowering plants, are distributed throughout the world and include 19,00,00 species. These are diverse family ranging from herbaceous annuals to woody perennials. Legumes are cheap sources of protein when animal protein is scarce. The word 'legume' basically originated from the Latin word 'legumen' which means 'seeds harvested in pods', and the alternative term for edible seeds of leguminous plants is 'pulse' which also came from Latin word 'puls', meaning 'pottage'. Legumes were among the earliest crops cultivated by human beings (Aykroyd et al. 1982). The nitrogen-fixing capacity of legumes makes it to withstand all kinds of terrestrial ecosystem and has a good nitrogen-fixing ability associated with symbiotic bacterium such as *Rhizobium* sp., and simultaneously it enriches the soil texture and fertility by improving nutrients. In many countries, it has been used as a cover crop to reduce soil erosion, and it is a cheap source of nutrient for animals to increase their fodder value.

The proteins from legume seeds could be used as natural antioxidant and therapeutics for the human health benefit. It also contains carbohydrates, dietary fibres, amino acids and minerals, etc. Legumes are not only domesticated for food but also for other uses like feed, forage, fibre, industrial and medicinal compounds. Asia shares more pulse production than other countries, and India is the largest pulse-producing country, and the production has been increased globally to 72.2 million tons in 2014. The intercellular/extracellular and degradative metabolites can degrade the plant secondary metabolites cells, and it can be transported in and out of the cell and reused in both primary and secondary metabolism. There are key regulatory nodes in the pathway that can control and alter the flux of metabolites (Cazzonelli and Pogson 2010).

Pulses are more nutritious and contained various phytochemicals such as alkaloids, flavonoids, tannins and phenols. Due to its nutritious state, edible pulses are widely consumed by humans in their diet. In the developing countries, the majority of new drugs and compounds have been generated from natural secondary metabolites (Rischer et al. 2006). Different techniques and methods are applied for the isolation and purification of compounds from medicinal plants. These plant extracts are concentrated, fractionated and purified biologically for active compound.

14.2 Targets for Genetic Modification of Legumes

Plant genetic transformation is a vital and major core research area in plant biotechnological studies, and it is a practical tool for improvement of the cultivar. Over 3000 field trials of transformed transgenic plants are in progress throughout the world. The first genetic transformation was done in tobacco plant in 1984, and it has been extended over many economically useful crops such as vegetables, ornamental, medicinal, fruit trees, etc. The new genetic engineering technology for gene transfer has enormous potential for plant improvement by introducing foreign genes into plant cells or tissues. It has become a reliable source for crop improvement of transgenic plants and their products without compromise on the yield and nutrition quality. Plant biotechnology offers the possibility to improve agricultural traits such as improvement and alteration of nutrition content and biotic and abiotic stress tolerance of economically important crops, without modifying cultivar identity. *In planta Agrobacterium*-mediated procedures have been successful in some nonleguminous plants like *Arabidopsis thaliana* (Chang et al. 1994), and consequently, similar transformation success was reported in legume *Medicago sativa* (Trieu et al. 2000) that shows *in planta* method can be successfully adapted to legume species.

The development of robust transformation systems for several useful legumes is desirable for their use in various gene identification strategies, for crop improvement and for pursuing a number of important biological questions related to legumes. There is a need for genetic modification in legumes to increase its productivity and enhance the nutritional values of pulses and other large-seeded legume crops. Recently, the production of genetically modified crops is considerably high, in that legumes play important positions because the molecular basis of nitrogen fixation and its unique metabolic pathway help us to use legumes for scientific and economic necessity (David et al. 2003).

14.3 Legume Transformations

In general, by using modern molecular techniques, the crop improvement is implemented against both biotic and abiotic stress by introducing the specific resistant gene into the target host plants, and genetically identical plantlets are needed in this study. These are derived predominantly from three different methods as (1) the plantlets obtained from direct regeneration, (2) indirect regeneration through the callus and (3) somatic embryos. Depending upon the plant and their explant sources, the regenerative potential, growth and development of plantlets vary within the above three methods of *in vitro* propagation. The genetic modification studies within the plant system are carried out in two distinct phases. The first phase is to produce the genetically identical plantlets as mentioned earlier by *in vitro* propagation methods, and the second phase is to introduce the manipulated gene into the target plant cell through primary and character gene transfer methods. Indeed, both phases are very important in the crop improvement programmes. The expression of the introduced genes within the target tissue depends on the growth and developmental status of *in vitro*-derived plants. During the past few decades, there has been considerable success in obtaining transformation of forage and pasture legumes. The studies on plant transformation technology are considered as the true 'genetic engineering technology'. *Agrobacterium tumefaciens* (α-proteobacterium of the family Rhizobiaceae) is a gram-positive soil phytopathogen. In natural conditions, a portion of transfer DNA (T-DNA) present in plasmid of *Agrobacterium* induces tumour in plants. The tumour-inducing plasmid (Ti plasmid) is transferred into plant cells and stably integrated into plant genome and causes the crown gall tumour. *Agrobacterium* enters into the plant cells, and the T-DNA ultimately integrates with the nuclear genome and expresses that result in the tumour phenotype of transformed cells (Zhenying and Binns 2003).

Gene transfer technology is a reliable tool for crop improvement, and it depends on cost-effective production of transgenic plants and their products without compromise on yield and nutrition quality in a shorter duration, but in the case of conventional methods, it needs a long duration. Plant genetic transformation vectors and methodologies have been improved to increase the efficiency of plant transformation and to achieve stable expression of transgenes in plants. Due to the simplicity of the transformation and precise integration of transgenes, *Agrobacterium*

Ti plasmid-based vector continues to offer the best system for plant transformation (Veluthambi et al. 2003). Precision and predictability in the production of transgenic crops will become increasingly important towards increasing the quality of transgenic events using *Agrobacterium*. The selective advantage of using *Agrobacterium* mediated transformation typically results in a low copy number of transgene. The various parameters such as explant size, duration of pre-culture, inoculation and cocultivation time, acetosyringone concentration and media composition significantly influence the T-DNA delivery and regeneration process during the gene transformation (Wu et al. 2003).

The expression of introduced gene within the host or target cell is analysed by several methods. The identification by β-glucuronidase (GUS) and chloramphenicol acetyltransferase (CAT) expressions is the main biomarker to identify the expression of introduced genes. The transformed transgenic plants are evaluated histochemically and anatomically. The detection of most reporter proteins such as GUS and CAT relies on the enzymatic production of a coloured, fluorescent product which then allows an easy quantification or localization. The advantage of enzymatic reporters resides in their sensitivity, as a single protein can produce abundant product molecules. Glucuronidases are of special interest since their assays do not involve any radioactivity. However, the levels of transgenic expression are generally unpredictable and vary among the independent transformants (Finnegan and Mc Elroy 1994).

There are three sequential and critical aspects which limit the gene-transferring process: (1) the response and growth status of a target tissue which is involved in the transformation process, (2) a convenient method to introduce DNA into those regenerative cells and (3) a standard procedure to select and regenerate transformed plantlets at a satisfactory frequency. In general, the plant transformation is carried out by two methods such as primary gene transfer and character transfer. The primary gene transfer is an ideal method to introduce the foreign gene and find out the responses of target cell during regeneration. The expression of primary gene transfer is identified by reporter and selection marker genes. The commonly employed reporter genes are GUS and CAT, and the selection marker genes are *NPT*II, *HPT*II, etc. The reporter genes are easily identified either by simple histological sections provided with substrate, and selectable marker genes have been identified by selection of transgenics in antibiotic growth media, polymerase chain reaction (PCR) analysis using specific primers and Southern blot technique. The standardization of technique in primary gene transfer is more useful and a prerequisite for the character gene introduction into the target cells.

14.3.1 Factors Influencing the Genetic Transformation

Agrobacterium-mediated genetic transformation in normal conditions is determined by many factors including bacterial strains and cell density, plant species and genotype, culturing temperature and light, explant wounding, phenolic compounds, effect of virulence genes and host cell division (Srivatanakul et al. 2001). This transforming ability of the *Agrobacterium* is exploited for the gene-transferring studies indeed, the method of gene transfer is imitated in the lab conditions and the *Agrobacterium* strain carrying the binary plasmid is introduced with the desirable gene into the plant system. This technique is mainly used to develop the transgenics and crop improvement programmes. However, many experimentation works are needed to produce the stable transgenic plants (Lyznik et al. 1996).

Lack of efficient regeneration protocol in non-meristematic tissues was the main disadvantage to produce the stable transformation in common bean (Mukeshimana 2013). However, *Agrobacterium*-mediated genetic transformation is one of the best methods used to obtain several transgenic legumes. Progress in improving legume transformation has also been achieved by many methods. *Agrobacterium*-mediated T-DNA delivery reducing or overcoming factors inhibits the host-pathogen interaction. The development of super-binary strains with enhanced virulence and the addition of acetosyringone have increased transformation efficiency. The recalcitrance of many legumes to tissue culture initiation and plant regeneration has driven researchers to develop non-tissue culture method. Nowadays several non-tissue culture methods like *agroinfiltration*, vacuum infiltration, sonication-assisted *Agrobacterium*-mediated genetic transformation (SAAT) and floral dip methods are now gaining importance to the researchers for progress in legume genetic transformation.

14.3.2 Gene Transfer Studies

The *Agrobacterium*-mediated genetic gene transfer in plants is a sequential process, which includes the following steps: (a) determination of antibiotic sensitivity, (b) pre-culture of the explant, (c) *Agrobacterium* culture, (d) inoculation and cocultivation, (e) selection of co-cultivated explants, (f) callus induction, (g) shoot/root induction and (h) transplantation.

14.4 Drawback of In Vitro Transformation of Legume

In vitro plant regeneration methods such as micropropagation, organogenesis and embryogenesis have been developed for several economically important legumes such as soybean (Olhoft et al. 2003; Arun et al. 2014), chickpea (Krishnamurthy et al. 2000; Kadri et al. 2014; Sunil et al. 2015), cowpea (Romeis et al. 2004; Tang et al. 2012) black gram (Saini and Jaiwal 2007; Adlinge et al. 2014; Rajendiran et al. 2016) and forage legumes like barrel *Medicago truncatula* (Trieu et al. 2000), these methods have not provided enough robust regeneration for applying the transformation protocols. So, it is a serious limitation to the exploitation of gene transfer to its full potential. Therefore, an alternate strategy is required to overcoming the above reported problem.

Legumes are regarded as difficult to transform and the limitation mostly being their poor regeneration ability by *in vitro* method. Genetic gene transfer for these species can therefore be challenging. Even among few legumes that are apparently amenable to transformation, the process is often slow, and the number of plants generated from each explants following DNA transfer is often low (Somers et al. 2003). To overcome the above problems, a number of researchers have pursued for plant transformation methods that avoid tissue culture or *in vitro* regeneration protocols. For non-tissue culture approaches, *Agrobacterium* or tungsten particles have been used in a number of species to transform cells in or around the apical meristems that are subsequently allowed to grow into plants and produce seeds (Birch 1997). However, transformed sectors have typically not persisted into gametes at reasonable frequencies, or the methods have been difficult to reproduce (Birch 1997). Injection of naked DNA into ovaries has also been reported to produce transformed progeny (Zhou et al. 1983). Variations of this method and 'pollen tube pathway' delivery of DNA are still practised in China (Hu and Wang 1999). Electroporation-mediated-based gene transfer into intact meristem by *in planta* and a variety of pollen transformation procedures have also been reported (Chowrira et al. 1995). However, most of these methods have been difficult to reproduce and have not gained widespread acceptance. So, many other types of *in planta* transformation such as vacuum infiltration, floral dip and seed infiltration methods are now gaining importance.

The chemical diversity of plant-derived natural products allows them to function in a multitude of ways including flavour enhancers, agricultural chemicals and importantly human medicine. Supply of pharmaceutically active natural products is often a challenge due to the slow-growing nature of some species. Several production options are available including natural harvest, total chemical synthesis, semi-synthesis from isolated precursors and expression of plant pathways in microbial systems (Kolewe et al. 2008).

In recent years, secondary metabolic-related genes have overexpressed in the original plant or in other plant species for production of transgenic plants (Verpoorte and Memelink 2002), and a number of important breakthroughs in the engineering of specific and novel carotenoids have also been obtained (Jayaraj et al. 2008). *Carotenoid synthesis* gene has been overexpressed in transgenic tomato to target the phytoene synthase (PSY) enzyme for metabolic engineering. (Fray et al. 1995), and this compound has received attention for its significant antioxidant activities and for playing important roles in inhibiting the onset of chronic diseases (Rao and Rao 2007).

Current Method: The genetic manipulation of plant species has a profound impact on basic plant research and biotechnology; the global rift around genetic engineering is between medicinal plants and all other uses such as pharmaceuticals. The most widely used method of gene transfer into plant cells employs a reformed plant pathogen called *Agrobacterium*, which has a natural ability to put foreign genes into plant cells. Transient expression of foreign proteins in plants has been mediated via simple tissue infiltration using *Agrobacterium tumefaciens*, by gene vectors based on a variety of plant viruses, and the integration of viral vectors into modified *A. tumefaciens* Ti plasmids, so that a replicating vector is delivered and released into host tissues, infiltrated with recombinant *A. tumefaciens*. Currently, using different gene transfer techniques employed agroinfiltration, polyethylene glycol (PEG)-mediated DNA uptake, silicon carbide fibres, electroporation and microparticle bombardment.

Genetic engineering is applied in medicinal plants to increase the level of active phytochemical constituent production and is a well-established target bacterial gene *ipt*, which promotes the endogenous production of cytokinin growth hormones, expressed in *Artemisia*, and there is a coordinated increase in chlorophyll and artemisinin levels (Lee et al. 2004). The drawback, particularly metabolic pathways by which active compounds are biosynthesized, is poorly understood; relatively several genes for key enzymatic or regulatory steps have been isolated (Bradford et al. 2005). Optimization of culture conditions and production lines via targeted gene replacement helped to enhance product yields and safety (Decker and Reski 2007). The improved compounds and many other plant species like grapes, peanut and several berry varieties, which are the major dietary sources of resveratrol and phytoalexin production, reported especially resveratrol and its derivatives of stilbenes; pterostilbene has antifungal and pharmacological properties which contribute to the protection against various pathogens. Resveratrol stilbene synthase (STS) displays a wide range of biological effects (Siemann and Creasy 1992). In addition, pterostilbene is five to ten times more fungi toxic than resveratrol. *In vitro* studies showed that pterostilbene exhibits anticancer, hypolipidemic and antidiabetic properties (Schmidlin et al. 2008).

Over 10,000 structurally characterized members of plant alkaloids are significant and privileged compounds having many pharmacological activities. In addition, the products are free of microbe and insect contamination (Beghyn et al. 2008). Yun et al. (1992) also reported the enhancement of ninefold sedative compounds such as scopolamine in hairy root cultures of *Hyoscyamus niger,* overexpressing two genes encoding the rate-limiting upstream and downstream biosynthetic enzymes (Beghyn et al. 2008).

The importance of tissue culture and application for industry to commercialization has essentially been impeded by economic feasibility, arising from both biological and engineering considerations. Being essential elements of a system for rational engineering of the genetic makeup of medicinal plants, tissue culture and regeneration could make important contributions to commercial production of high-value phytochemicals by an *in vitro* system for growing plant organs, tissues or cells. The disadvantage is the limited yield of some bioactive compounds in plant tissues, which presents a significant challenge for large-scale drug development. Numerous drugs and drug precursors in the current pharmacopoeia originate from plant sources, and once a potent natural product is identified, the main limiting factor for drug discovery is the ability to produce enough material for clinical applications.

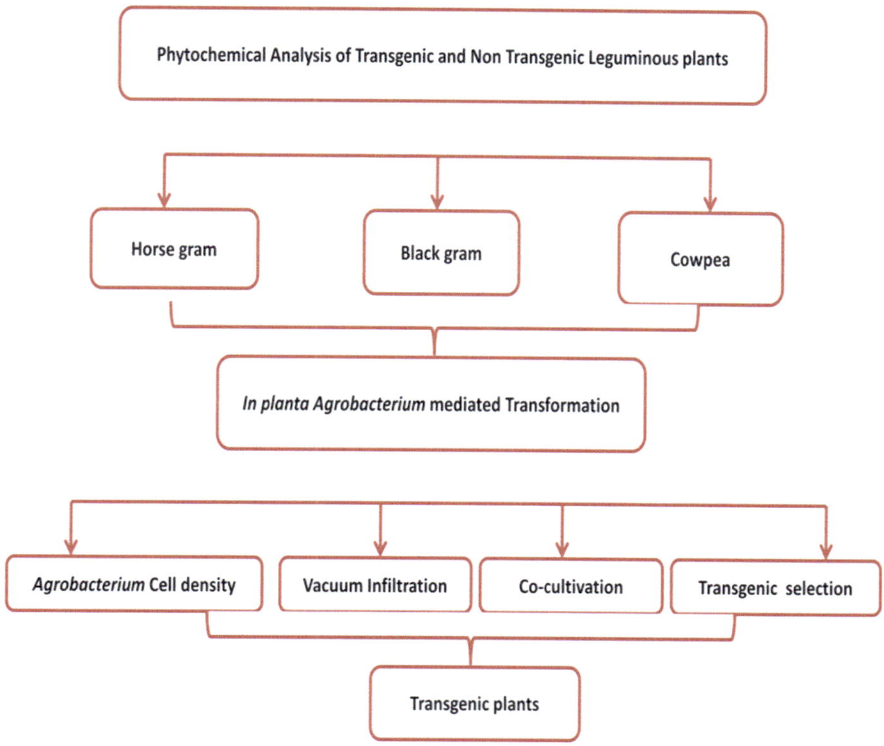

Fig. 14.1 The flow chart describes *in planta Agrobacterium*-mediated transformation

Genetic transformation for the resource of plant tissue culture offers promising alternatives for the production of chemicals because totipotency enables plant cells and organs to produce useful secondary metabolites by *in vitro* conditions to alter the biosynthesis of secondary metabolites; medicinally important plant biosynthetic pathways have been characterized, particularly the pathways that are responsible for carotenoid, alkaloid and proanthocynidine biosynthesis (Bradford et al. 2005) (Fig. 14.1).

Three economically important pulse crops such as horse gram, black gram and cowpea were selected for the present study. The seeds were subjected to *Agrobacterium*-mediated transformation by using a super-virulent *A. tumefaciens* strain EHA105 harbouring the binary vector pCAMBIA2301. Different parameters such as *Agrobacterium* cell density, vacuum infiltration, cocultivation and primary and secondary selection were analysed to obtain the maximum transformation efficiency (Fig. 14.2).

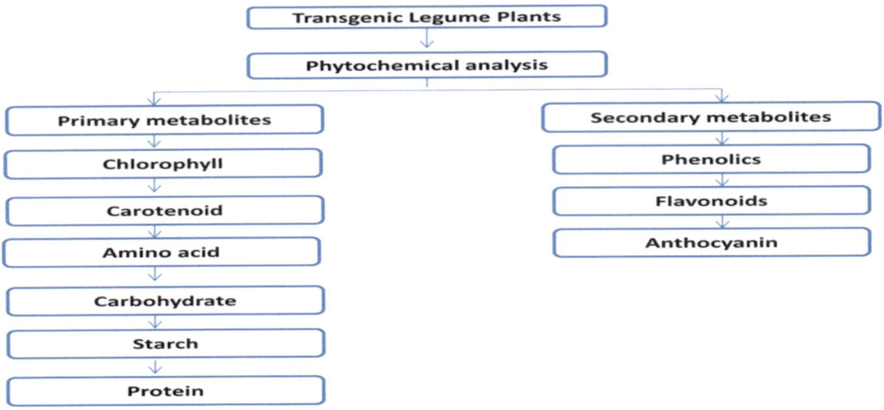

Fig. 14.2 The flow chart shows various primary and secondary metabolites screened in transgenic leguminous and non-transgenic control plants, respectively. Primary metabolites such as chlorophyll, carotenoid, amino acid, carbohydrate, starch and protein were analysed, whereas secondary metabolites like phenolics, flavonoids and anthocyanin were quantified

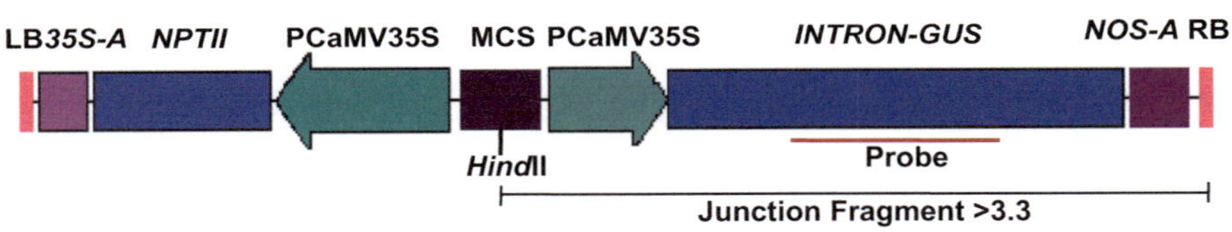

Fig. 14.3 The linear T-DNA map of pCAMBIA2301

14.5 Methods

14.5.1 Agrobacterium *Strain and Binary Vector*

Transformation studies were carried out using super-virulent *A. tumefaciens* strain EHA105 harbouring the binary vector pCAMBIA2301, carrying *uid*A or GUS (ß-glucuronidase) reporter gene driven by the CaMV35S promoter and kanamycin resistance gene (*NPT*II) as selectable marker driven by the CaMV35S promoter. The chimeric GUS gene within the vector is a useful tool for study and analyses of the early events of transformation in plant tissue by histochemical staining (Fig. 14.3).

14.5.2 *Preparation of* Agrobacterium *Culture*

Glycerol stock of *A. tumefaciens* strain EHA105 stored at –80 °C was streaked on yeast extract mannitol (YEM) agar medium containing rifampicin (20 mg/l) and kanamycin (50 mg/l) (1.4%). Streaked *Agrobacterium* cultures were incubated in darkness at 28 °C until the formation of bacterial colony formation. A loop of colony was inoculated into 2 ml of liquid YEM broth containing the above antibiotics and incubated in an orbital shaker at 220 rpm for 16–18 h at 28 °C. Afterwards, 200 μl of *Agrobacterium* culture was transferred to a 100 ml of YEM broth containing the above antibiotics and incubated overnight at 28 °C at 220 rpm. The cultures were monitored by measuring the absorbance at OD_{600}nm. A culture that reads 0, 0.3, 0.6 and 0.8 OD was used to infect the germinated seeds. The *Agrobacterium* cells from the overnight culture were harvested by centrifuge at 5000 rpm for 10 min. The supernatant was discarded and the pellet was resuspended in a known volume of infiltration medium and further used for the seed infection.

14.5.3 *Explant Preparation, Infection and Sonication Combined with Vacuum Infiltration*

Seeds were germinated on sterilized moistened paper towels in petri dishes at 25 °C for 24 h in darkness. Once the plumule has come out from the seeds to around 1 cm in length, then the seeds were used for the infection. Explants were gently wounded by sterile syringe needle (Dispovan, New Delhi, India). The suspension of *Agrobacterium* culture was diluted with half strength MS medium to obtain 0, 0.3, 0.6 and 0.8 OD. Further, the beaker was placed in the centre of ultrasonic homogenizer (SKL-150DN model, frequency 45 KHZ). Germinated seeds were inoculated through sonication at various intervals from 0, 2.5, 5.0 to 7.5 min (temperature 27^{0}C; 8 sec on, 2 sec off; 17% power). After sonication treatment, seeds were placed in a vacuum system (GAST DOA-P704-AA) for further treatment of vacuum infiltration for 0, 2.5, 5.0 and 7.5 min (300 mmHg). Seeds were taken out from the beaker and blotted on filter paper and transferred to petri dishes for coculture in the dark at 25^{0}C for 24 h. Following the coculture, the inoculated seeds were rinsed thrice in sterilized distilled water containing 300 mg/l^{-1} of cefotaxime sodium salt, and then the seedlings were transferred to soil pots to attain the maturity of plants so that the seeds could be harvested for further analysis. Seeds without *agroinfection* were used as control. Transformation efficiency was calculated using the formula:

$$\text{Transformation efficiency}\,(\%) = \frac{\text{No. of GUS positive explants} \times 100}{\text{No. of explants infected}}$$

14.5.4 *Biochemical Analysis*

14.5.4.1 Chlorophyll (Arnon 1949)

500 mg of fresh leaf material was grounded with a mortar and pestle with 10 ml of 80% acetone. The homogenate was centrifuged at 6000 rpm for 15 min at 4 °C. The supernatant was saved and the residue was re-extracted with 10 ml of 80% acetone. The resulting supernatant was saved and pooled together, and absorbance values were read at 645 and 663 nm in a UV spectrophotometer. Chlorophyll 'a', chlorophyll 'b' and the total chlorophyll content were calculated using the following formula, and the results are expressed as mg/g fresh weight:

$$\text{Chlorophyll 'a'} = (0.0127) \times (\text{O.D } 663) - (0.00269) \times (\text{O.D } 645)$$

$$\text{Chlorophyll 'b'} = (0.0229) \times (\text{O.D } 645) - (0.00488) \times (\text{O.D } 663)$$

$$\text{Total chlorophyll} = (0.0202) \times (\text{O.D } 645) + (0.00802) \times (\text{O.D } 663)$$

14.5.4.2 Carotenoid (Kirk and Allen 1965)

The 80% acetone extract used for chlorophyll estimation was further used for carotenoid estimation. The acetone extract (500 mg/ 20 ml) was read at 480 nm in a UV spectrophotometer (Shimadzu UV-1800). The carotenoid content was calculated using the following formula, and results are expressed as mg/g^{-1} fresh weight:

$$\text{Carotenoid} = (\text{O.D } 480) - (0.114) \times (\text{O.D } 663) - (0.638) \times (\text{O.D } 645)$$

14.5.4.3 Estimation of Phenolics (Siddhuraju and Becker 2003)

Aliquots of appropriate concentration of the extracts were taken in a test tube and made up to the volume of 1 ml with distilled water. Then 0.5 ml of Folin-Ciocalteu phenol reagent (1:1 with water) and 2.5 ml of sodium carbonate solution (20%) were added sequentially in each tube. Soon after vortexing the reaction mixture, the test tube was placed in dark for 40 min, and the absorbance was recorded at 725 nm against reagent blank, and the result was expressed as gallic acid equivalent (GAE)/g extract.

14.5.4.4 Estimation of Amino Acid (Yasuma and Ichikawa 1953)

10 ml of aliquots of the extract (500 mg/10 ml) was taken in a test tube and made up to 1 ml using distilled water, and 1 ml of ninhydrin solution was added and heated for 20 min in a water bath. An equal volume of water and n-propanol was added and incubated for 15 min, and absorption was read at 570 nm. The amount of amino acids (mg/g sample) present in the plant materials was calculated by using a standard graph using leucine as a standard.

14.5.4.5 Estimation of Carbohydrate (Hodge and Hofreiter 1962)

500 mg of fresh material was hydrolysed using 2.5 N HCL and neutralized with solid sodium carbonate until the effervescence ceased. The volume was made up to 100 ml and centrifuged. To 1 ml of supernatant, 4 ml of anthrone reagent was added and heated for 8 min in a boiling water bath. The absorption was read at 630 nm, and the amount of carbohydrate (mg/g of sample) present in the material was calculated using a standard graph, glucose as a standard.

14.5.4.6 Estimation of Starch (Hodge and Hofreiter 1962)

500 mg of fresh material was homogenized using 10 ml of 80% ethanol and centrifuged. Five per cent of perchloric acid was added to the residue and centrifuged, and then the supernatant was collected. To 1 ml of supernatant, 4 ml of anthrone reagent was added and heated for 8 min in boiling water bath. The absorbance was read at 630 nm. The starch present in the extract was determined using a standard graph, glucose as a standard.

14.5.4.7 Estimation of Protein (Lowry et al. 1951)

500 mg of the sample was grounded well in 10 ml of phosphate buffer (0.2 M, pH 7.3) with the help of pestle and mortar and centrifuged. 5 ml of alkaline copper solution was added to 0.1 ml of supernatant and incubated for 10 min. After that, 0.5 ml of Folin-Ciocalteau reagent was added and incubated in dark for 30 min and the absorbance was read at 660 nm. The protein present in the extract was determined from a standard graph using BSA as standard.

The results are expressed as amounts of protein in mg/g of sample.

14.5.4.8 Estimation of Total Flavonoids (Zhishen et al. 1999)

The flavonoid content of the sample extract was determined by the use of a slightly modified colorimetric method. 0.5 ml aliquot of appropriately diluted sample solution was mixed with 2 ml of distilled water and subsequently with 0.15 ml of 5% $NaNO_2$ solution. After 6 min of incubation, 0.15 ml of 10% $AlCl_3$ solution was added and allowed to stand for 6 min, and then 2 ml of 4% NaOH solution was added to the mixture. Immediately, distilled water was added to bring the final volume to 5 ml, and then the mixture was thoroughly mixed and allowed to stand for another 15 min. Absorbance of the mixture was determined at 510 nm. Quercetin was used as standard for quantification of total flavonoids, and the results were expressed as quercetin equivalents (QE)/g of extract.

14.5.4.9 Estimation of Anthocyanin (Moyer et al. 2002)

Total anthocyanin content (AC) was performed by differential methods of samples that were diluted 1:150 in 0.025 M KCL buffer (pH 1.0) and 0.4 M sodium acetate buffer (pH 4.5) and absorbance measured at 520 and 700 nm. The results were calculated using the following formula and expressed as mg of malvidin 3-glucoside equivalent mg/g extract:

$$\mathrm{AC} = (\mathrm{A520\ pH\ 1.0} - \mathrm{A700nm\ pH\ 1.0}) - (\mathrm{A520nm\ pH\ 4.5} - \mathrm{A700nm\ pH\ 4.5})$$

14.5.4.10 Free Radical-Scavenging Activity Using DPPH (Blios 1958)

The antioxidant activity of the extract was determined in terms of hydrogen-donating or radical-scavenging ability using the stable radical DPPH. The sample extract at various concentrations was adjusted to 100 µl with methanol. 5 ml of 0.1 mM methanolic solution of DPPH was added and shaken vigorously. The tubes were allowed to stand for 20 min at 27 °C. The absorbance of the sample was measured at 517 nm. Radical scavenging percentage of free radical by the sample was calculated using the formula.

$$\%\ \mathrm{DPPH\ radical\ scavenging\ activity} = \mathrm{control\ OD} - \mathrm{sample\ OD/control\ OD} \times 100$$

14.5.5 Phytochemical Screening of Transgenic and Non-transgenic Leguminous Plants of Horse Gram

Horse gram seed is considered to be the poor man's pulse crop particularly in South India. It is native to Southeast Asian subcontinent and tropical Africa. They are extensively cultivated especially in dry areas of Australia, Burma, India and Sri Lanka (Duke and Reed 1981). In peninsular India, the maximum cultivated area belongs to Andhra Pradesh, Karnataka and Tamil Nadu. In Tamil Nadu, they were cultivated on 47,231 hectares, and yield per hectares was 462 kg during the years 2009–2010 (Government of Tamil Nadu Report 2010). Horse gram is grown as a main crop in kharif and rabbi seasons with annual rainfall of 300–600 mm. However, the major limitation of this crop is that it doesn't tolerate flooding and water logging. The favourable average temperature is 18–27 °C with wide adaptability to drought stress and thrives under adverse climatic conditions with poor soils (Bolbhat and Dhumal 2010).

Nitrogen-fixing capacity is an important phenomenon of this crop. As the name implies, it is also used as horse and cattle feed. The seed sprouts of this crop are used by large populations in rural areas (Kanaka 2012). They are considered a potential remedy for common cold, throat infection and fever. The soups made from the seeds were used to generate heat and help to dilute renal stones (Siddhuraju and Manian 2007). Earlier studies showed that it forms a good source of protein, carbohydrates, essential amino acids, iron and molybdenum (Kadam and Salunkhe 1985; Sudha et al. 1995; Bravo et al. 1999).

14.5.5.1 Effect of *Agrobacterium* Cell Density on Biochemical Parameters on Horse Gram

The biochemical components play an important role in the process of plant growth and development. *Agrobacterium*-mediated transformation induces the stress for the plants and subsequently changes their metabolites. The metabolites present in the plants have been altered due to the insertion of foreign genes. The biochemical changes are responses to various *Agrobacterium* cell densities ranging from 0, 0.3, 0.6 to 0.8 OD and at the same time correlated with the results of the control plants. The leaf protein content of *Agrobacterium* transmitted plant was found to be significantly higher (62.83 ± 0.67^{a}) at 0.8 OD than control plants (59.47 ± 0.5^{b}). Protein content was moreover similar in control (59.47 ± 0.5^{b}) and 0.6 OD ($60.67 \pm 0.61^{a,b}$), whereas it decreased in 0.3 OD (56.2 ± 0.87^{c}) when compared with control plants (59.47 ± 0.5^{b}). The results revealed that optimum protein content in leaf increased with increase in *Agrobacterium* cell density up to 0.8 OD (62.83 ± 0.67^{a}). The total phenolic content was found to be increased in 0.8 OD levels (31.13 ± 0.65^{a}) than in control plants (22.95 ± 0.97^{d}). The total phenolic content was found to increase in 0.3 and 0.6 *Agrobacterium* cell densities (26.82 ± 0.62^{c} and 29.47 ± 0.75^{b}) than in control plants (22.95 ± 0.97^{d}), respectively.

Carbohydrate content in transgenic leaf was decreased when the cell density was increased as compared to control plants. The highest level of carbohydrate content (49.36 ± 0.99^{a}) was observed in control plants. However, according to DMRT range test at $P < 0.05$, significance level showed that the carbohydrate content in 0.3 OD (48.21 ± 0.99^{a}) and control plants (49.36 ± 0.99^{a}) was moreover less similar in condition, whereas carbohydrate content in 0.6 OD ($46.79 \pm 0.51^{b,c}$) and 0.8 OD (45.89 ± 0.69^{c}) decreased in nature than control plants. The starch activities in the leaves were found to be significantly higher (44.43 ± 0.90^{c}) in control plants. The *Agrobacterium* cell densities 0.3 and 0.6 showed that starch content in leaves was $43.39 \pm 0.89^{a,\ b}$ and $42.11 \pm 0.46^{b,\ c}$, respectively. Among these values, the lowest starch content was observed in 0.8 OD (41.30 ± 0.62^{c}). These results revealed that there is a gradual decrease of starch content with increase in cell density.

The level of amino acids was decreased in the transgenic with increase in *Agrobacterium* cell density. This may be due to the inverse relationship between protein and amino acid. The decreased activity may be due to the *Agrobacterium* stress on protease activity. The levels of total free amino acids contents in 0.3, 0.6 and 0.8 OD were 37.72 ± 0.25^{b}, 34.02 ± 0.78^{c} and 33.19 ± 1.03^{c}, respectively. The control plant showed higher total free amino acid (40.09 ± 0.32^{a}) than other cell densities.

Chlorophyll estimation is one of the important plant parameters which is used as an index of production capacity of the plants. Chlorophyll content is positively correlated to net photosynthetic rate, and hence it rates a major role in controlling grain growth and grain-filling process (Liu 1980). The leaves of chlorophyll-a, chlorophyll-b and carotenoid contents were recorded. Chlorophyll-a content increased in 0.3 OD (0.0137 ± 0.000^{a}), whereas in 0.6 and 0.8 OD showed $0.0129 \pm 0.001^{a,\ b}$ and 0.0102 ± 0.000^{b}, respectively. According to DMRT at $P < 0.05$, the level of significance showed in control ($0.0121 \pm 0.003^{a,b}$) and in 0.3 OD (0.0102 ± 0.000^{b}). Chlorophyll-b content in plant leaf showed higher activity in control plants (3.77 ± 0.09^{a}), whereas, cell density, such as 0, 0.3, 0.6 and 0.8, showed the OD 3.28 ± 0.03^{b}, 2.54 ± 0.17^{d} and 2.82 ± 0.13^{c}, respectively. The level of carotenoid content was 0.3 OD (0.94 ± 0.02^{b}), 0.6 OD (0.77 ± 0.01^{d}) and 0.8 OD (0.85 ± 0.01^{c}), respectively. The control plant showed the highest carotenoid content (1.11 ± 0.01^{a}) among other cell densities. The results showed that reduction in chlorophyll content induced by *Agrobacterium* stress on different cell density also affects the growth parameters in plants. This may be due to the formation of enzyme chlorophyllase which is responsible for chlorophyll degradation (Neelam and Sahai 1988) (Table 14.1).

14.5.6 Phytochemical Screening of Transgenic and Non-transgenic Leguminous Plants on Black Gram

Black gram is being cultivated during monsoons, post monsoon and summer season as a sole crop or intercrop in India. Although better agricultural and breeding practices have significantly improved the yield of black gram over the last decades, it is

Table 14.1 Effect of *Agrobacterium* cell density on biochemical parameters in horse gram

			Agrobacterium cell density (OD600nm)		
S.No	Biochemical estimation	Control	0.3	0.6	0.8
1	Protein (mg/g)	59.47 ± 0.5^{b}	56.2 ± 0.87^{c}	60.67 ± 0.61^{b}	62.83 ± 0.67^{a}
2	Phenol (mg/g)	22.95 ± 0.97^{d}	26.82 ± 0.62^{c}	29.47 ± 0.75^{b}	31.13 ± 0.65^{a}
3	Carbohydrate (mg/g)	49.36 ± 0.99^{a}	$48.21 \pm 0.99^{a,b}$	$46.79 \pm 0.51^{b,c}$	45.89 ± 0.69^{c}
4	Starch (mg/g)	44.43 ± 0.90^{a}	$43.39 \pm 0.89^{a,b}$	$42.11 \pm 0.46^{b,c}$	41.30 ± 0.62^{c}
5	Amino acid (mg/g)	40.09 ± 0.32^{a}	37.72 ± 0.25^{b}	34.02 ± 0.78^{c}	33.19 ± 1.03^{c}
6	Chlorophyll-a (mg/g)	$0.0121 \pm 0.003^{a,b}$	0.0137 ± 0.000^{a}	$0.0129 \pm 0.001^{a,b}$	0.0102 ± 0.000^{b}
7	Chlorophyll- b (mg/g)	3.77 ± 0.09^{a}	3.28 ± 0.03^{b}	2.54 ± 0.17^{d}	2.82 ± 0.13^{c}
8	Carotenoids (mg/g)	1.11 ± 0.01^{a}	0.94 ± 0.02^{b}	0.77 ± 0.01^{d}	0.85 ± 0.01^{c}

limited and could not fulfil the domestic consumption demand of the country (Jaiwal and Gulati 1995). As a result, the developing nation requires tremendous scientific efforts to meet the requirements.

The crop improvement was done by breeding methods in early days. However, breeding is difficult due to the fact that *Vigna mungo* is a self-pollinating crop, and the genetic variation among the black gram variation is narrow. Recently, the genetic improvement of black gram has been carried out through traditional plant breeding methods, but the limited genetic variability in black gram germplasm has considerably slowed the pace of breeding varieties for agronomically important traits. *Agrobacterium*-mediated genetic transformation is more efficient as it results in integration of well-defined DNA sequence, potentially low copy number, high co-expression of the introduced genes and preparential integration into actively transcribed regions (Gheysen et al. 1991). Though genes conferring resistance to biotic and abiotic stresses have been reported in many wild and related species, these are sexually incompatible with the cultivated ones. Some of such genes have been isolated, characterized and cloned from other plant species and microorganisms (Grove et al. 2004). Therefore, genetic engineering allows these genes to be transferred to cultivated varieties in order to enhance the tolerance to various stresses and hence stabilize yield (Kaur and Murphy 2012).

Mechanical wounding of explants prior to inoculation with *Agrobacterium*, time lag in regeneration due to removal of cotyledons from explants and a second round of selection on kanamycin at rooting stage were found to be critical for transformation. However, the efficiency of transformation of meristematic cells in the axil of cotyledonary node was low (1%) which may be attributed to the presence of limited number of regenerable cells whose capacity for regeneration was short lived in explants and inefficient T-DNA delivery to regenerable cells (Saini and Jaiwal 2005). An *Agrobacterium*-mediated transformation of shoot apex explants was developed to overcome these limitations. A significant improvement in transformation efficiency from an average of 1 to 6.5% was obtained. Evidence for stable integration of transgene and their inheritance to progeny was presented. Shoot apex explants are preferred because of their high regeneration potential with minimal tissue culture manipulations in a less genotype-dependent fashion. Saini and Jaiwal (2007) optimized the conditions for enhanced transformation of cotyledonary node and generated stable transgenic plants at a frequency of 4.3%.

Muruganantham et al. (2010) developed herbicide-tolerant *V. mungo* plants using cotyledonary node and shoot tip explants (from seedlings germinated by *in vitro* from immature seeds) and *A. tumefaciens* harbouring a binary vector carrying bar and *uidA* genes. Bhomkar et al. (2008) engineered salt stress tolerance in *V. mungo* by introducing glyoxalase I (*gly*I) gene under a novel constitutive *Cestrum* yellow leaf curling virus promoter (CmYLCV) into regenerable cells present at cotyledonary node of embryonic axis using *A. tumefaciens*. The T_1 plants expressing glyoxalase I activity survived and set seeds under NaCl stress (100 mM). Varalaxmi et al. (2013) optimized parameters for *Agrobacterium*-mediated transformation of black gram using cotyledons as explants which are used to mobilize genes conferring tolerance to various biotic and abiotic stresses from diverse sources into *V. mungo* with more precision. Expanding the range of genotypes within a species that undergo transformation will allow faster varietal improvement for enhancing crop yield. Although it possesses easy manipulation under *in vitro* condition, the lack of reproducibility of regeneration protocols and highly problem in transplantation of the *in vitro*-regenerated shoots is a major limiting factor for obtaining complete transgenic plants and their progeny. Tissue culture is labour intensive and can be difficult to masters (Birch 1997). Numerous critical factors are involved in these approaches. In transformation by *in vitro*, key factors include choice of vectors and bacterial strains, types of plant tissues to be infected, procedures of preparing the tissues, protocols of infection and cocultivation, methods for subsequent culture and selection of transformed cells, antibiotics to remove infecting bacteria, techniques for regeneration of transgenic plants and genotypes of plants (Komari et al. 1996).

14.5.6.1 Effect of Vacuum Infiltration Duration on Biochemical Parameters of Black Gram

There was a variation among all the major secondary metabolites by the impact of vacuum infiltration. Among all treated/untreated plants were recorded which total phenolic contents were varied between 23.566 and 40.333. However, the effect of vacuum infiltration treatment on following order, 5 min vacuum infiltrated (40.333) >3 min vacuum infiltrated (37) > 1 min vacuum infiltration (32.466) > untreated control (23.566). The effect of vacuum infiltration on flavonoid content showed that all the samples contained significant amount of flavonoids which varied between 20.33 and 31 mg/g of fresh weight in the sample tested. The highest amount of flavonoids (31 mg/g of frwt) was recorded at 5 min of vacuum infiltration-treated plants. The lowest amount of flavonoid content (20.33 mg/g of frwt) was recorded in control plants, whereas, the highest amount of protein content (38 mg/g of frwt) at 5-min-vacuum-infiltrated plant was recorded. Followed by 3 min (35.6 mg/g of frwt), 1 min (31.96 mg/g of frwt) and control (29.66 mg/g of frwt) plants was noted. The influence of vacuum infiltration on starch content (mg/g of frwt) of black gram (83.63 mg/g frwt) showed that the highest starch content was obtained in control plants, while the least total amount of starch content was recorded in 5-min-vacuum-infiltrated plants (54.33 mg/g frwt). The highest amount of carbohydrate (90.66 mg/g frwt) was recorded in control plants, whereas the lowest amount (61.85 mg/g frwt) was recorded in 5-min-vacuum-infiltrated plants. The effect of vacuum infiltration on amino acid contents resulted that the highest amount of amino acid content (12.46 mg/g frwt) was recorded in control plant, while the 5-min-vacuum-treated plants recorded the lowest amino acid content (11.13 mg/g frwt). The effect of various time durations of vacuum infiltration on chlorophyll ‘a’, chlorophyll ‘b’, total chlorophyll and carotenoid was analysed. The highest content of chlorophyll ‘a’ (0.0312 mg/g frwt), chlorophyll ‘b’ (0.048 mg/g frwt), total chlorophyll (0.0791 mg/g frwt) and carotenoid content (3.179 mg/g frwt) was recorded in control plants. On the contrary, the lowest value of chlorophyll ‘a’ (0.013 mg/g frwt), chlorophyll ‘b’ (0.023 mg/g frwt), total chlorophyll (0.0399 mg/g frwt) and carotenoid (0.7506 mg/g frwt) was recorded in 5-min-vacuum-infiltrated plants. In the present study, vacuum infiltration-treated plants that were observed for the biochemical content showed that the chlorophyll, carbohydrate, starch and amino acid contents gradually decreased, whereas protein, total phenolics, flavonoids and anthocyanin gradually increased. The biotic stress in plant might have lowered the levels of starch and photosynthetic pigments and increase total phenolic content (Sreedevi et al. 2013), which is also similarly observed in tobacco and petunia plant (Loake et al. 1988).

The free radical-scavenging activity assessed in the seeds is subjected to vacuum infiltration-treated/untreated plants. All the plant extracts showed radical-scavenging ability which varied widely. However, the control plant extract exhibited 43.66% scavenging, the vacuum infiltration-treated plants compared to control treatment (77, 72 and 56 at 5 min, 3 min and 1 min vacuum infiltration, respectively). However, the standard antioxidant ascorbic acids recorded the highest value (89.6). The 1,1-diphenyl-2-picryl hydrazyl (DPPH) radical is widely used as the model system to investigate the scavenging activities of several natural compounds such as phenolics or crude extracts of plants. DPPH is a relatively stable radical at room temperature and accepts an electron or hydrogen radical to become a stable diamagnetic molecule. The assay is based on the measurement of the scavenging ability of antioxidants towards the stable radical DPPH which reacts with suitable reducing agent (Halliwell and Gutteridge 2007). In the present study, 5-min-vacuum infiltration-treated plants showed strongest scavenging activity than other treated or untreated plants. The radical-scavenging activity was increased with increasing of percentage of the free radical inhibition. The degrees of discoloration indicated that free radical scavenging potentials of the sample by influencing hydrogen-donating ability. The electrons become paired off, and solution loses colour stochiometrically depending on the number of electrons taken up. Therefore, from the results, it could be interpreted that the phenolic content increased might have favoured an enhanced antioxidant activity (Table 14.2, Fig. 14.4).

14.5.7 Phytochemical Screening of Transgenic and Non-transgenic Leguminous Plants of Cowpea

Vigna unguiculata (L.) Walp is commonly known as cowpea, southern pea and black-eyed pea. It originated from the African continent. Now, it is cultivated in semiarid regions of Asia and America. It is a warm weather and drought-resistant crop. It is well adapted to low rainfall, heat and wide range of soil conditions (Kochhar 2009). It is an annual, herbaceous legume. It matures from 3 to 5 months. Worldwide, it is cultivated on 10.5 million hectares with a total production of 3.8 million tons per year. Nigeria is the topmost producer (2.61 million tons per year) of cowpea than other countries (FAOSTAT 2012).

Table 14.2 Vacuum infiltration treatment of black gram and biochemical changes

Biochemical contents	Vacuum treatment			
	Control	1 min	3 min	5 min
Chlorophyll 'a'	0.03 ± 0.08^{c}	0.02 ± 0.05^{b}	0.02 ± 0.05^{b}	0.01 ± 0.06^{a}
Chlorophyll 'b'	0.04 ± 0.03^{c}	0.04 ± 0.02^{b}	0.04 ± 0.03^{b}	0.02 ± 0.01^{a}
Total chlorophyll	0.07 ± 0.032^{c}	0.06 ± 0.001^{b}	0.06 ± 0.002^{b}	0.03 ± 0.003^{a}
Carotenoid	3.17 ± 0.03^{d}	2.20 ± 0.01^{c}	1.77 ± 0.01^{b}	0.75 ± 0.03^{a}
Phenolics	23.56 ± 0.29^{d}	32.46 ± 0.29^{c}	37.00 ± 0.57^{b}	40.33 ± 0.33^{a}
Protein	29.66 ± 0.33^{d}	31.96 ± 0.54^{c}	35.60 ± 0.30^{b}	38.00 ± 0.51^{a}
Amino acid	12.46 ± 0.06^{d}	11.80 ± 0.11^{c}	11.50 ± 0.05^{b}	11.33 ± 0.13^{a}
Carbohydrate	90.66 ± 1.20^{d}	79.23 ± 0.39^{c}	75.13 ± 0.59^{a}	61.83 ± 1.01^{a}
Starch	83.63 ± 2.3^{d}	73.13 ± 0.59^{c}	67.33 ± 1.2^{b}	54.33 ± 0.88^{a}
Flavonoids	20.33 ± 0.33^{d}	21.66 ± 0.88^{c}	26.66 ± 0.2^{b}	31 ± 0.57^{a}
Anthocyanin	0.04 ± 0.02^{d}	0.07 ± 0.04^{c}	0.17 ± 0.09^{b}	0.18 ± 0.1^{a}

It is able to grow in poor soil and improves the soil quality by fixing the atmospheric nitrogen (Adesoye et al. 2010). It is mainly cultivated for their pods or as forage crop. The young leaves and shoots are eaten like spinach. The seeds are rich in protein (24.6%) and carbohydrate (55.7%). The crop is used as a hay, silage, pasture, soil cover and green manure. The fodder is highly palatable to all types of livestock (Kochhar 2009).

In cowpea, the first evidence shows the *Agrobacterium*-mediated genetic transformation successfully performed on the cotyledon explants by tissue culture method. Approximate bacterial cell density is 5×10^{8} cells ml^{-1}. Hygromycin phosphotransferase genes performed as reporter gene. Transformation frequency was improved by the factors like pre-culture and coculture period. But only 6 T_0 plants survived out of 17 T_0 plants established in pots, and further studies are required on heritability of introduced genes (Muthukumar et al. 1996).

The seed explants produced multiple shoots. The cotyledonary nodes of multiple shoots are introduced to *Agrobacterium* infiltration. Addition of thiol compounds during infection and coculture with *Agrobacterium* induces the transformation rate. In a total of 73 tissue culture plants produced, only 20 plants were successfully transferred. Transgenic cowpeas that transmit the transgenes to their progeny recovered at the rate of one fertile plant per thousands explants. But these transgenic plants show low survival rate (Popelka et al. 2006).

1×10^{9} cells ml^{-1} bacterial cell density was cocultured with *Agrobacterium* strain EHA105 harbouring a binary vector pCAMBIA2301 that carried GUS and neomycin phosphotransferase (*NPT*II) genes as a reporter and selectable marker genes, respectively. Explants allow forming multiple shoots. From 395 explants, a total of 160 shoot tips are GUS positive. But only 44 plants survived in pots. The transformation efficiency was 0.76%, and 20–23 weeks were required from infection to seed generation (Chaudhury et al. 2007).

Fig. 14.4 *In planta Agrobacterium*-mediated transformation (**a**) horse gram seeds, (**b**) black gram seeds, (**c**) cowpea seeds, (**d**) horse gram transgenic plants, (**e**) black gram transgenic plants, (**f**) cowpea transgenic plants, (**g**) GUS-positive leaves of horse gram, (**h**) GUS-positive seeds of black gram, (**i**) GUS-positive leaf section of cowpea

Using binary plasmid pCAMBIA1301, influence of inoculation and cocultivation media composition on transient gene expression was determined. Embryos were inoculated on MS solutions supplemented with various concentrations of acetosyringone. The highest transformation rate (55.3%) was recorded. A total of 32 embryos survived from 200 infected embryo explants (Adesoye et al. 2010).

Four-day-old cotyledonary nodes were used as explants. Explants were vacuum infiltrated at 500 mm Hg for 5 min, and sonication was given for 20 s. The GUS genes act as reporter and used for histochemical staining and resulted in 93% transient transformation efficiency. The stable transformation rate is increased by 88.4% using both sonication-assisted *Agrobacterium*-mediated gene transfer in cowpea (Bakshi et al. 2011).

A rapid and efficient regeneration system via organogenesis from cotyledonary node explants of cowpea (*Vigna unguiculata* L. Walp) has been established. The cotyledonary node explants excised from 4-day-old seedlings is placed *in vitro* medium containing salts of Murashige and Skoog and vitamins of Gamborg's media (MSB5). Adventitious shoots occurred at the basal end of the initiated axillary buds that pre-existed at the node regions. BAP at 1.25 mg/l was the optimum for shoot induction. The combination of BAP with IBA had worthless effect on shoot proliferation. The number of adventitious buds was promoted when the seeds were preconditioned with appropriate concentrations of BAP (2 to 3 mg/l), whereas it was depressed with higher concentrations of BAP (5 to 15 mg/l). The regeneration system was further optimized due to the presence of cotyledons attaching to the cotyledonary node explants. Explants with two entire cotyledons from 4-day-old seedlings produced greater number of shoots (7.83) after 3 weeks on MSB5 medium supplemented with 1.25 mg/l BAP. Regenerated shoots could well elongate on regulator-free basal medium and well root with 100% of success on the half strength medium supplemented with various concentrations of IBA (0, 0.1, 0.3 and 0.5 mg/l). The regenerated plantlets were cultured on the pots containing sterilized vermiculite and soil (1:1) with 27% of survival (Tang et al. 2012).

14.5.7.1 Effect of *Agrobacterium* Cell Density and Biochemical Parameters in Cowpea

The carbohydrate content of the plant reduced with the increase in the cell density of *Agrobacterium*. Reduction in carbohydrate content can be attributed to increased respiration and decreased CO_2 fixation due to biotic stress (Pandey et al. 2006). The protein content was found to be increased in the increased cell density. The enhanced protein denaturation and breakdown of existing protein to amino acid is the main cause of reduction in protein content (Pandey et al. 2006). The reducing sugars are increased with the decreased cell density. Amino acid is the monomer of protein, the common reserve food material manufactured by plant system. An increased the amino acid content of cowpea with decreased amount of cell density of *Agrobacterium*. The changes in the amino acid concentration could be due to the breakdown of proteins (Pandey et al. 2006). The flavonoids are increased in the decreased cell density. Phenols are the secondary metabolites; it is decreased with the increased amount of cell density. Chlorophyll is the plant pigment and mainly receives the solar energy in plants. The content of chlorophyll is increased in 0.6 amount of cell density. Similar result was noted in tobacco (Hansen et al. 1993). The biochemical contents such as carbohydrates, phenolics, proteins, amino acids, starch and total chlorophyll of control and transformed cowpea plants were estimated. These parameters increased with minimum level of cell density.

In the present study, sonication-treated plant observed for biochemical content showed that phenolic, protein, amino acid, chlorophyll and carotenoid contents were gradually increased, and starch, carbohydrates and reducing sugars were gradually decreased. The highest carbohydrate (8.03 mg/g of sample at control), protein

(1.88 mg/g of sample at 0.3 OD), phenolic (0.94 mg/g of sample at 0.6 OD), amino acid (8.99 mg/g of sample at 0.3 OD) and starch (26 mg/g of sample at control) contents were analysed. Chlorophyll-a (66 mg/g of sample at 0.3 OD), chlorophyll-b (54 mg/g of sample at 0.3 OD), carotenoids (57 mg/g of sample at 0.3 OD), reducing sugars (20.97 mg/g of sample at control) and secondary metabolites were analysed in nematode plant phenolics (30.13 mg/g) and flavonoids (37.06 mg/g).

14.6 Conclusions

In planta Agrobacterium-mediated genetic transformation has become one of the most important technologies to produce stable transgenic plants in the short time. Over the last decade, significant progress has been made in the development of new and efficient transformation methods in plants. Despite of a variety of reports available, there have been no reports in the phytochemical analysis between transgenic and non-transgenic leguminous plant species. There was a significant metabolic change in plants by the insertion of new foreign genes into the plant genome. It affected the primary and secondary metabolic changes to overcome the barriers. Hence, improvement in the analysis of various metabolic profilings and antioxidant analyses should be applied to create new improved transformation technology.

Acknowledgements The authors would like to thank DST-SERB (Department of Science and Technology-Science and Engineering Research Board) for the research funds to carry out the work and Dr. P. Ananda Kumar, Indian Institute Rice Research (IIRR), Hyderabad, for providing the *A. tumefaciens* strain harbouring the binary vector pCAMBIA2301 for this study. We also acknowledge Bharathiar University, Coimbatore, for providing necessary facilities and support to carry out the research.

References

Adesoye AI, Togun AO, Machuka J (2010) Transformation of cowpea (*Vigna unguiculata* L. Walp.) by *Agrobacterium* infiltration. J Appl Biosci 30:1845–1860

Adlinge PM, Samal KC, Kumara Swamy RV, Rout GR (2014) Rapid in vitro plant regeneration of black gram (Vigna mungo L. Hepper) Var. Sarala, an important legume crop. Proc Natl Acad Sci India Sect B Biol Sci 84:823–827

Arnon DI (1949) Copper enzyme in isolated chloroplast polyphenoloxidase in *Beta vulgaris* L. Plant Physiol 24:1–15

Arun M, Subramanyam K, Theboral J, Ganapathi A, Manickavasagam M (2014) Optimized shoot regeneration for Indian soybean: the influence of exogenous polyamines. Plant Cell Tissue Organ Cult 117:305–309

Aykroyd WR, Doughty J, Walker A (1982) Legumes in human nutrition. FAO Food and Nutrition Paper, Rome

Bakshi S, Sadhukhan A, Mishra S, Sahoo L (2011) Improved *Agrobacterium*-mediated transformation of cowpea via sonication and vacuum infiltration. Plant Cell Rep 30:2281–2292

Beghyn T, Deprez-Poulain R, Willand N, Folleas B, Deprez B (2008) Natural compounds: leads or ideas? Bioinspired molecules for drug discovery. Chem Biol Drug Des 72:3–15
Bhomkar P, Upadhyay CP, Saxena M, Muthusamy A, Shia Parkash N, Sarin NB (2008) Salt stress alleviation in transgenic *Vigna mungo L.* Hepper by overexpression of glyoxalase 1 gene using a novel Cestrum yellow leaf curling virus (CmYLCV) promoter. Mol Breeding 22:169–181
Birch RG (1997) Plant transformation: problems and strategies for practical application. Annu Rev Plant Physiol Plant Mol Biol 48:297–326
Blios MS (1958) Antioxidant determinations by the use of a stable free radical. Nature 181:1199–1200
Bolbhat SN, Dhumal KN (2010) Desirable mutants for pod and maturity characteristics in M2 generation of horse gram (*Macrotyloma uniflorum* lam) Verdc. Res on Crops 11:437–440
Bradford KJ, Van Deynze A, Gutterson N, Parrott W, Strauss SH (2005) Regulating transgenic crops sensibly lessons from plant breeding biotechnology and genomics. Nat Biotechnol 23:439–444
Bravo L, Siddhuraju P, Saura-Calixto F (1999) Composition of underexploited Indian pulses. Comparison with common legumes. Food Chem 64:185–192
Cazzonelli CI, Pogson BJ (2010) Source to sink: regulation of carotenoid biosynthesis in plants. Trends Plant Sci 15:266–274
Chang SS, Park SK, Kim BC, Kang BJ, Kim DU, Nam HG (1994) Stable genetic transformation of *Arabidopsis thaliana* by *Agrobacterium* inoculation *in planta*. Plant J 5:551–558
Chaudhury D, Madanpotra S, Jaiwal R, Saini R, Kumar PA, Jaiwal PK (2007) *Agrobacterium tumefaciens*-mediated high frequency genetic transformation of an Indian cowpea cutivar and transmission of transgenes into progeny. Plant Sci 172:692–700
Chowrira GM, Akella V, Lurquin PF (1995) Electroporation-mediated gene transfer into intact nodal meristems in planta: generating transgenic plants without in vitro tissue culture. Mol Biotechnol 3:17–23
Chrispeels MJ, Sadava DE (1994) Plants. Genes and agriculture. Jones & Bartlett Publishers, London
David AS, Deborah AS, Paula MO (2003) Recent advances in legume transformation. Plant Physiol 131:892–899
Decker EL, Reski R (2007) Moss bioreactors producing improved biopharmaceuticals. Curr Opin Biotechnol 18:393–398
Duke JA, Reed CF (1981) *Macrotyloma uniflorum* (Lam.) Verdc. In: Duke JA (ed) Handbook of legumes of world economic importance. Plenum Press, New York
Faostat (2012) FAO agriculture database. FAO, Rome. http://faostat3.fao.org/home/index.html#visualize
Finnegan J, McElroy D (1994) Transgene inactivation: plants fight back. Nature Biotechnol 12:883–888
Fray RG, Wallace A, Fraser PD, Valero D, Hedden P, Bramley PM, Grierson D (1995) Constitutive expression of a fruit phytoene synthase gene in transgenic tomatoes causes dwarfism by redirecting metabolites from the gibberellin pathway. Plant J 8:693–701
Gheysen G, Villarroel R, Van Montagu M (1991) Illegitimate recombination in plants: a model for T-DNA integration. Genes Dev 5:287–297
Government of Tamil Nadu (2010) Department of economics and statistics, seasons and crop report 2009–2010
Grove TJ, Hendrickson JW, Sidell BD (2004) Two species of Antarctic ice fishes (genus *Champsocephalus*) share a common genetic lesion leading to the loss of myoglobin expression. Polar Biol 27:579–585
Halliwell B, Gutteridge JMC (2007) Free radicals in biology and medicine. Oxford University Press, Oxford
Hansen G, Vaubert D, Heron JN, Clerot D, Tempe J, Brevet J (1993) Phenotypic effects of over expression of agrobacterium T-DNA ORF13 in transgenic tobacco plants are mediated by diffusible factor(s). Plant J 4:581–585

Hodge JE, Hofreiter BT (1962) Determination of reducing sugars and carbohydrates. In: Whistler RL, Wolfrom ML (eds) Methods in carbohydrate chemistry. Academic Press, New York, pp 380–394

Hu CY, Wang L (1999) In planta transformation technologies developed in China: procedure confirmation and field performance. In Vitro Cell Dev Biol Plant 35:417–420

Jaiwal PK, Gulati A (1995) Current status and future strategies of in vitro culture technique for genetic improvement of mung bean (Vigna radiata). Euphytica 86:167–181

Jayaraj J, Devlin R, Punja Z (2008) Metabolic engineering of novel ketocarotenoid production in carrot plants. Transgenic Res 17:489–501

Kadam SS, Salunkhe DK (1985) Nutritional composition, processing, and utilization of horse gram and moth bean. Crit Rev Food Sci Nutri 22:1–26

Kadri A, Chalak L, El Bitar A, Nicolas N, Mroué S, Grenier De March G (2014) In vitro plant regeneration system for two middle east cultivars of chickpea (Cicer arietinum L.) Adv Crop Sci Tech 2:1–4

Kanaka KD (2012) Variability and divergence in horse gram (*Dolichos uniflorus*). J Arid Land 4:71–76

Kaur N, Murphy JB (2012) Enhanced isoflavone biosynthesis in transgenic cowpea (*Vigna unguiculata* l.) Callus. Plant Mol Biol Biotechnol 3:1–8

Kirk JTO, Allen RL (1965) Dependence of chloroplast pigment synthesis on protein synthesis: effect of actidione. Biochem Biophys Res Commun 21:523–530

Kochhar SL (2009) Economic botany in the tropics, 3rd edn. Macmillan publishers, Delhi, pp 152–153

Kolewe ME, Gaurav V, Roberts SC (2008) Pharmaceutically active natural product synthesis and supply via plant cell culture technology. Mol Pharm 5:243–256

Komari T, Hiei Y, Saito Y, Murai N, Kumashiro T (1996) Vectors carrying two separate T-DNAs for co-transformation of higher plants mediated by *Agrobacterium tumefaciens* and segregation of transformants free from selection markers. Plant J 10:165–174

Krishnamurthy KV, Suhasini K, Sagare AP, Meixner M, Kathen A, Pickardt T, Schieder O, Kathen de A (2000) *Agrobacterium* mediated transformation of chickpea (*Cicer arietinum* L.) embryo axes. Plant Cell Rep 19:235–240

Lee MH, Yoon ES, Jeong JH, Choi YE (2004) *Agrobacterium rhizogenes*-mediated transformation of *Taraxacum platycarpum* and changes of morphological characters. Plant Cell Rep 22:822–827

Liu ZC (1980) A photo synthetic characteristic of different plant types in rice. Sci Agric Sin 3:3–10

Loake GJ, Ashby AM, Shaw CH (1988) Attraction of *Agrobacterium tumefaciens* C58C1 towards sugars involves a highly sensitive chemotaxis system. J Gen Microbiol 134:1427–1432

Lowry OH, Rosebrough NJ, Farr AL, Randall RJ (1951) Protein measurement with the Folin phenol reagent. J Biol Chem 193:265–275

Lyznik LA, Rao KV, Hodges TK (1996) FLP-mediated recombination of FRT sites in the maize genome. Nucleic Acids Res 24:3784–3789

Moyer RA, Hummer KE, Finn CE, Frei B, Wrolstad RE (2002) Anthocyanins phenolics and antioxidant capacity in diverse small fruits: vaccinium rubus ribes. J Agric Food Chem 50:519–525

Mukeshimana G (2013) Dissecting the genetic complexity of drought tolerance mechanisms in common bean (*Phaseolus vulgaris* L.). Ph.D dissertation. Michigan State University, East Lansing, Michigan, USA, p 211

Muruganantham M, Amutha S, Ganapathi A (2010) Somatic embryo productions by liquid shake culture of embryogenic calluses in *Vigna mungo* (L.) Hepper. In Vitro Cell Dev Biol Plant 46:34–40

Muthukumar B, Mariamma M, Veluthambi K (1996) Genetic transformation of cotyledon explants of cowpea (*Vigna unguiculata* L. Walp) using *Agrobacterium tumefaciens*. Plant Cell Rep 15:980–985

Neelam R, Sahai J (1988) Effect of textile waste water on seed germination seedling growth pigment content and biomass of *Sesamum indicum*. Linn. J Environ Biol 9:45–50

Olhoft PM, Flagel LE, Donovan CM, Somers DA (2003) Efficient soybean transformation using hygromycin B selection in the cotyledonary-node method. Planta 216:723–735

Pandey AK, Gopinath KA, Chatiacharya P, Hooda KS, Hooda S, Sushil SN, Kundu SKG, Gupta HS (2006) Effect of source and rate of organic manures on yield attributes, pod yield and economics of garden pea (*Pisum sativum* subsp. hortense) in North-East Himalaya. Indian J Agric Sci 76:230–234

Popelka JC, Gollasch S, Moore A, Molvig L, Higgins TJV (2006) Genetic transformation of cowpea (*Vigna unguiculata* L.) and stable transmission of the transgenes to progeny. Plant Cell Rep 25:304–312

Rajendiran K, Thiruvarasan K, Vijayalakshmi R (2016) In vitro callus induction in leaf explants of black gram varieties grown under in situ uv-b radiation. Int J Agric Sci Vet Med 4:53–64

Rao AV, Rao LG (2007) Carotenoids and human health. Pharmacol Res 55:207–216

Rischer H, Orešič M, Seppänen-Laakso T, Katajamaa M, Lammertyn F, Ardiles-Diaz W, Van Montagu MCE, Inzé D, Oksman- Caldentey KM, Goossens A (2006) Gene-to-metabolite networks for the terpenoid indole alkaloid biosynthesis in *Catharanthus roseus* cells. Proc Natl Acad Sci U S A 143:5614–5619

Romeis J, Sharma HC, Sharma KK, Das S, Sarmah BK (2004) The potential of transgenic chickpeas for pest control and possible effects on non-target arthropods. Crop Prot 23:923–938

Saini R, Jaiwal KJ (2005) Transformation of a recalcitrant grain legume, *Vigna mungo* L. Hepper using *Agrobacterium tumefaciens*-mediated gene transfer to shoot apical meristem cultures. Plant Cell Rep 24:164–171

Saini R, Jaiwal PK (2007) *Agrobacterium tumefaciens*-mediated transformation of blackgram: an assessment of factors influencing the efficiency of *uidA* gene transfer. Biol Plant 51:69–74

Schmidlin L, Poutaraud A, Claudel P, Mestre P, Prado E, Santos-Rosa M, Wiedemann –MS, Karst F, Merdinoglu D, Hugueney P (2008) A stress-inducible resveratrol O-methyltransferase involved in the biosynthesis of Pterostilbene in grapevine. Plant Physiol 148:1630–1639

Siddhuraju P, Becker K (2003) Antioxidant properties of various solvent extracts of total phenolic constituents from three different agroclimatic origins of drumstick tree (*Moringa oleifera* Lam.) leaves. J Agric Food Chem 51:2144–2155

Siddhuraju P, Manian S (2007) The antioxidant activity and free radical-scavenging capacity of dietary phenolic extracts from horse gram (*Macrotyloma uniflorum* (Lam.) Verdc.) seeds. Food Chem 105:950–958

Siemann EH, Creasy LL (1992) Concentration of the phytoalexyn resveratrol in wine. Am J Enol Vitic 43:49–52

Somers DA, Samac DA, Olhoft PM (2003) Recent advances in legume transformation. Plant Physiol 131:892–899

Sreedevi S, Remani NK, Benjamin S (2013) Biotic stress induced biochemical and isozyme variations in ginger and tomato by *Ralstonia solanacearum*. Am J Plant Sci 4:1601–1610

Srivatanakul M, Park SH, Salas MG, Smith RH (2001) Transformation of parameters enhancing T-DNA expression in kenaf (*Hibiscus cannabinus*). J Plant Physiol 158:255–260

Sudha N, Mushtari Begum J, Shambulingappa KG, Babu CK (1995) Nutrients and some antinutrients in horse gram (*Macrotyloma uniflorum* (Lam.) Verdc.) Food and Nutritional Bulletin 16:81–83

Sunil SP, Robinson JP, Karthick Balan SS, Anandhaprabhakaran M, Balakrishnan V (2015) *In vitro* regeneration and induction of multiple shooting in *Cicer arietinum* L. using cotyledonary nodal explants. Afr J Biotechnol 14:1129–1138

Tang Y, Chen L, Li XM, Li J, Luo Q, Lai J, Li HX (2012) Effect of culture conditions on the plant regeneration via organogenesis from cotyledonary node of cowpea (*Vigna unguiculata* L. Walp). Afr J Biotechnol 11:3270–3275

Trieu AT, Burleigh SH, Kardailsky IV, Maldonado-Mendoza IE, Versaw WK, Blaylock LA, Shin H, Chiou TJ, Katagi H, Dewbre GR, Weigel D, Harrison MJ (2000) Transformation of *Medicago truncatula* via infiltration of seedlings or flowering plants with *Agrobacterium*. Plant J 22:531–541

Varalaxmi Y, Prasanna A, Sushil KY, Venkateswarlu B, Maheswari M (2013) Optimization of parameters for *Agrobacterium* mediated transformation of black gram (*Vigna mungo* L. Hepper) using cotyledon explants. Afr Biotech 12:1209–1215

Veluthambi K, Aditya KG, Arun S (2003) The current status of plant transformation technologies. Curr Sci 84:3

Verpoorte R, Memelink J (2002) Engineering secondary metabolite production in plants. Curr Opin Biotechnol 13:181–187

Wu H, Sparks C, Amoah B, Jones HD (2003) Factors influencing successful *Agrobacterium*-mediated genetic transformation of wheat. Plant Cell Rep 21:659–668

Yasuma A, Ichikawa T (1953) Ninhydrin-Schiff and alloxanschiff staining: a new histochemical staining method for protein. J Lab Clin Med 41:296–299

Yun DJ, Hashimoto T, Yamada Y (1992) Metabolic engineering of medicinal plants: transgenic Atropa belladonna with an improved alkaloid composition. Proc Natl Acad Sci U S A 89:11799–11803

Zhenying L, Andrew NB (2003) Functional subsets of the VirB type IV transport complex proteins involved in the capacity of *Agrobacterium tumefaciens* to serve as a recipient in virB-mediated conjugal transfer of plasmid RSF1010. J Bacteriol 185:3259–3269

Zhishen J, Mengcheng T, Jianming W (1999) The determination of flavonoid contents in mulberry and their scavenging effects on superoxide radicals. Food Chem 64:555–559

Zhou GY, Weng J, Zhen YS, Huang JG, Qian SY, Liu GL (1983) Introduction of exogenous DNA into cotton embryos in methods in enzymology. In: Wu R, Grossman L, Molddave K (eds) Recombination DNA part C, vol 101. Academic, New York, pp 433–481

Chapter 15
Somatic Embryogenesis from Immature Anther Explants: Toward the Development of an Efficient Protocol Production of Grapevine

Krishnan Vasanth and Melané A. Vivier

Abstract This paper reports on successful establishment of friable embryogenic callus (FEc) initiated from immature anther explants collected from the vineyard within 1 month. The anther collection, filament attachment, filament length, stages, anther size, color, and maturation are the multiple factors involved in the establishment of embryogenic callus. Anthers with immature filaments having a translucent white color, a length that was between 0.02 mm and 0.05 mm, and at either stage II or stage III gave a high frequency of embryogenic callus. In contrast, mature anthers with green translucent or yellow color filaments attached with a length between 0.08 mm and 0.2 mm and at either stage V or stage VI produced non-embryogenic callus (NEc), with browning of the anther in the medium. Inflorescences collected from bud initiation onward could be cultured as immature anthers, and a good response was recorded; FEc and NFc were separated from initiation in the form of a callus as early as 9 weeks. The initiation of proembryogenic masses (PEMs) and the subsequent stages of development were analyzed; the maximum percentage of embryogenic callus obtained was 56% for Sultana, 48% for Red Globe, and 39% for Merlot at stage III. In the modified medium as described in Franks et al. (Mol Breed 4:321–333, 1998), subculturing was carried out on NN medium containing 2,4-D (1.5 mg dm-3), BA (1.5 mg dm^{-3}), sucrose (60 g dm^{-3}), and PVP (25 mg dm^{-3}). The excretion of polyphenols into the subculture was reduced by culturing the callus in the same medium in the presence of 25 mg dm^{-3} PVP, and this reduced the browning of the culture medium. In conclusion, many factors are involved in determining

K. Vasanth (✉)
Molecular Biology Laboratory, Department of Botany, Bharathiar University, Coimbatore, Tamil Nadu 641 046, India

Institute for Wine Biotechnology, Department of Viticulture and Oenology, Stellenbosch University, Stellenbosch ZA-7600, South Africa
e-mail: vasanthlabbu@gmail.com

M.A. Vivier
Institute for Wine Biotechnology, Department of Viticulture and Oenology, Stellenbosch University, Stellenbosch ZA-7600, South Africa

A. Dhanarajan (ed.), *Sustainable Agriculture towards Food Security*,
https://doi.org/10.1007/978-981-10-6647-4_15

embryogenic callus initiation, and these are, in turn, linked to subsequent embryo initiation, development, and regeneration during somatic embryogenesis.

Abbreviations

2, 4-D	2, 4-Dichlorophenoxyacetic acid
FEc	Friable embryogenic callus
NAA	α-Naphthyl acid
NFc	Non-friable callus
PEMs	Pre-embryogenic masses
PVP	Polyvinylpyrrolidone

15.1 Introduction

Grapevine is a woody perennial crop plant species of ancient origins. From the Caucasian area, its cradle of origin, it spreads first in the Mediterranean area and later across the world, such as, Australia CSIRO working in the Cooperative Research Centre for Viticulture reported (extremely rare and independent mutations in two genes [*VvMYBA1* and *VvMYBA2*] [of red grapes] produced a single white grapevine that was the parent of almost all of the world's white grape varieties). One gene had been mutated, most grapes would still be red, and we would not have more than 3000 white grape cultivars available today (Amanda et al. 2007).

Grapevines grow in temperate regions that generally require two consecutive growing seasons to complete their reproductive developmental cycle. The flowering spreads in two seasons; flowering is induced during the first season in latent summer buds in which the shoot apical meristem produces two to three lateral meristems that become inflorescence meristems. The inflorescence meristems proliferate within the bud to give rise to inflorescence branch meristems with a spiral phyllotaxis and generate an immature raceme structure before the bud enters dormancy at the end of the summer. In the next spring (second season), additional inflorescence branch meristems can be formed before each one gives rise to a cluster of three to four flower meristems that develop into flowers arranged in a dichasium (Mullins et al. 1992; Carmona et al. 2008). The interactions between environmental factors influenced genetic mechanisms controlling the induction, and development of inflorescences, flowers, and berries is also an important area that requires increased emphasis, especially given the large seasonal fluctuations in yield of wine-producing regions (Carmona et al. 2008).

Currently 102,146 hectares of vine-producing wine grapes are under cultivation in South Africa over an area of some 800 kilometers in length. White varietals constitute 55% of the plantings for wine, with Chenin blanc plantings comprising 19%

of the total. Red varietals account for 45% of the national vineyard in South Africa. Grapes are used in wine making and fresh fruit (table grapes) and dried fruit (raisins) production. Other products derived from grapes or wine-making waste include grape juice, jelly products, ethanol, vinegar, grape seed oil, tartaric acid, and fertilizer.

There is also an increasing interest in the health benefits of certain grape-derived antioxidant compounds (e.g., polyphenols, resveratrol), and these compounds are being investigated and used in the food additive, cosmetic, and pharmaceutical industries. Grapevine plants annually undergo a major developmental transition from vegetative to reproductive development, that is, flowering.

Physiological and genetic analysis of flowering has shown that multiple environmental and endogenous inputs can influence the timing of this switch (Boss et al. 2004; Lebon et al. 2005). And also floral transition is one of the most drastic changes occurring during the life cycle (Boss et al. 2002). Furthermore, flowering occurs only once a year for a few weeks, to identify the correct developmental stage of explants and to initiate for the callus, but cuttings of cane material were used as explants (Dhekney et al. 2009; Kikkert et al. 2005; Perrin et al. 2004; Rajasekaran and Mullins 1979; Rajasekaran and Mullins 1983).

Moreover, the correspondence between the developments of the inflorescences in in vitro cutting is able to mimic sexual reproduction in the field in this context; vineyard is not well explained (Lebon et al. 2005). Grapevine biotechnology is one of the most promising developments in the global wine industry, which is increasingly faced with conflicting demands from markets, consumers, and environmentalists.

Wine industries support the establishment of stress-tolerant and disease-resistant varieties of *Vitis vinifera* (Vivier and Pretorius 2000, 2002). During the past two decades, there has been a great scientific effort aimed at obtaining somatic embryos (SEs) from various grapevine genotypes, and a number of different tissues including anthers and the stages of classification (I–VI) have been used (Dhekney et al. 2009; Gribaudo et al. 2004; Kikkert et al. 2001, 2005).

The anther color and separation from calyx have been explored in in vitro conditions. However, any great extent in the literature in the field conditions not well established the establishment of somatic embryogenesis and plant regeneration.

It appears to be dependent on interaction between the genotype, explant source, medium composition, plant growth regulators, and explant stage at the beginning of culture. Nevertheless, success with the same cultivar in different years still may differ because many factors influence the differentiation of embryogenic callus.

Thus, it has become necessary to develop specific regeneration protocols for each grapevine species or cultivar (Emershad and Ramming 1994; Gribaudo et al. 2004; Jayasankar et al. 2003; Lápez-Pérez et al. 2005; Vidal et al. 2009). Anther explants seem to be the best responding material; however, other factors seem to act to inhibit embryo development under conventional culture conditions.

Embryogenic callus can be maintained, although factors like lack of torpedo conversion, erratic embryogenesis, and arrest of embryo meristematic development are frequently observed for long period of culture (Coutos-Thévenot et al. 1992).

Embryogenesis and subsequent plant development of this cultivar are very inefficient (Wang et al. 2005). In addition to this, after a transformation event, regeneration ability is considerably reduced (Marsoni et al. 2008).

This study investigates the appropriate culture condition and various factors like flower maturation, collection, and medium response in combination in order to establish a mass production system for the establishment of friable callus of South African conditions. This is then expanded to investigate the different parameters influencing somatic embryogenesis.

15.2 Materials and Methods

15.2.1 Plant Material

Inflorescences were collected from vineyards in Stellenbosch in the early morning and at night; there were daily fluctuations in temperatures at the time of collection. However, the cuttings in the vineyard were all collected at temperatures between 20 °C and 25 °C. The presence of inflorescence on grapevine cultivars depends on the local environmental conditions with flowering usually starting from the first week of October. The varieties Sultana, Red Globe, and Merlot, were collected before anthesis until the anther color had turned to yellow, was more than 0.08 mm in length, and was at stage according to Gribaudo et al. (2004) and Goussard et al. (1991).

The anthers were collected at about 2-day intervals for 4 weeks and surface sterilized in 5.0% (w/v) calcium hypochlorite solution containing three drops of Tween-20 for 10 min and then rinsed three times with sterile distilled water. Anthers were carefully removed from the inflorescence with the attachment portion of the filaments at 0.03 mm in order to obtain calyx (Fig. 15.1a.) initial stage I.

15.2.2 Establishment of Pre-embryogenic Masses and Somatic Embryos

The anthers were placed on a (90 mm x 15 mm) Petri plate with 30 ml of callus induction NN medium (Nitsch and Nitsch (1969) containing NN macroelements, MS microelements, B_5 vitamin, 6% (w/v) sucrose with 2,4-dichlorophenoxyacetic acid (2,4-D, 1.5 mg dm^{-3}), and benzylaminopurine (BA 1.5 mg dm^{-3}) in 0.3% Phytagel. The pH of the medium had been adjusted to 5.7 and autoclaved at 121 °C and 1.2 kg/cm pressure for 15 min. Independent experiments were carried out, and each plate contained 50 anthers, and at least 5000–5500 anthers every year were cultured from each of the cultivars for 3 years. After 2 weeks, embryogenic callus (EC) tissue was subcultured on the same medium for 8 weeks at 25 ± 2 °C under complete darkness.

Fig. 15.1 Anther culture response during embryogenic callus initiation and regeneration. (**a**) Anther plating green and white florescent color, (**a**) immature anther attachment with filament white color (0.03 mm), (**b**) embryogenic callus initiation, friable texture, and white from first month response, (**c**) pre-embryogenic masses and high embryogenic capacity during cultivation on proliferation media containing PVP at 3 months, (**d**) *PEMs* – proembryogenic mass conversion at various stages of development, (**e**) pre-embryogenic mass stages that are normally induced in somatic embryos at the globular to torpedo stages, (**f**) third week advanced somatic embryos at different stages of development on embryo maturation media with activated charcoal – *em* embryo maturation, (**g**) developing somatic embryos on maturation media, the early cotyledonary stage, (**h**) cotyledonary portion with roots after 2 weeks on maturation media, greened somatic embryos ready for germination, (**i**) converted plantlets after germination on MS medium followed by 3 weeks without hormone under light. (**j**) Plant regeneration
Non-embryogenic callus from Sultana, Red Globe, and Merlot. Mature anther at first month response (Sultana) (**j**). Anther with green and white florescent color, *ma* – mature anther attachment green color length (0.1 mm) from before anthesis. (**k**) *ap* – Filament attachment portion. (**l**) *nec*, nodular embryogenic callus; *snesc*, non-embryogenic spongy callus – (Merlot) (**m**) nodular non-embryogenic callus (Sultana) (**n**) nodular non-embryogenic callus (Red Globe). (**o**) Watery non-embryogenic callus (wnec) (all three cultivars)

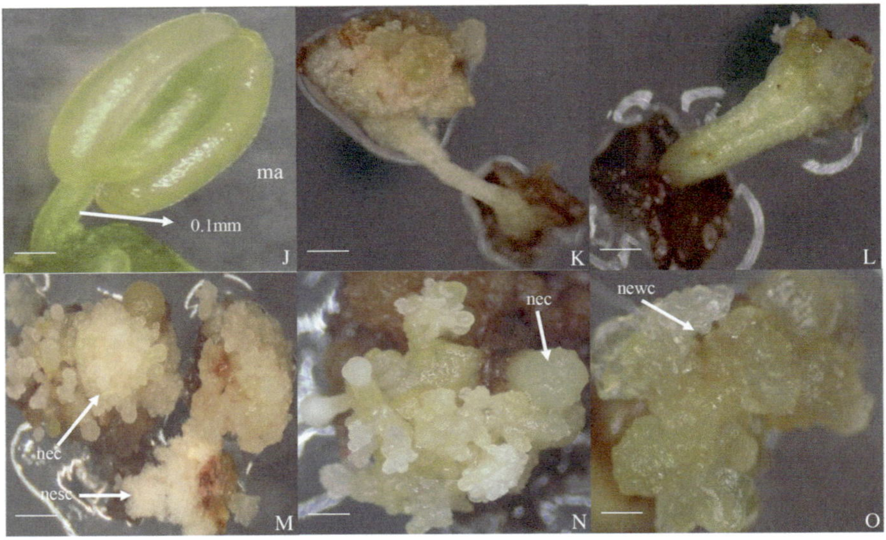

Fig. 15.1 (continued)

The anthers were dissected under a stereomicroscope after 75 days in culture and their length recorded. Subculture medium containing plant growth regulators with PVP (25 mg $dm^{-3)}$ was used for the induction of EC. Non-embryogenic callus (NEC) and EC tissue from the anthers were separated on a morphological basis under a stereomicroscope and then transferred into PIV for 4 weeks until the production of proembryogenic masses.

The embryogenic cultures were maintained for multiplication using a medium (Franks et al. 1998; Perl 1995). The cell lines have been maintained in our laboratory for 5 years. Finally, the embryo germination callus was transferred to NN medium with 3% sucrose and 2.5% activated charcoal without any plant growth regulators for 2 weeks in order to study the various stages of development under dark conditions.

Different colchicine concentrations treated on the anther and ovule culture doubling by colchicine treatment of embryogenic cell aggregates growing in grapevine suspension cultures on the ploidy level of somatic embryo-derived plantlets, 20 plantlets derived from ESCs treated with each of the colchicine concentrations, and 20 plantlets derived from nontreated ESCs were analyzed by flow cytometry method reported by Acanda et al. (2015).

15.2.3 Plant Regeneration and Acclimatization

The late cotyledonary embryos' radical portion was cut to 0.5 cm; transferred to NN medium containing macroelements, sucrose (30 g dm^{-3}), and activated charcoal (2.5 g dm^{-3}) with and without plant growth regulators; and then maintained in the dark (Vasanth and Vivier 2011). Once normal leaflets were obtained, they were transferred to the rooting MS basal medium with activated charcoal (Murashige and Skoog 1962). Well-rooted plants were thoroughly washed in sterile water to remove agar and transferred to plastic pots containing a mixture of autoclaved sandy soil and vermiculite (1:1). For acclimatization, the potted plants were grown in a growth chamber at 85% relative humidity for 3 weeks and in a greenhouse (initially covered with polythene bag to maintain high humidity) for 4 weeks before final transfer. The plants were watered once every 3 days with Hoagland nutrient solution (Hoagland and Arnon 1950).

15.3 Results and Discussion

15.3.1 Effect of Plant Growth Regulators (PGR) and Polyvinylpyrrolidone (PVP) on Anther Response

Anthers were carefully removed from the inflorescence with the attachment portion of filament length of 0.03 mm to obtain calyx (Fig. 15.1a.). The maturation of the anther, the calyx with attachment of the filaments, the anther size, and the anther color influenced the efficiency of the callus induction. The three different cultivars gave different culture responses. Anthers were then detached from the calyx and tissue arising from the same anther plated in bulk (5). The tissue needed to be translucent green with a white anther for the induction of embryogenic callus (Fig. 15.1b).

Explants were transferred to a fresh medium of the same formulation every 2 weeks during the first month and every 4 weeks thereafter for five additional months. At each subculture, explants were examined under a dissecting microscope for callus initiation. The response of anthers reached 80% in the first month, and in the second month, they were subcultured and were separated from the embryogenic callus.

Inflorescence was collected from the vineyard during the first week on stage I, first week on stage II, and second week on stages III and IV responded well. However, colection of influrescence later 14th day, most of the cultivars in stage V failed to initiate callus, and the matured anther's color turned black. The right stage and the size of the anther clearly influenced the percentage response of the three different cultivars.

The best type of inflorescence was cultivar specific, and the optimum numbers of days of anther maturation also varied depending on the specific cultivars. Callus formation was observed 15 days after the initiation of the cultures on NN medium

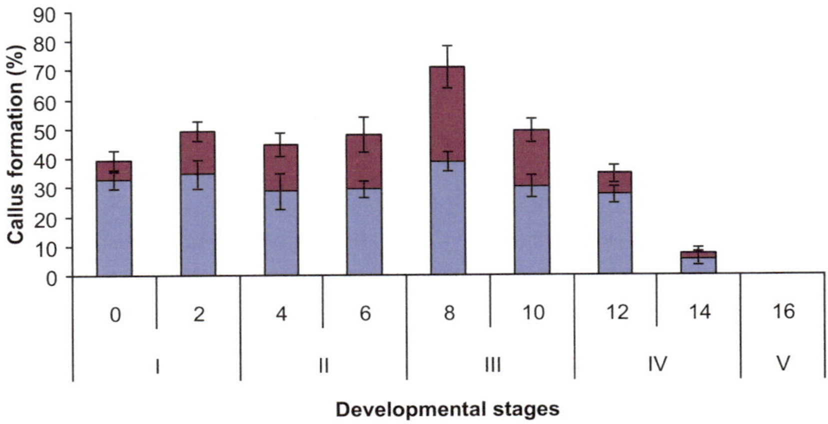

Fig. 15.2 Anther collection date, stages, and response after 3 months in terms of producing embryogenic callus

and was then further maintained. After 2 months, the callus appeared to be friable and surrounded the explants. The majority of EC was initiated up to the fourth week, and the culture was maintained for 9 weeks on the same medium. The anther collection date, stages, and response in terms of producing embryogenic callus over 3 months are shown in Fig. 15.2.

The initiated embryogenic callus was friable yellow in color and contained pre-embryogenic masses that could be maintained on PIV medium supplemented with PVP (25 mg dm^{-3}) in darkness by regular subculture. The highest percentage of response in terms of embryogenic callus during the first month was recorded for Sultana at 56%, followed by Red Globe at 48% and Merlot at 39% (Fig. 15.3). The first month produced embryogenic and non-embryogenic callus; nonetheless, embryogenic masses could be separated and subcultured on the same medium.

A semi-compact pre-embryogenic callus was favorable to allowing embryogenic culture and plant regeneration via somatic embryogenesis (Fig. 15.4). Anther color was usually a translucent green-yellow and was either attached or could be cut away from the attachment site on the calyx; the detached or without attachment anthers could be dead or non-embryogenic callus. Friable embryogenic callus from the different cultivars was subcultured and maintained on the medium for multiplication (Franks et al. 1998, Perrin et al. 2004).

This should be compared with the use of optimized cane material for inflorescence production and then anther cutting and a three-step culture media approach that allowed the induction of embryogenic callus with 19 genotypes at 84 days after anther plating; this gave efficiency rates of 33–42% (Perrin and Pin 2004), Sultanina at 47% (Vidal et al. 2009), Sultanina 27%, and Merlot 35% at PIV medium (Dhekney et al. 2009). In addition, inflorescence and flower development have been studied in Gewurztraminer, Pinot Noir, and other cultivars (Dhekney et al. 2009; Gribaudo et al. 2004; Perrin et al. 2004).

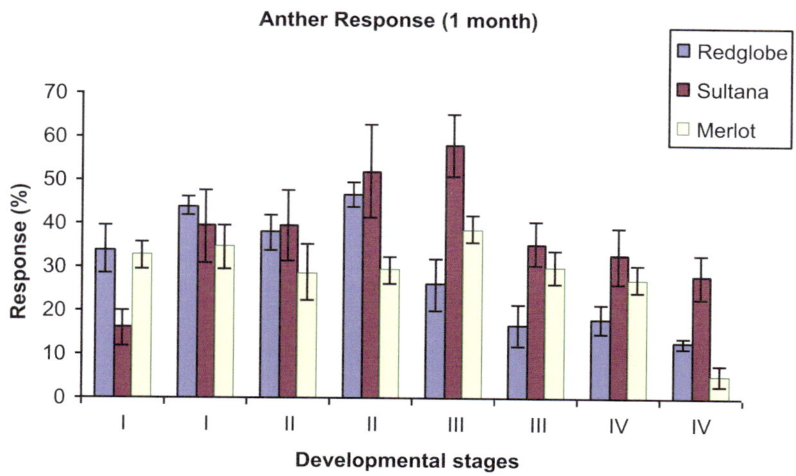

Fig. 15.3 Embryogenic cells derived from the anther culture system at 25 days after incubation in the medium, the response of the callus

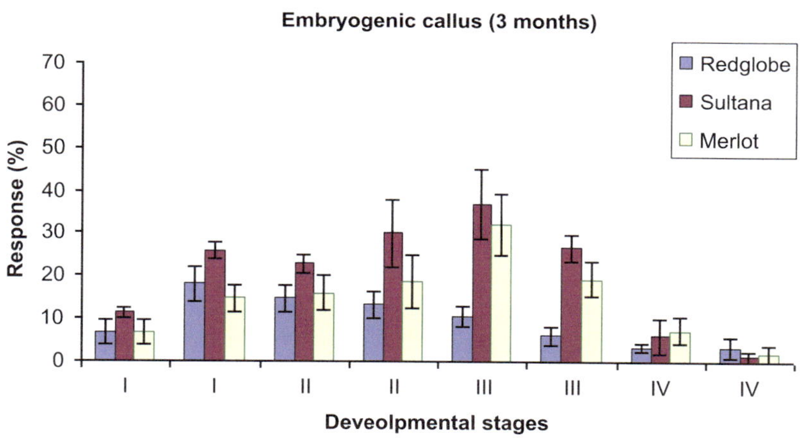

Fig. 15.4 Three-month record of embryogenic callus proliferation and multiplication (Sultana, Red Globe, and Merlot)

Callus induction for most anthers in the PIV medium occurred in the anther attachment portion of filaments (0.02–0.09 mm) and at the stages of anther development that are recorded in Table 15.1. The maturation of the flowers and other factors like color, size, growth medium, and plant growth regulators affected the embryogenic callus induction and the percentage response. As a model of flower development in a woody species, our study will assist in efforts to understand the physiological traits that favor inflorescence initiation and flower development. The twice monthly subculturing of the anthers increased the cell density of the embryogenic callus. The

Table 15.1 Anther stage, filament length, color, and callus morphology

Anther no. of days	Stage of development	Attachment portion length (mm)	Collection of anther color	Morphology of callus
0	I	0.02	Green	Friable, yellow
2	I	0.02	Green with white	Soft friable, light yellowish
4	II	0.03	Green with white	Soft friable, light brown
6	II	0.03	Green with white	Soft friable, white
8	III	0.04	Florescence	Compact, white
10	III	0.05	Green with white florescence	Nodular friable, green
12	IV	0.06	Green with white	Soft friable, spongy white
14	IV	0.07	White with yellow	Compact, greenish
16	V	0.08	Green with yellow	Green, creamish watery
18	VI	0.09	Yellow	Black brown

Data recorded after 5 weeks of culture. Each treatment was replicated twice and each replicate consisted of 50 anther explants

white synchronous anthers that were able to initiate callus have embryogenic potential; however, matured anthers are non-embryogenic and turn brown/dark, which results in death. Similar results have been found with other genotypes, where anther to the calyx of 1–1.5 mm length and translucent green anthers have been reported as optimal (Dhekney et al. 2009; Joly et al. 2004; Lápez-Pérez et al. 2005).

Furthermore, anther classification of a few cultivars by Gribaudo et al. (2004) suggested that the highest percentage of anther response was in the first month (52.5%). In another study, callus induction by Sultana was higher than previously reported (Perl et al. 1995; Vidal et al. 2009). Embryogenic callus cells arise from diploid cells and success mostly depends on the genotype used. Some agronomically important cultivars are still not amenable to tissue culture and the production of embryogenic callus (Wang et al. 2005).

In this study, the phenotype of the EC was observed, and it was found that the color of the callus differed between genotypes. Red Globe and Merlot callus tissues needed to be white, yellow, and brown to produce embryos from a friable callus when compared to the Sultana. Whitish semi-compact embryogenic callus was transferred to a proliferation medium for long-term expansion.

During the long-term maintenance of the grapevine embryogenic cultures, the cell line cultures that gave rise to well-developed somatic embryos were infrequent. The increase in weight of the callus always depends on the density of the cells. The density of the solid culture PEMs started to increase from the 9th day onward up to the 12th day of every subculture. When subculturing was stopped, the cell density did not increase, and moreover, polyphenol accumulation occurred along with degradation of nitrogen contents of the cell lines (data not shown).

Many reports have shown that treatment with a high concentration of growth regulator 2,4-D can be regarded as a trigger for induction of somatic embryogenesis (Perrin et al. 2004; Jayasankar et al. 2003). In the majority of plant species, the synthetic auxin 2,4-D is the most suitable plant growth regulator for inducing somatic embryogenesis. Subculturing explants under the influence of 2,4-D results in an increase in the endogenous auxin levels in the explants (Michalczuk et al. 1992), thus increasing cell division, which allows the development of the hormonal gradient necessary for embryogenesis.

Once the somatic cells are embryogenically induced, further regeneration of somatic embryogenesis may be initiated by a number of factors, depending on plant species, cultivar type, and physiological condition of the donor plant. Marsoni et al. (2008) reported that callus formation decreased as the concentration of 2,4-D decreased and increased with an increasing concentration of 2,4-D.

When the differentially expressed proteins between embryogenic callus and non-embryogenic calluses are compared, it suggests that grapevine embryogenic status is related to better control of oxidative stress, both by regulation of the ROS-scavenging system and by the preservation of protein structure by heat shock proteins.

The totipotency in cultured somatic cells is thus part of a general stress adaptation process. An oxidative burst, probably caused by 2,4-D, may be the key event resulting in the embryogenic competence of some of the callus cells but leads to the programmed death of other cells. Embryogenic callus from recalcitrant genotypes shows a variety of responses including the release of polyphenols, which is manifested as tissue browning and glycoproteins, both of which are frequently accompanied by cell death (Franks et al. 1998; Perl et al. 1995).

15.3.2 Embryogenic and Non-embryogenic Callus Differentiation

Sultana, Red Globe, and Merlot embryogenic calluses were obtained from PIV medium containing 2,4-dichlorophenoxyacetic acid (2,4-D) and benzylaminopurine (BA) that had been kept under dark conditions for 8 weeks. After 2 weeks of culture, many anthers started to become brown due to the presence of excreted phenolic compounds, which also turned the medium brown. As a result of this, it was necessary to subculture into fresh medium with the same composition (NN medium with PVP 25 mg dm^{-3}).

Embryogenic callus was white with a little yellow tinge as well as being friable and compact. This contrasted with non-embryogenic callus, which was white, soft and spongy, or else brown watery and quick-growing loose aggregating.

The undifferentiated mass of cells could be differentiated into (i) friable callus (Fig. 15.1b) and (ii) non-friable callus (Fig. 15.1m, n). The friable callus could be further differentiated into (ia) friable white callus (Fig. 15.1c) and (ib) friable nodular callus (Fig. 15.1k). Non-embryogenic callus could be (a) friable watery loose callus (Fig. 15.1o), (b) friable white nodular callus (Fig. 15.1l), and (c) black callus (Fig. 15.1m). Most of them remained as non-embryogenic calluses that would not convert into somatic embryos after prolonged subculture. For non-embryogenic callus, there is a 35-kDa polypeptide that was not detected in embryogenic callus reported (Marsoni et al. 2008). The embryogenic potential of each callus was recorded, and it was then transferred to the same medium. Friable embryogenic callus was able to maintain proembryogenic mass proliferation during long-term maintenance on the medium (Ben Amar et al. 2007; Perl et al. 1995).

In grapevines, initiation is a critical life history trait, function of flowering remaining unknown however; in vivo conditions uncommitted primordium in the latent bud develops into inflorescence primordium and then into a potential fully developed inflorescence in the next spring, depending on the environmental conditions and soil conditions (Boss and Thomas 2002).

Rapid shoot growth can produce tendrils rather than inflorescences with stress delaying or stopping reproductive development. For example, low temperatures near flowering affect ovule development and pollen tube growth (Lebon et al. (2005). Further development of the flowers is then influenced by the plant variety and environment. Flower abscission is poorly described in grapevine but can dramatically diminish the vine's yield (Boss et al. 2004).

It is known that the growth conditions of the grapevine canopy, mainly the climate and the pruning, also influence the reserves stored in the roots, the rootstock, and the hardwood and thus may play an important role in the flowering ability of the latent bud in the year after its initiation. All of the above factors may influence the culture response of somatic cells that are derived from anthers. Under our conditions, the inflorescence development relied essentially on the hardwood reserves of cutting, which consists of two internodes (around 30 cm) as described in Perrin et al. (2004).

It has also been reported that significant modification of media composition, plant growth regulators, and culture conditions can lead to a design with five to six culture steps, each better adapted to the physiological stage of the cells.

15.3.3 Embryo Germination and Regeneration

For the germination of mature somatic embryos, these were transferred to conversion medium (MS basal medium supplemented with 3% sucrose and 2.5% activated charcoal). Next, the late cotyledonary stage was transferred to an MS medium under light (Vasanth and Vivier 2011). At this stage, embryo germination was defined as embryos showing root elongation and root development, whereas plantlet conversion was considered to be embryos showing normally developed cotyledons under the light. This agrees with studies of previous cultivars reported in the literature (Lápez-Pérez et al. 2005; Perrin et al. 2004).

An embryogenic culture typical of grapevines developed when the PEMs were transferred to and grown on solidified medium. The embryogenic cultures grew nonsynchronously, and therefore somatic embryos at different stages of development were present at any given time. Specifically, mature embryos and small plantlets were cultured on solid culture on NN medium containing 0.25% activated charcoal without plant growth regulators under dark conditions.

It is known that the addition of activated charcoal to the medium (AC) for root development is a common practice (Perrin et al. (2009), Ben Amar et al. (2007)), but the exact mechanism whereby AC promotes root development is still unknown. In this study, when the regenerated plantlets were cultured on MS medium with AC, the roots were more vigorous, and more lateral roots were developed.

Sanchez et al. (1996) suggested that the stimulatory effect of AC on root development may involve firstly the reduction of light intensity at the base of shoots, which provides an environment conducive to the accumulation of auxin or cofactors or both, and secondly the absorption of substances such as inhibitory phenolics and any excess auxin or cytokinin is carried over from previous media.

The efficiency of the embryo germination and conversion into plantlets did not involve a problem with shoot development, but it was found that the addition of activated charcoal to the medium used for the callus culture was an essential prerequisite to obtain somatic embryos, and similar results have been reported (Lápez-Pérez 2005; Perl et al. 1995; Zhu et al. 1997).

In previous studies, plant regeneration rates have often been low, and embryos did not survive when placed in CPM alone. For example, 6% of embryos germinated and 2% regenerated into plants (Jayasankar et al. 1999). For instance, in some cultivars, only 3% of somatic embryos were capable of developing into complete plants, although 27% had shoot and root apices (Faure 1990; Goussard et al. 1991).

The highest percentage of plant regeneration is reported (Dhekney et al. 2009; Vidal et al. 2009). This progressive reduction in the capacity for maturation in the supplied medium finally resulted in the continued maintenance of uniform somatic PEM culture. A plant *defensin gene* Vv-AMP-1 and PGIP has been reported in tobacco (De Beer and Vivier 2008; Joubert et al. 2007); these genes transformed into grapevine cell line using established protocol and at the analyzing data. Somatic embryogenesis has been projected as an effective tool for large number of plant multiplication of elite germplasm. This approach has an added advantage that it will reduce the complexities, the consumable requirements, and the number of subcultures needed, which are significant limitations to the commercial applications of this potentially useful technology.

Acknowledgments This work is financially supported by the National Research Foundation, THRIP, and Winetech.

References

Acanda Y, Martınez O, González MV et al (2015) Highly efficient in vitro tetraploid plant production via colchicine treatment using embryogenic suspension cultures in grapevine (Vitis vinifera cv. Mencıa). Plant Cell Tiss Org 123:547–555

Amanda R, Walker EL et al (2007) White grapes arose through the mutation of two similar and adjacent regulatory genes. Plant J 49(5):772–785

Ben Amar A, Cobano P, Boonrod K et al (2007) Efficient procedure for grapevine embryogenic suspension establishment and plant regeneration: role of conditioned medium for cell proliferation. Plant Cell Rep 26:1439–1447

Boss PK, Thomas MR (2002) Association of dwarfism and floral induction with a grape 'green revolution'mutation. Nature 416(6883):847–850

Boss PK, Bastow RM, Mylne JS et al (2004) Multiple pathways in the decision to flower: enabling, promoting, and resetting. Plant Cell 16:18–31

Carmona MJ, Chaïb J, Martínez-Zapater JM et al (2008) A molecular genetic perspective of reproductive development in grapevine. J Exp Bot 59:2579–2596

Coutos-Thévenot P, Goebel-Tourand I, Mauro MC et al (1992) Somatic embryogenesis from grapevine cells. I-improvement of embryo development by changes in culture conditions. Plant Cell Tiss Org 29:125–133

De Beer A, Vivier MA (2008) Vv-AMP1, a ripening induced peptide from Vitis vinifera shows strong antifungal activity. BMC Plant Biol 8(1):75

Dhekney SA, Zhijian T, Li Compton ME et al (2009) Optimizing initiation and maintenance of *Vitis* embryogenic cultures. Hort Sci 44(5):1400–1406

Emershad RL, Ramming DW (1994) Somatic embryogenesis and plant development from immature zygotic embryos of seedless grapes (*Vitis vinifera* L). Plant Cell Rep 14:6–12

Faure O (1990) Embryons somatiques de *Vitis rupestris* et embryons zygotiques de *Vitis* sp: morphologie, histologie, histochimie et développement. Can J Bot 68:2305–2315

Franks T, Gang He D, Thomas M (1998) Regeneration of transgenic shape *Vitis vinifera* L. Sultana plants: genotypic and phenotypic analysis. Mol Breed 4:321–333

Goussard PG, Wiid J, GGF K (1991) The effect of *in vitro* somatic embryogenesis in eliminating fanleaf virus and leafroll associated viruses from grapevines. S Afr J Enol Vitic 12:77–81

Gribaudo I, Gambino G, Vallania R (2004) Somatic embryogenesis from grapevine anthers: the optimal developmental stage for collecting explants. Am J Enol Vitic 55:427–430

Hoagland DR, Arnon DI (1950) The water culture method for growing plants without soil. California Agricultural experiment station circular 347. College of Agriculture, University of California, Berkeley, p 32

Jayasankar S, Gray DJ, Litz RE (1999) High-efficiency somatic embryogenesis and plant regeneration from suspension culture of grapevine. Plant Cell Rep 18:533–537

Jayasankar S, Bonbada BR, Li Z et al (2003) Comparative anatomy and morphology of *Vitis vinifera* (Vitaceae) somatic embryos from solid- and liquid-culture-derived proembryogenic masses. Am J Bot 90:973–979

Joly D, Perrin M, Gertz C et al (2004) Expression analysis of flowering genes from seedling-stage to vineyard life of grapevine cv. Riesling. Plant Sci 166:1427–1436

Joubert DA, Kars I, Wagemakers L et al (2007) The grapevine polygalacturonase-inhibiting protein (VvPGIP1) reduces *Botrytis cinerea* susceptibility in transgenic tobacco and differentially inhibits fungal polygalacturonases. Mol Plant Microb Interac 20(4):392–402

Kikkert JR, Thomas MR, Reisch BI (2001) Grapevine genetic engineering. In: Roubelakis-Angelakis KA (ed) Molecular biology and biotechnology of the grapevine. Kluwer Academic Publishers, Dordrecht, pp 393–410

Kikkert JR, Striem MJ, Vidal JR et al (2005) Long-term study of somatic embryogenesis from anthers and ovaries of 12 grapevine (*Vitis* sp) genotypes. In Vitro Cell Dev Biol-Plant 41(3):232–239

Lápez-Pérez J, Carreño J, Martínez-Cutillas A et al (2005) High embryogenic ability and plant regeneration of table grapevine cultivars (*Vitis vinifera* L) induced by activated charcoal. Vitis 44:79–85

Lebon G, Duchene E, Brun O et al (2005) Phenology of flowering and starch accumulation in grape (*Vitis vinifera* L.) cuttings and vines. Ann Bot 95:943–948

Marsoni M, Bracale M, Espen L et al (2008) Proteomic analysis of somatic embryogenesis in Vitis vinifera. Plant Cell Rep 27(2):347–356

Michalczuk L, Ribnicky DM, Cooke TJ et al (1992) Regulation of indole-3-acetic acid biosynthetic pathways in carrot cell cultures. Plant Physiol 100:1346–1353
Mullins MG, Bouquet A, Williams LE (1992) Biology of the grapevine. Cambridge University Press, Cambridge
Murashige T, Skoog F (1962) A revised medium for rapid growth and bioassays with tobacco tissue cultures. Physiol Plant 15:473–497
Nitsch JP, Nitsch C (1969) Haploid plants from pollen grains. Science 163:85–87
Perl A (1995) Cryopreservation of grapevine (*Vitis* spp.) embryogenic cell suspensions by encapsulation-vitrification. Plant Cell Tiss Org 77:267–275
Perl TM, Dvorak L, Hwang T et al (1995) Long-term survival and function after suspected gram-negative sepsis. JAMA 274(4):338–345
Perrin D, Pin J-É (2004) Infinite words: automata, semigroups, logic and games, vol 141. Academic Press, Amsterdam
Perrin M, Martin D, Joly D, Demangeat G, This P, Masson JE (2004) Medium-dependent response of grapevine somatic embryogenic cells. Plant Sci 161:107–116
Perrin M, Gertz C, Masson JE (2009) High efficiency initiation of regenerable embryogenic callus from anther filaments of 19-grapevine genotypes grown worldwide. Plant Sci 167:1343–1349
Rajasekaran K, Mullins MG (1979) Embryos and plantlets from cultured anthers of hybrid grapevines. J Exp Bot 30:399–407
Rajasekaran K, Mullins MG (1983) The origin of embryos and plantlets from cultured anthers of hybrid grapevines. Am J Enol Vitic 34:108–113
Sanchez MC, San-Jose A, Ballester A (1996) Requirements for in vitro rooting of Quercus robur and Q. rubra shoots derived from mature trees. Tree Physiol 16:673–680
Vasanth K, Vivier MA (2011) Optimisation of improved cryopreservation procedure for long term storage synchronised culture of grapevine (*Vitis vinifera* L.) Biol Plant 55:365–369
Vidal JR, Rama J, Taboada L et al (2009) Improved somatic embryogenesis of grapevine (*Vitis vinifera*) with focus on induction parameters and efficient plant regeneration. Plant Cell Tissue Organ Cult 96:85–94
Vivier MA, Pretorius IS (2000) Genetic improvement of grapevine: tailoring grape varieties for the third millennium – a review. S A J Enol Viticul 21:5–26
Vivier MA, Pretorius IS (2002) Genetically tailored grapevines for the wine industry. Trends Biotechnol 20:472–478
Wang Q, Li P, Hanani U (2005) Improvement of agrobacterium-mediated transformation efficiency and transgenic plant regeneration of Vitis vinifera L. by optimizing selection regimes and utilizing cryopreserved cell suspensions. Plant Sci 168:565–571
Zhu HX, Mills NJ, Knott JF (1997) Analysis of the high strain compression of open-cell foams. J Mech Phys Solids 45(11):1875–1904

Part VI
Soil Health

Chapter 16
Soil Security: A Key Role for Sustainable Food Productivity

Palaniswamy Thangavel and Ganapathi Sridevi

Abstract Soil health, soil quality, and soil security, all the three mainly focused on the status of the soil fertility that is essential for all living things in the terrestrial environment. Due to several anthropogenic and natural sources, soil degradation is one of the major constraints for agricultural productivity. Around 40% of the arable land is already degraded for various factors including urbanization and soil sealing, soil acidification, salinization, soil erosion, soil contamination, etc. Despite there is a link between soil quality and food productivity, the status of global food production has been updated regularly rather than the status of world soil resources. Both the UN Millennium Development Goals (MDGs) and Sustainable Development Goals (SDGs) emphasized to ensure the food security globally as we need to serve more than 9 billion people by 2050. Further, the soil acts as a carbon sink efficiently rather than aquatic ecosystems in many parts of the world, which will help to mitigate the climate change. Hence, the soil protection is of the utmost importance for all the securities especially water, food, and energy. Sustainable soil management and agricultural practices such as efficient water utilization, climate-smart agriculture, AeroFarms, and organic farming are the key aspects to ensure the food security globally.

16.1 Introduction

The theme for World Environment Day 2012 encompasses various aspects of human living, ranging from transport to energy to food to sustainable livelihood. Green technology, an eco-friendly clean technology, contributes to sustainable development to conserve the natural resources and environment which will meet

P. Thangavel (✉)
Department of Environmental Science, School of Life Sciences, Periyar University, Periyar Palkalai Nagar, Salem 636011, Tamil Nadu, India
e-mail: thanvel@gmail.com

G. Sridevi
Department of Plant Biotechnology, School of Biotechnology, Madurai Kamaraj University, Madurai 625021, Tamil Nadu, India

A. Dhanarajan (ed.), *Sustainable Agriculture towards Food Security*,
https://doi.org/10.1007/978-981-10-6647-4_16

the demands of the present and future generations. The implementation of the principles of sustainability is a world's challenge in managing a life-sustaining and environmentally sound global ecosystem. We are currently facing a lot of global environmental issues during the twenty-first century such as: (1) population of >7.0 billion with an increase at the rate of 1.14% year^{-1}, (2) per capita arable land area of 0.22 ha and decreasing to <0.07 ha for 30 countries by 2025, (3) soil degradation of 2 billion ha (Bha) and increasing at the rate of 5–10 million hectares (Mha) year^{-1}, (4) renewable freshwater supply of <1000 m^3 for 30 countries and increasing to 58 countries by 2050, (5) atmospheric CO_2 concentration of 378 ppm and increasing at 0.46% year^{-1}, (6) energy use of 435 quads (1015 BTU) year^{-1} and increasing at the rate of 2.2% year^{-1}, and (7) decrease in per capita grain consumption of 300 Kg year^{-1}. Most of these issues are mainly due to anthropogenic activities under Anthropocene era.

The world population has doubled at least ten times during the last 10,000 years, and it will project 11 billion people by 2100. Further, the future increase in population (expected to a combined population of 9 billion by 2150) will mostly occur in developing counties of Asia and Africa, and those regions are more prone to soil and water stresses. In addition, increasing standards of living and change in food habits of people from emerging economies especially China and India will remain a major challenge until 2050 and perhaps beyond. Agricultural land is occupying 40% of the land surface (Tilman et al. 2001, 2002) and 70% of all the water withdrawn from aquifers, streams, and lakes only for the purpose of agriculture (FAO 2011). Soil security, an overarching concept of soil motivated by sustainable development, is concerned with the maintenance and improvement of the global soil resource to produce food, fiber, and freshwater, to contribute to energy and climate sustainability, and to maintain the biodiversity and the overall protection of the ecosystem (Mc Bratney et al. 2014). In this review, soil security has been discussed toward sustainable food production integrated with soil health, hindering of soil fertility, and strategies of improving soil nutrients.

16.2 Food Productions and Demand in India

India remains the second largest producer of rice and wheat among the cereals and the top producer of pulses globally. The rice production is estimated to be 106.54 Mt, a new record and higher by 1.30 Mt than the production of rice during 2012–2013. Similarly, the wheat production is estimated to be 95.91 Mt which is a record and higher than the production of 93.51Mt during 2012–2013. There will be an increase in the food consumption levels in India from the current level of 2400 kcal/capita per day to about 3000 kcal/capita per day in 2050. Also, there will be a demand for cereals to 243 Mt in 2050 (Singh 2009). Similarly, there will be an increase in the rainfed crops to 1.8 t/ha in 2030 and 2.0 t/ha in 2050. During the above period, the irrigated crops are expected to increase from 3.5 to 4.6 t/ha. In India, the projected increase in the cereal production was 0.9% per year from 1999 to 2001 and is expected to exceed the demand by 2050.

16.2.1 Current Status of Agriculture in India (as on March 2017)

The food grain production is estimated to be 271.98 million tonnes in 2016–2017 as per the estimate by the Ministry of Agriculture (Second Advance Estimates of Production of Food Grains 2016–2017). The requirement of food grains to meet population demand is projected to be 300 million tonnes by 2025 (29th Report, Standing Committee on Agriculture, August 11 2016). There is a need for the crop output to grow at an annual average of 2%, which is closer to the present growth trend. In spite of all these high levels of grain production, India's agriculture yield is still lower than other larger grain-producing countries such as China, Brazil, and the USA. Further, many issues including yield reduction, population pressure, water supply, climate change, markets, and policies have coalesced to determine agricultural trends in recent decades (Alexandratos 1999; Hazell and Wood 2008; Foley et al. 2011; Alexandratos and Bruinsma 2012; Conway 2012; Ray et al. 2012).

16.3 Global Food Scarcity

Nearly 1.4 million children are at "imminent risk" of death in famines in Nigeria, Somalia, South Sudan, and Yemen according to recent report in UNICEF. Globally some 795 million people are chronically undernourished; 125 million under 5-year-olds (one out of four) are estimated to be stunted in spite of considerable progress. Also, 2.1 billion adults are overweight and obese, and many are deficient in key micronutrients, particularly iron, zinc, and vitamin A (Global Nutrition Report 2016).

16.4 Challenges for Food Productivity

As expected from the population explosion and changing in the food consumption pattern globally, we need to increase the agricultural areas as well as yield potential of staple food crops to meet the future projected demand of food that is challenging nowadays. The "Status of the World's Soil Resources" report is released by the UN's Food and Agriculture Organization (FAO 2015).

16.4.1 Land Crisis

Among the 13.2 billion ha of global land mass, 12% (1.6 billion ha) is currently in used for the cultivation of agricultural crops, 35% (4.6 billion ha) includes grasslands and woodland ecosystems, and 28% (3.7 billion ha) are covered with forests

(FAO 2011). The demand for new agricultural land (due to population pressure, diet change, and demand for biofuels) is expected to increase by about 50% by 2050. Utilization of tropical forests for agricultural purposes will lead to an increase in the extent of soil degradation. According to FAO (2015), the arable area in developing countries will have to increase by almost 13%, or 120 million ha, over the years from 1997–1999 to 2030 to meet the food demand.

16.4.2 Urbanization and Soil Sealing

The imminent population increase will be accompanied by rapid expansion of cities (urbanization). In the past 200 years, the proportion of the world's population living in cities has grown from about 5% to more than 50% (McMichael 2000). The urbanization process affecting soils was coined as "land take" and "soil sealing" (European Commission 2012). Increases in soil productivity will not compensate reductions in the extent of agricultural areas and thus will not be sufficient to buffer overall losses in crop yields (Gardi et al. 2015).

16.4.3 Water Scarcity

About 40% of the world's food depends on irrigation, which draws largely from stores of underground water, called aquifers, which make up 30% of the world's freshwater. The impact of agriculture expansion into marginal areas will also be felt through increased competition for irrigation water (Matuschke 2009).

16.4.4 Climate Change

Although several reasons have mentioned for the soil degradation, global climate change issues are the major threat to the agricultural system. Due to deforestation, transportation sectors, and industrialization especially emission from thermal power stations are the major sources of greenhouse gases released to the atmosphere. According to US National Oceanic and Atmospheric Administration (NOAA), the CO_2 concentration is 410.28 ppm as on April 25, 2017.

16.4.4.1 Erratic Rainfall Pattern

According to India Meteorological Department (IMD) data, India received normal rainfall as 2% less than the 100-year average by the end of August 2016 after two consecutive droughts; however, more than one third (221 out of 610 districts) of

India suffered deficit rainfall pattern. Also, the IMD data revealed that as much as 16% of the country's area is now rain deficient. The monsoon deficit was greater (between 30 and 40%) in northeast India which was repeating the situation in 2013.

16.4.4.2 Relation Between El Niño and Drought in India

The El Niño – characterized by surface waters of the equatorial Pacific warming up more than half a degree – is known to dry up monsoon rains every 6 out of 10 years. The most prominent droughts in India, eight of them, since 1871 have been El Niño-triggered droughts, including the recent ones that occurred in 2002, 2009, 2014, and 2015. Nevertheless, it is important to note that all El Niño years do not lead to drought in India.

16.4.4.3 Declining Agricultural Productivity

The warming trend in India over the past 100 years has indicated an increase of 0.6 °C, which is likely to impact many crops, negatively impacting food and livelihood security of millions of farmers.

Climate change could result in global crop yield losses as large as 5% in 2030 and 30% in 2080 (Hunger Report 2017). In 2015, 42.2 million Americans were food insecure, including 29.1 million adults and 13.1 million children (Hunger Report 2017).

16.4.4.4 Loss of Food Stability

Increased climate variability will decrease food stability especially in semiarid and subhumid areas such as sub-Saharan Africa, Southeast Asia, and parts of the Americas (Brown and Funk 2008; Schmidhuber and Tubiello 2007). Climate change will increase the occurrence of water- and food-borne diseases and especially increase the frequency of food and reef-fish poisoning (ciguatera and salmonellosis) in temperate and tropical regions, respectively (Schmidhuber and Tubiello 2007).

16.4.5 Soil Nutrient Deficiency/Nutrient Imbalances

Soil nutrient availability is the prevalent soil limitation in current cultivated land in most regions, particularly in tropical developing countries. The lower nutrient availability was evident in several countries including sub-Saharan Africa, Southern America, East Asia, Southeast Asia, and Australia and New Zealand. The highest nutrient constraint was reported in high-income countries (76%) as compared to low-income countries (68%). However, only 44% of cultivated soils (about 196 Mha) have no or only minor constraints in low-income countries. On the other hand,

24% of the soil (247 Mha) contains poor nutrient availability at different degrees ranged from light to severe. In addition, the natural fertility status of some soils has deteriorated over time through "nutrient mining" (FAO 2011). The human malnutrition is related to food scarcity, and the consumption of nutrient-deficient crops especially Fe, Mg, Zn, Cu, and Mn is directly attributable to nutrient impoverished soils (St Clair and Lynch 2010).

16.4.6 Modern Agronomic Practices

Due to increase in agricultural productivity, 70% of the agricultural land followed several intensification agricultural practices such as new varieties, irrigation, and the use of inputs, and the remaining 30% is the conversion of tropical forests into agricultural land (Conway 2012; Alexandratos 1999; Gibbs et al. 2010). The modern agricultural practices such as continual plowing of fields coupled with heavy input of fertilizer applications have degraded soils globally. Industrial farming contributes more greenhouse gas emissions than the transportation sector. Industrial agriculture also depends on massive phosphorus fertilizer application – another dead end on the horizon. Almost 75% of the world's reserve of phosphate rock, mined to supply industrial agriculture, is in a politically unstable area of northern Africa centered in Morocco and Western Sahara.

16.5 Soil Degradation

Approximately 40% of soil used for agriculture around the world is either degraded or seriously degraded – meaning, among other things, that 70% of the topsoil, the layer allowing plants to grow, is gone (WEF 2012). Soil degradation mainly occurs due to several anthropogenic activities such as overgrazing of farm animals (~35%), agricultural activities (~28%), deforestation (~30%), overexploitation of land to produce fuel wood (~7%), and industrialization (~ < 1%). The European Commission has also identified five threats classified as erosion, compaction, contamination, organic matter decline, salinization, landslides, and surface sealing. The mitigation of soil degradation is utmost important as the human beings are obtained more than 99.7% of food resources from land and less than 0.3% from the aquatic ecosystems. Soils are the second largest active store of carbon after the oceans; more carbon is stored in soil than in the atmosphere (760 billion tonnes) and in vegetation (560 billion tonnes) combined (Global Opportunity Report 2017). In this view, soil degradation increases the intensity of global warming scenarios as the capability of C sequestration in soil is less. Soil degradation such as erosion, fertility loss, salinity, acidification, soil carbon decline, and compaction have long been reported, and it directly and indirectly impacts food security through changes in soil functions.

These have detrimental consequences for agricultural productivity, provision of water, increased greenhouse gases, and loss of biodiversity (Koch et al. 2013).

16.5.1 Soil Erosion and Soil Compaction

Soil erosion is a form of land degradation leading to the removal of topsoil (typically the layer with the most effect on plant production and thus food production) by water and/or wind. Topsoil contains organic matter, provides micro- and macronutrients to plants, and is responsible for soil structural stability – the determinant factor for the provision of water to plants (Rojas et al. 2016). Thus, soil erosion affects soil health by reducing the thickness of the topsoil, altering soil properties, and depleting SOM and nutrients. Blanco and Lal (2010) identified Asia, Sahel, Central America, and Africa as the regions predominately affected by soil erosion. The authors also reported that 60% of the rural population in the African tropics and subtropics affected food insecurity due to 40% of soil erosion. Effective land management practices with sustainable utilization of soil resources are the only option to achieve the food security in these affected regions. The African losses cropland of about 100,000 km^2 (approximately the size of Iceland) due to soil erosion, which is the type of soil degradation that refers to ultimate soil losses in terms of topsoil and nutrients. If the scenario will persist in the same trajectory, the agricultural productivity will be lesser (about 30%) over the next 20–50 years. In addition to soil erosion, inappropriate land management may cause surface crusting and soil compaction, which hampers seed germination.

16.5.2 Soil Acidification

Soil acidification ($pH < 5.5$) is one of the major constraints for the agricultural productivity and nearly 50% of the arable land affected globally. The problem of soil acidification has been intensified in many agricultural systems (Sullivan et al. 2013; Martins et al. 2014; Li et al. 2016; Goulding 2016). Soil acidification is mainly caused by acid deposition (Lu et al. 2014), removal of farm products (Bolan et al. 1991), and application of ammonium-based fertilizers (Cai et al. 2014). Soil acidification can lead to increased toxicities of aluminum and manganese to crops and cause deficiency of phosphorus, base cations (calcium, potassium, and magnesium), and molybdenum (Fenn et al. 2006). The synergistic impact of nutrient deficiencies and increasing the availability of toxic metals can subsequently decrease soil quality and reduce yield potential of several crops.

16.5.3 Soil Salinization

Soil salinity is a severe abiotic stress caused primarily by an abundance of sodium chloride (NaCl), from both natural accumulations and from irrigation and crop evapotranspiration (Flowers and Flowers 2005). Saline soils are characterized by having an electrical conductivity higher than 4 dS m^{-1} (where 4 dS m^{-1} ~ 40 mM NaCl), which many crops are unable to tolerate (Qadir et al. 2000). Poor-quality irrigation water, overuse of fertilizers, and seawater intrusion onto land also contribute to salt accumulation in soil (Rengasamy 2006). More than 75 countries around the world are struggling with salinity problems (Alaghmand et al. 2016). Saline soils are distinguished by the large content of soluble salts, sodic soils with higher levels of sodium ions, and saline-sodic soils with an excess of salts and exchangeable sodium (Sastre-Conde et al. 2015). It is estimated that at least 20% of the irrigated lands in the world are affected by varying levels of salt (Qadir et al. 2008). Almost 40% of salt-affected soils in the world are saline and 60% are sodic (Qadir et al. 2006).

16.5.4 Soil Contamination

Heavy metals are ubiquitous environmental pollutant throughout the world. Atmospheric deposition (mainly from mining, smelting, and fly ash) and livestock manures are the main sources of trace elements contaminating arable soil (Luo et al. 2009). By obstructing the breakdown of soil organic matter and altering nutrient cycling, soil contamination is considered responsible for decreasing soil biodiversity and fertility and decreasing soil health (Edwards 2002).

According to the data provided by the European Environmental Agency (EEA 2014), total potentially contaminated sites in Europe are estimated to be more than 2.5 million, of which 340,000 are thought to be actually contaminated. Approximately one third of the high-risk sites have been positively identified as contaminated, and of these only 15% have so far been successfully remediated (EEA 2014).

Asian countries experience considerable contamination of agricultural soil and crops by trace elements, and this contamination is becoming a threat to human health and the long-term sustainability of food production in the contaminated areas. In China, approximately one fifth of China's total farmland (nearly 20 million ha) is contaminated by heavy metals which may result in a reduction of more than 10 million tonnes of food supplies each year (Wei and Chen 2001). Among the different trace elements contaminating Chinese agricultural soils, Cd is the biggest concern. Due to its high mobility in the soil (except in poorly drained soil where sulfides are present), it can be easily transferred to the food chain and so poses risks to human health. Arsenic is also naturally present in groundwater in many regions of Southeast Asia especially in Bangladesh and West Bengal in India. This represents a threat to agriculture, particularly in rice paddy fields where anaerobic conditions prevail

(Smedley 2003; Hugh and Ravenscroft 2009). Asia is also the largest contributor to the atmosphere of anthropogenic Hg, which originates from the chemical industry, from Hg mining and from gold mining (Li et al. 2009). All across Asia, areas under rapid economic development are experiencing moderate to severe contamination by heavy metals (Ng 2010).

16.6 Soil Securities: Prerequisite for Sustainable Food Security

Soil is one of the fundamental production factors determining food production stability, food nutrient quality, and yield quantity (FAO 2008). The decrease in soil quality at a pace of up to 100 times greater than the rate of soil formation. It takes around 500 years for just 2.5 cm of topsoil to be created amid unimpeded ecological changes. The measured average soil production rates are ca. 0.036 ± 0.04 mm $year^{-1}$ (Montgomery 2007) and are even lower on most agricultural soils because agricultural soils have a certain thickness and soil production rates decrease with increasing soil depth (Stockmann et al. 2013). Like atmosphere and oceans, soils of both the agricultural and nonagricultural lands are vital for humanity life support system. Hence, we need to ensure the soil quality is fundamental to fulfill other global essential needs including water and food.

Soil security is a new, wide-ranging, multidimensional, multidisciplinary concept that has antecedence in notions such as soil conservation, care, quality, health, and protection. Soil security as in McBratney et al. (2014) as being concerned with the maintenance and improvement of the world's soil resource to produce food, fiber, and freshwater contributes to energy and climate sustainability and maintains the biodiversity and the overall protection of the ecosystem. Based on the concept of soil security, the seven functions of the soil to (1) produce food and other biomass would be related to soil capability and soil condition, while soil capital would relate to (2) storing, filtering, and transformation and (3) the provision for a habitat and gene pool. The cultural environment for mankind (4) is related to soil connectivity and valued through the soil capital, where (6) acting as a carbon pool is related to soil condition and capital, and being an archive for archeological heritage (7) is covered by soil condition and its connectivity. Although described as a function, we would consider (5) source for raw materials, as a threat (McBratney et al. 2014).

In 2012, the Food and Agriculture Organization (FAO) of the United Nations established the "Global Soil Partnership" to highlight effective and concerted actions against soil degradation and to advocate healthy soils for ensuring food security globally. Food security has four dimensions: (1) food production and availability through agronomic management of soil resources, (2) stability of food production and availability at all times, (3) food access through economic and physical capacity of households or communities, and (4) food safety and utilization through nutritious and biological quality (FAO 2011; Schmidhuber and Tubiello 2007).

The success of food security concept is entirely dependent on availability, access, and utilization; these are the three pillars of food security. Essentially, food availability is referring to having sufficient quantity and a reliable source of food supply, and access focuses on having the resources to obtain high-quality and nutritious food (Pinstrup-Andersen 2009; Godfray et al. 2010). Food use describes having the knowledge of basic nutrition, as well as access to nonfood inputs of adequate water and sanitation or lack of contamination (World Health Organization 2012). Based on these three concerns, stimulated efforts have been taken to ensure food security by improving yield and quality, minimizing loss of productivity by land degradation, pollution, and urbanization, as well as the need for water supply and storage (Chen 2007; Godfray et al. 2010).

16.7 Sustainable Soil Management: A Way Toward Global Food Security

The world could produce up to 58% more food for restoring and protecting living soils through sustainable soil management. The soil is the habitat for the largest gene pool and diversity of species, and these organisms participate actively in soil processes that affect its formation and function (Lavelle et al. 2006). This gene pool will also contribute to the future development of products that sustain human health (Brevik and Burgess 2013). Soil microorganisms mainly contributed to the maintenance of the organic matter and energy transfer in terrestrial environment (Filip 2002) especially for contribution to nutrient and water efficiencies, improve soil structure, and protect against soilborne diseases (Brussaard et al. 2007).

The promotion of sustainable soil management should aim to: (a) manage soil erosion through soil conservation and soil rehabilitation/restoration, (b) increase or maintain soil organic matter content which constitutes the key factor in effective soil management and resilience, (c) limit soil sealing (land take) especially on the most suitable and productive agricultural soils, (d) enhance soil water storage through improved soil structure (i.e., increase soil organic matter content), (e) boost long-term soil fertility through integrated management approaches, (f) maintain and enhance soil biodiversity as a foundation for soil health, and (g) increase soil carbon sequestration. Ultimately, sustainable soil management is an important precondition to address the increasing pressures on soils posed by rapid global population growth.

The Global Soil Partnership (GSP), established by members of the FAO in 2012, aims to enhance synergies and to promote coordination and partnerships through numerous initiatives, actions, and actors worldwide aimed at promoting sustainable soil management. To achieve sustainable soil management and ensure food security, the established GSP identified five pillars of action: (1) adoption of soil protection and conservation strategies for sustainable food production; (2) encouragement of investment, technical cooperation, policy, education awareness, and extension in soil; (3) promotion of soil research and development on identified gaps and priorities;

(4) enhancement of the quantity and quality of soil data and information; and (5) harmonization of methods, measurements, and indicators for sustainable soil management (Montanarella and Vargas 2012).

The Government of India established All India Soil Survey Organization in 1956 with headquarters at the Indian Agricultural Research Institute with five Regional Soil Correlation Centres at Bangalore, Delhi, Kolkata, Jorhat, and Udaipur. Later in 1958, this scheme was integrated with the Land Use Planning Scheme of the Central Soil Conservation Board primarily to carry out detailed soil surveys in the catchment areas of major River Valley projects, with setting up the organization, "All India Soil and Land Use Survey." Recently, the Government of India launched the *Soil Health Card Scheme* in February 2015 to assess the soil quality parameters periodically. The objective of this scheme is to issue the soil cards to about 14 crore farmers spread all over India by 2017. The process of this scheme as the various soil testing laboratories in the country will carry out the testing of the soil samples, the results of which will be analyzed by the experts. The results are related to the strength and weaknesses of the soil. The experts also suggest methods to improve the soil quality. These results and suggestions are displayed in the soil health cards.

16.8 Sustainable Food Productivity

Sustainable agricultural practices include many farming practices that can both maintain crop production and improve soil quality, such as uses of organic fertilizers, no-till or minimum tillage, polyculture, and biological pest management. These approaches aimed at enhancing farming systems resilience and spreading the risks to agricultural practices through high levels of adaptation and mitigation approaches to climate change. These practices should increase soil health by improving soil water storage capacity and soil structure with organic matter, controlling soil erosion, managing soil organic matter (SOM) for soil carbon sequestration, and boosting nutrient management (FAO 2013).

16.8.1 Agroecology or Organic Farming

Organic farming, commonly understood as farming with no synthetic pesticides and fertilizers, is a key dimension of agroecology. Among the 1.24 million tonnes of organic production in the country, around 80,000 million is supplied from Sikkim. In India, the other leading states for organic farming practices are Kerala, Mizoram, and Arunachal Pradesh. Sustainable farming goes a step ahead as it provides environmental protection, biodiversity conservation, and better agricultural production with enhanced soil quality by crop rotation, intercropping, polyculture, covering crops, and mulching. According to the recent report on global data for organic farming worldwide by the Research Institute of Organic Agriculture (FiBL) and

IFOAM – Organics International, a total of 50.9 million hectares were organically managed at the end of 2015, representing a growth of 6.5 million hectares over 2014, the largest growth ever recorded. Australia is the country with the largest organic agricultural area (22.7 million hectares), followed by Argentina (3.1 million hectares), and the USA (2 million hectares). Further, 2.4 million organic producers were reported in 2015. India continues to be the country with the highest number of producers (585’200), followed by Ethiopia (203’602) and Mexico (200’039) (The World of Organic Agriculture 2017). Further, the report also highlighted that the countries with the largest share of organic agricultural land of their total farmland are Liechtenstein (30.2%), Austria (21.3%), and Sweden (16.9%). In eleven countries, 10% or more of all agricultural land is organic. Several authors (e.g., Pretty and Hine 2001; FAO 2002, 2008; Halberg et al. 2006; Badgley and Perfecto 2007; Badgley et al. 2007; El-Hage Scialabba 2007; Niggli et al. 2007, 2008) argued that organic agriculture could benefit developing countries because organic practices contribute considerably to increasing soil stability and resilience, an important factor in food supply stability, and also save water, another critical resource in many areas.

16.8.2 Climate-Smart Agriculture (CSA) or Climate Resilient Agriculture (CRA)

According to the Food and Agriculture Organization (FAO), climate-smart agriculture (CSA) is an integrated approach that addresses the interlinked challenges of climate change and food security with the objectives of (1) sustainably increasing productivity to support equitable increases in farm incomes; (2) adapting and building resilience of food production systems to climate change at multiple levels such as drought, flood, heat/cold wave, erratic rainfall pattern, pest outbreaks, and other threats caused by changing climate; and (3) reducing GHG emission from agriculture (including crops, livestock, and fisheries). In India, the estimated countrywide agricultural loss in 2030 is expected to be over $7 billion that will severely affect the income of at least 10% of the population. However, this could be reduced by 80%, if cost-effective climate resilient measures are implemented.

16.9 Ensuring Global Food Security: Role of MDGs and SDGs

> Food security exists when all people, at all times, have physical and economic access to sufficient, safe and nutritious food that meets their dietary needs and food preferences for an active and healthy life. (World Food Program 2009)

The above definition focuses on four dimensions of food security such as (1) availability of food, (2) accessibility (economically and physically), (3) utilization (the way it is used and assimilated by the human body), and (4) stability of these three dimensions. The countries China, Indonesia, India, Bangladesh, and East and West Africa (particularly Nigeria) have been identified and forecasted as food security hotspots in both 2005 and 2050 (Matuschke 2009). According to Global Hunger Index 2016 released by the International Food Policy Research Institute (IFPRI), India ranked 97 among 118 countries; 15% of Indian population is undernourished – lacking in adequate food intake, both in quantity and quality. The GHI is calculated by taking into account four key parameters: shares of undernourished population, wasted and stunted children aged under 5, and infant mortality rate of the same age group.

The Millennium Development Goals (MDGs) provided a framework to mobilize global action against hunger and poverty and other development objectives. The Asia-Pacific region has achieved the Millennium Development Goals' hunger target (MDG-1c) of halving the proportion (236 million) of undernourished people in 2015. However, this was not sufficient to meet the target set by the World Food Summit of halving the number of undernourished people by 2015 (FAO 2015).

16.10 Future Perspectives

Innovative strategies are needed to strengthen and to attain the sustainable soil management, which plays a crucial role in food security. These are:

1. A multidisciplinary approach and/or experts with extensive knowledge from geology, geography, environmental science, and agricultural science are involved to solve the various issues on soil and food security.
2. Initiative and reports to be updated regularly including the areas of various categories of soil degradation especially contaminated sites, and these informations will be helpful for remediation and rehabilitation of such contaminated sites.
3. Similar to Global Strategy for Plant Conservation, identification of different sustainable soil management and agricultural productivity strategies depending upon the site/land condition and other influential parameters for increased food productivity (e.g., a diversified production system, with soil conservation practices, incorporation of trees, and reduced energy use, provides both adaptation and mitigation benefits for more adaptation; monocrop with reduced tillage and reduced use of agrochemicals reduced GHG emissions for more mitigation).
4. Extensive utilization of multigene approach for various staple food crops rather than rice and wheat.
5. To find the possibility for many plants through AeroFarms as the productivity increased 75 times per square feet than the commercial farm.
6. Identification for alternative source of phosphatic fertilizers – possibility for exploration of mechanisms in several thousand species of AM fungi use to

extract phosphorus from the environment or how these processes work in degraded soils and in various crop types.

7. To ensure proper price for farmers products without any intermediates especially for smallholder farms, which use less than 2 ha of land; 83% of such farms are existed in Asia and sub-Saharan Africa.

Acknowledgments P.T. is highly thankful to Ministry of Environment, Forest and Climate Change (MoEFCC), New Delhi, for the financial support of the project (File No. 19-150/2014-RE).

References

29th Report: Impact of chemical fertilizers and pesticides on agriculture and allied sectors in the country, standing committee on agriculture, August 11, 2016

Alaghmand S, Beecham S, Woods JA et al (2016) Quantifying the impacts of artificial flooding as a salt interception measure on a river-floodplain interaction in a semi-arid saline floodplain. Environ Model Softw 79:167–183

Alexandratos N (1999) World food and agriculture: outlook for the medium and longer term. Proc Natl Acad Sci U S A 96:5908–5914

Alexandratos N, Bruinsma J (2012) World agriculture: towards 2030/2050: the 2012 revision. FAO, Rome. Available online: http://www.fao.org/docrep/016/ap106e/ap106e.pdf. Accessed 10 Oct 2015

Badgley C, Perfecto I (2007) Can organic agriculture feed the world? Ren Agric Food Syst 22:80–86

Badgley C, Moghtader J, Quintero E et al (2007) Organic agriculture and the global food supply. Ren Agric Food Syst 22:86–108

Blanco H, Lal R (2010) Soil erosion and food security. Principles of soil conservation and management. Springer, Dordrecht, pp 493–512

Bolan N, Hedley M, White R (1991) Processes of soil acidification during nitrogen cycling with emphasis on legume based pastures. Plant Soil 134:53–63

Brevik EC, Burgess LC (2013) Soils and human health. CRC Press, Boca Raton

Brown ME, Funk CC (2008) Food security under climate change. Science 319:580–581

Brussaard L, de Ruiter PC, Brown GG (2007) Soil biodiversity for agricultural sustainability. Agric Ecosyst Environ 121:233–244

Cai Z, Wang B, Xu M et al (2014) Nitrification and acidification from urea application in red soil (Ferralic Cambisol) after different long-term fertilization treatments. J Soils Sediment 14:1526–1536

Chen J (2007) Rapid urbanization in China: a real challenge to soil protection and food security. Catena 69(1):1–15

Conway G (2012) One billion hungry: can we feed the world. Cornell University Press, Ithaca

Edwards CA (2002) Assessing the effects of environmental pollutants on soil organisms, communities, processes and ecosystems. Eur J Soil Biol 38:225–231

El-Hage Scialabba N (2007) Organic agriculture and food security. International conference on agriculture and food security. FAO, Rome

European Commission (2012) Commission staff working document. Guidelines on best practice to limit, mitigate or compensate soil sealing, Brussels

European Environment Agency (EEA) (2014) Progress in Management of Contaminated Sites, European Environment Agency

FAO (2002) Organic Agriculture, Environment and Food Security. Environment and Natural Resources Service Sustainable Development Department
FAO (2008) The state of food insecurity in the developing world. FAO, Rome. http://www.fao.org/docrep/012/i0876e/i0876e00.HTM. Accessed 20 Jan 2010
FAO (2011) The state of the World's land and water resources for food and agriculture (SOLAW) – managing systems at risk; Food and Agriculture Organization of the United Nations: Rome, Italy; Earthscan: London, UK, 2011. Available online: http://www.fao.org/docrep/017/i1688e/i1688e.pdf. Accessed 15 Sept 2015
FAO (2013) State of Food and Agriculture (SOFA) 2013 – food systems for better nutrition. FAO, Rome
FAO (2015) The State of Food Insecurity in the World 2015: meeting the 2015 international hunger targets: taking stock of uneven progress. FAO, Rome, 8, 44. http://www.fao.org/3/a-i4646e.pdf
Fenn ME, Huntington T, McLaughlin S, Eagar C, Gomez A, Cook R (2006) Status of soil acidification in North America. J For Sci 52:3–13
Filip Z (2002) International approach to assessing soil quality by ecologically-related biological parameters. Agric Ecosyst Environ 88:169–174
Flowers TJ, Flowers SA (2005) Why does salinity pose such a difficult problem for plant breeders? Agric Water Manag 78:15–24
Foley JA, Ramankutty N, Brauman KA et al (2011) Solutions for a cultivated planet. Nature 478:337–342
Fourth Advance Estimates of Production of Food grains for 2015–16, Directorate of Economics and Statistics, Ministry of Agriculture, August 17, 2015
Gardi C, Panagos P, Van Liedekerke M et al (2015) Land take and food security: assessment of land take on the agricultural production in Europe. J Environ Plan Manag 58(5):898–912
Gibbs HK, Ruesch AS, Achard F et al (2010) Tropical forests were the primary sources of new agricultural land in the 1980s and 1990s. Proc Natl Acad Sci U S A 107(38):16732–16737
Global Nutrition Report (2016) IFPRI, Washington, DC: International Food Policy Research Institute, 2016
Global Opportunity Report (2017) First edition, DNV GL AS, Hovik, Oslo
Godfray C, Beddington J, Crute I et al (2010) Food security: the challenge of feeding nine billion people. Science 27:812–818
Goulding K (2016) Soil acidification and the importance of liming agricultural soils with particular reference to the United Kingdom. Soil Use Manag 32:390–399
GSP (2014) World soil day. 5 December. http://www.fao.org/globalsoilpartnership/world-soil-da/it/. Accessed 16 Dec 2014
Halberg N, Alroe HF, Knudsen MT et al (2006) Global development of organic agriculture: challenges and prospects. CABI Publishing
Hazell P, Wood S (2008) Drivers of change in global agriculture. Philos Trans R Soc Lond B 363:495–515
Hugh B, Ravenscroft P (2009) Arsenic in groundwater: a threat to sustainable agriculture in South and South-east Asia. Environ Int 35(3):647–654
Hunger Report (2017) Fragile Environments, Resilent Communities. Bread for the World Institute, Washington, DC
Koch A, McBratney AB, Adams M et al (2013) Soil security: solving the global soil crisis. Glob Policy J 4:434–441
Lavelle P, Decaens T, Auberts M et al (2006) Soil invertebrates and ecosystem services. Eur J Soil Biol 42:S3–S15
Li P, Feng X, Qiu G, Shang L et al (2009) Atmospheric mercury emission from artisanal mercury mining in Guizhou Province, Southwestern China. Atmos Environ 43:2247–2251
Li S, Li H, Yang C et al (2016) Rates of soil acidification in tea plantations and possible causes. Agric Ecosyst Environ 233:60–66
Lu X, Mao Q, Gilliam FS et al (2014) Nitrogen deposition contributes to soil acidification in tropical ecosystems. Glob Chang Biol 20:3790–3801

Luo Y, Wu L, Liu L et al (2009) Heavy metal contamination and remediation in Asian agricultural land. MARCO symposium 2009: challenges for agro-environmental research in monsoon Asia
Martins AP, de Andrade Costa SEV et al (2014) Soil acidification and basic cation use efficiency in an integrated no-till crop–livestock system under different grazing intensities. Agric Ecosyst Environ 195:18–28
Matuschke I (2009) Rapid urbanization and food security: using food density maps to identify future food security hotspots
McBratney A, Field DJ, Koch A (2014) The dimensions of soil security. Geoderma 213:203–213
McMichael AJ (2000) The urban environment and health in a world of increasing globalization: issues for developing countries. Bull World Health Organ 78(9):1117–1126
Montanarella L, Vargas R (2012) Global governance of soil resources as a necessary condition for sustainable development. CurrOpin Environ Sustain 4:559–564
Montgomery DR (2007) Soil erosion and agricultural sustainability. Proc Natl Acad Sci U S A 104:13268–13272
Ng H (2010) Regional assessment: soil contamination policy and practice in Asia. Asian Environmental Compliance and Enforcement Network (AECEN)
Niggli U, Earley J Ogorzalek K (2007) Organic agriculture and environmental stability of the food supply. International conference on Agriculture and Food Security. FAO, Rome
Niggli U, Slabe A, Schmid O et al (2008) Vision for an organic food and farming research agenda to 2025 organic knowledge for the future. IFOAM
Pinstrup-Andersen P (2009) Food security: definition and measurement. Food Sec 1(1):5–7
Pretty J, Hine R (2001) Reducing food poverty with sustainable agriculture: a summary of new evidence. Final report from the 'SAFE World' research project, University of Essex. http://www.essex.ac.uk/ces/esu/occasionalpapers/SAFE%20FINAL%20-%20Pages1-22.pdf. Accessed 12 Feb 2010
Qadir M, Ghafoor A, Murtaza G (2000) Amelioration strategies for saline soils: a review. Land Degrad Dev 11:501–521
Qadir M, Noble AD, Schubert S et al (2006) Sodicity-induced land degradation and its sustainable management: problems and prospects. Land Degrad Dev 17(6):661–676
Qadir M, Tubeileh A, Akhtar J et al (2008) Productivity enhancement of salt affected environments through crop diversification. Land Degrad Dev 19:429–453
Ray DK, Ramankutty N, Mueller ND et al (2012) Recent patterns of crop yield growth and stagnation. Nat Commun:1–19
Rengasamy P (2006) World salinization with emphasis on Australia. J Exp Bot 57:1017–1023
Rojas RV, Achouri M, Maroulis J et al (2016) Healthy soils: a prerequisite for sustainable food security. Environ Earth Sci 75:180
Sastre-Conde I, Carmen Lobo M, Icela Beltrán-Hernández R et al (2015) Remediation of saline soils by a two-step process: washing and amendment with sludge. Geoderma 247–248:140–150
Schmidhuber J, Tubiello FN (2007) Global food security under climate change. Proc Natl Acad Sci U S A 104:19703–19708
Second Advance Estimates of Production of Food Grains for 2016–17, Directorate of Economics and Statistics, Ministry of Agriculture and Farmers Welfare, February 15, 2017. http://eands.dacnet.nic.in/Advance_Estimate/2nd_Advance_Estimate_ENG.pdf
Singh RB (2009) Towards a food secure India and South Asia. Making hunger a history. http://www.apaari.org/wp-content/uploads/2009/08/towards-a-food-secure-india-makinghunger-history.pdf
Smedley PL (2003) Arsenic in groundwater – south and east Asia. In: Welch AH, Stollenwerk KG (eds) Arsenic in ground water. Kluwer Academic Publishers, Boston, pp 179–209
St Clair SB, Lynch JP (2010) The opening of pandora's box: climate change impacts on soil fertility and crop nutrition in developing countries. Plant Soil 335:101–115
Stockmann U, Adams M, Crawford JW et al (2013) The knowns, known unknowns and unknowns of sequestration of soil organic carbon. Agric Ecosyst Environ 164:80–99

Sullivan TJ, Lawrence GB, Bailey SW, McDonnell TC, Beier CM, Weathers K, McPherson G, Bishop DA (2013) Effects of acidic deposition and soil acidification on sugar maple trees in the Adirondack Mountains, New York. Environ Sci Technol 47:12687–12694

The World of Organic Agriculture (2017) Research Institute of Organic Agriculture (FiBL)

Tilman D, Fargione J, Wolff B et al (2001) Forecasting agriculturally driven global environmental change. Science 292:281–284

Tilman D, Cassman KG, Matson PA et al (2002) Agricultural sustainability and intensive production practices. Nature 418:671–677

Wei CY, Chen TB (2001) Hyperaccumulators and phytoremediation of heavy metal contaminated soil: a review of studies in China and abroad. Acta Ecol Sinica 21:1196–1203

World Economic Forum (2012) What if the world's soil runs out? WEF in collaboration with TIME talked to University of Sydney Professor John Crawford on December 14, 2012

World Food Program (2009) Emergency food security assessment handbook, 2nd edn. World Food Programme, Rome

World Health Organization (2012) Trade, foreign policy, diplomacy and health: food security. (http://www.who.int/trade/glossary/story028/en/. Accessed 25 Oct 2012

Chapter 17
Amelioration of Environmental Stress for Sustainable Crop Productivity

K. Jothimani, D. Arulbalachandran, and K. Yasmin

Abstract Provision of adequate and quality food supply is essential in human population in the world. However, present-day, crop production and future demand of food toward people are challenges due to the loss of production by various biotic and abiotic stresses. Environmental stresses play pivotal role in the production, survival, and reproductive biology of plants as well as crops. Crop plants are exposed and always impacted with harsh environmental stress which threatens to fulfill the demand of agricultural production for food to the world in the future. Inevitable environmental stress factors, such as irregular and insufficient rainfall, excessive temperatures, drought, salinity, alkalinity, heavy metal toxicity, acidity, and others, would limit the yield and productivity of cultivated crop plants. With the increasing world population, many efforts and innovative approaches have been taken in order to increase the agricultural production and to improve the global supply of food and reduce their demands. In this context, the sustainable production of crop plants by overcoming the environmental stresses has been reviewed and discussed various strategies exist to crop improvement for biotic and abiotic stress tolerance. Besides, in this chapter, the current advances to improve the yield were investigated in all the levels of analyses and focused on potential leads that employed in understanding the full mechanisms behind the plant's tolerance to the environmental stress.

Abbreviation

ABA	Abscisic acid
AFLP	Amplified fragment length polymorphism
CAPS	Cleavable amplified polymorphic sequences
Cd	Cadmium
CK	Cytokinin
GA	Gibberellin

K. Jothimani (✉) • D. Arulbalachandran • K. Yasmin
Division of Crop Stress and Molecular Biology, Department of Botany, School of Life Sciences, Periyar University, Salem 636 011, Tamil Nadu, India
e-mail: jophd14@gmail.com

A. Dhanarajan (ed.), *Sustainable Agriculture towards Food Security*,
https://doi.org/10.1007/978-981-10-6647-4_17

H_2O_2	Hydrogen peroxide
Hg	Mercury
IAA	Indole acetic acid
IBA	Indole butyric acid
JA	Jasmonic acid
K	Potassium
NaCl	Sodium chloride
O^{-2}	Superoxide radicals
Pb	Lead
RAPD	Random amplified polymorphic DNA
RFLP	Restriction fragment length polymorphism
ROS	Reactive oxygen species
RWC	Relative water content
SA	Salicylic acid
SSCP	Single-strand conformation polymorphism

17.1 Introduction

Plants are generally exposed to several forms of environmental stress because they are immobile. The environmental conditions are affected in the current situations, where an alternate option is needed and to be emphasized by increasing the crop productivity for the overwhelming populations and ensure the food security. Environmental conditions differ based on the biotic and abiotic stress like drought, salinity, cold and heavy metal stress and biotic (pathogen, herbivore) are playing a major role of losses to agricultural production worldwide (Wafaa et al. 2015). Abiotic stresses are ultimate to the agricultural yield losses due to diversity in the dramatic changes of environment which are unpredictable. The knowledge of the rising environmental impact has led to worldwide efforts to increase the agricultural production in such difficult environmental conditions (Lobell et al. 2008; Ortiz et al. 2008). The universal food arrangement has responded to the doubling of world population by more than doubling food production for the duration of the past five decades. Feeding such wealthier population poses significant challenges for food security and environmental sustainability in the near future (Abdullah 2006). Food and Agriculture Organization of the United Nations (FAO) reported that the global grain production and consumption will be projected to increase by approximately 70% by 2050 and to almost double in developing countries and considered to be begun by abiotic stresses reduced the 70% of yield losses in the worldwide (Turral et al. 2009).

17.2 Types of Environmental Stress

Due to immobility, plants cannot escape from sessile and environmental stress. They are always exposed to different abiotic stress factor without any defense. On the other hand, animals are moveable and escape through harsh conditions. Tolerance of plant refers to its capacity to survive and reproduce under environmental stresses.

The schematic flow of the biotic and abiotic stresses in the environment is depicted in Fig. 17.1.

17.2.1 Drought Stress

Drought is one of the major abiotic stresses constraining agricultural production. Accessibility of water is the most significant factor in the atmosphere which determines the production of commercial crops. Water is gradually required for human populations and major agricultural lands for irrigation, where the availability of water could create a greater impact to produce crop plants (Boyer 1982). Drought stress affects different parts of plant that leads to reduced plant growth rate, photosynthetic pigment, CO_2 concentration, seed growth, and changes in molecular metabolism in plant (Fig. 17.2).

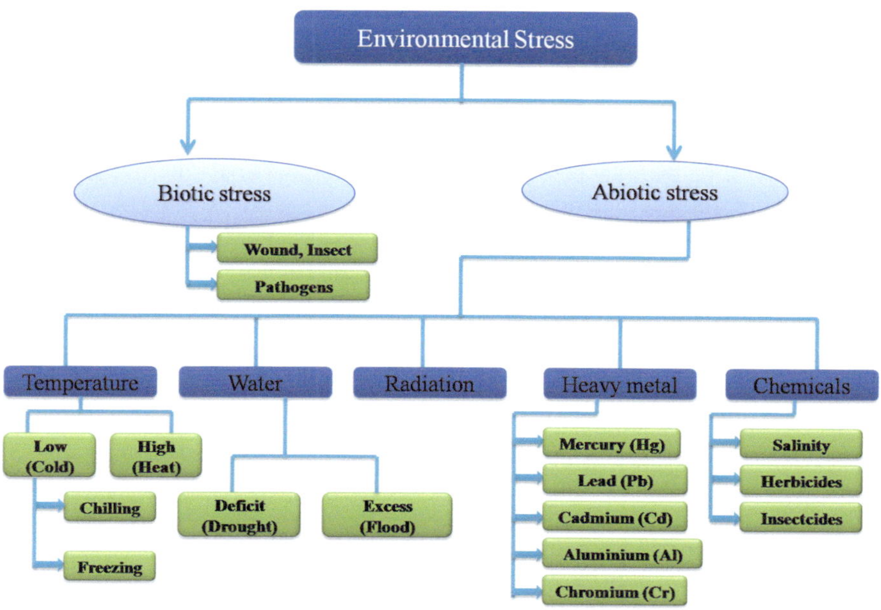

Fig. 17.1 Biotic and abiotic stress in the environment

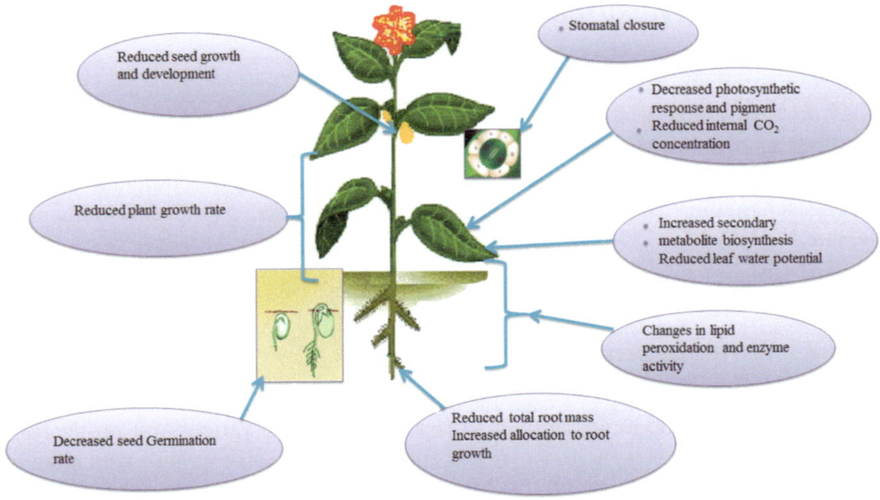

Fig. 17.2 Drought stress and plant growth

Nowadays, global water shortage is one of the major challenges in agriculture and food security caused by population growth and climate change. Environment changes, especially global warming caused by rising concentrations of atmospheric CO_2 produced by burning fossil fuels and deforestation, lead to increased aridity and desertification in several areas worldwide (Petit et al. 1999; Lu et al. 2007). Atmospheric CO_2 concentration has risen from ~260 parts per million (ppm) to ~380 ppm over the past 150 years indicated the Intergovernmental Panel on Climate Change (Griggs and Noguer 2002). The earth surface temperature has been increased by approximately 0.8 °C over the past 100 years, with about two-thirds of the increase occurring over the past three decades (Carnesale et al. 2011). Increase in every 1 °C of surface temperature, the global barley yield decreases up to 10% throughout the world (Adams et al. 1998).

Water deficit in the plant disrupts many cellular and molecular function of whole plant, having a negative impact on plant growth and reproduction. On average, 69% of crop yields are reduced by water deficit when plants are exposed to unfavorable conditions in the field (Boyer 1982). Water deficit is identified as the shortage of adequate moisture necessary for plants to grow and develop normally and complete their life period (Zhu 2002). The plant relative water content is one of the main physiological tolerances to plant directly related to soil water content (Sarker et al. 1999; Lawlor and Cornic 2002). Plant–water relations have some important characteristics that influence relative water content (RWC), leaf water potential, stomatal resistance, and the rate of transpiration leaf.

RWC is a measure of plant water status which reflecting the metabolic activity in tissues and used as a greatest expressive index for dehydration tolerance. In the early stages of leaf development, the RWC of leaves is higher and declines as the dry matter accumulates and leaf matures. The RWC is related to water uptake by the roots as well as water loss by transpiration. The RWC decreases in plant leaves are due to the response by drought stress (Nayyar and Gupta 2006). Water deficit directly affects the plant and reduces the potential for CO_2 fixation and chemical signaling and closed the stomata. Closure is caused by both chemical signaling and hydraulic effects by the latter being an adaptive function that increases transpiration efficiency (Davies et al. 2005).

17.2.2 Salinity Stress

The increasing human population of the twenty-first century is noticeable through the universal insufficiency of water resources, environmental pollution, and increased salinization of soil and water, and decrease in land available for cultivation is a threat to agricultural sustainability (Shahbaz and Ashraf 2013). Soil salinity is a universal abiotic stress that severely limits crop growth and development and productivity and causes the continuous loss of arable land, which results in desertification in arid and semiarid regions of the world, and soil salinization is a very harmful effect in semi-arid regions that decreases crop productivity (Pons et al. 2011).

Approximately, more than 970 million hectares and about 8% of the land in the world are adversely affected by high salinity (Munns and Tester 2008). Salinity stress induces toxicity as a consequence of the accumulation of ions' initial osmotic stress and subsequent matter. Consequece of excessive reactive oxygen species (ROS) such as superoxide radicals (O^{-2}), hydrogen peroxide (H_2O_2), and hydroxyl radicals (OH) produced at a high rate is usually accumulated in plant tissues due to ion imbalance and hyperosmotic stresses. ROS accumulation leads to lipid oxidation and has a harmful effect on cellular metabolism and physiology, therefore adversely ruining the membrane integrity (Munns James 2006).

Osmotic stress factors affect the first stage of reduced growth than after the next stage where toxic ions increase, specifically in leaves. Salt-induced water deficit leads to Ca^{2+} deficiency symptoms in fully expanded leaves which often show necrotic leaf tissue and their low nutrient availability (Fortmeier and Schubert 1995; Inal and Gunes 2008). Salt stress affects the seedling germination percentage, shoot and root length, fresh and dry weight, and many physiological parameters, and also salt stress reduced K^+ and Ca^{2+} level in stress condition (Lang et al. 2017). Water deficits in the soil environment arise due to the salt concentration around root surfaces and increase the osmotic pressure of root cells (Eapen et al. 2005).

The gene expression was reduced by the effects of salinity which arrests the cell cycle and transiently by the activity of cyclins and cyclin-dependent kinases. Such arrest results in fewer cells in the meristem, thus limiting the plant growth. The cyclin-dependent kinase activity was diminished also by posttranslational inhibition during salt stress. Recent reports showed that salinity harmfully affects the plant growth and development, hindering seed germination, seedling growth, as well as enzyme activity (Seckin et al. 2009).

17.2.3 *Heavy Metal Stress*

A series of metals and metalloids are harmful and toxic for plants and animals even though at lowest concentration. Heavy metals, such as As, Cd, Hg, Pb, or Se, are not essential heavy metals for plant growth, since they do not perform any identified physiological function in plants. Other known micronutrients such as Co, Cu, Fe, Mn, Mo, Ni, and Zn are essential elements, required for normal plant growth and metabolism. Heavy metal alters various physiological processes at cellular/molecular level by inactivating enzymes, blocking functional groups of metabolically important molecules, displacing or substituting for essential elements, and disrupting membrane integrity. Moreover, other common moment of heavy metal damaging is the enhanced production of reactive oxygen species (ROS) due to interference with electron transport activities in chloroplast membranes (Pagliano et al. 2006; La Rocca et al. 2009).

Heavy metal stress induces a range of potentials in plant physiological and biochemical syndromes, which in turn cause a reduced in crop productivity. This could be one of the main difficulties in crop production. Along with the naturally occurring elements, 53 are heavy metals, and the mass of these metals does not have any essential role in plants. Heavy metals contain some differences with respect to their biological significance and effects on crop productivity (Kavamura and Esposito 2010). Potentially, toxic and harmful effects by the heavy metals such as arsenic, cadmium, chromium, mercury, and lead have not been clearly studied in the biological function of plants (Peralta-Videa et al. 2009). The heavy metals encountered in the first organs of the plant are in roots that have been widely studied to assess the impact of a stressor. The heavy metal stress reduces the water and nutrient uptake capability of roots and nitrogen fixation (Poschenrieder and Barcelo 2004). Hence, the increased heavy metal in soil could always be decreased in both plant growth and yield (Keunen et al. 2011).

The seed germination is a very susceptible physiological phenomenon in the plant life cycle. Seed germination was being affected by various biotic and abiotic environmental factors and hormonal interactions as well as metals that hinder various developmental processes with flowering, embryogenesis, seed formation, and the changes in functioning of root and leaf; also in leguminous plants, the metal stress reduced the seed germination percentage (Moosavi et al. 2012; Guala et al. 2010).

17.2.3.1 Mercury Stress (Hg)

Heavy metal contamination is one of the most significant environmental difficulties worldwide. Among them, mercury (Hg) is one of the main toxic heavy metals commonly released into the environment in both developed and developing nations. The presence of extreme heavy metals in contaminated soil can't decrease the soil microbial activity, soil fertility, and heavy losses in agricultural yield production.

Their presence in the environment is highly dangerous, and also the major sources of contamination include mining of gold and silver, the coal industry, unprocessed discarded batteries, and industrial waste disposals (Du et al. 2005; Pilon-Smits and Pilon 2000). Oxidative stress is also induced by high amount of mercury ions, formation of reactive oxygen species (ROS) like superoxide radical (O^{2-}), singlet oxygen ($^1O_2$), hydroxyl radical (OH), and hydrogen peroxide (H_2O_2) in plant cells, which are responsible for peroxidase damages to fatty acids, nucleic acids, protein, and chlorophyll contents (Gallego et al. 2002).

17.2.3.2 Lead Stress (Pb)

Lead stress commonly affects the agroecosystem in much environmental stress, including heavy metal stress; lead toxicity decreased in plant development commonly in plant growth, changing physiological and biochemical metabolism (Ashraf et al. 2017). Lead is one of the highly toxic elements in the soil worldwide and is a bright silver metal. The reduction of toxic level of lead in soil is the consequence through the removal of municipal manure sludge, mining and smelting activities, lead-containing paints, paper and pulp, gasoline, and explosives, which exert an adverse effect on morphology, growth, and photosynthetic processes of plants. Also, high level of lead causes inhibition of enzyme activities, water imbalance, and alterations in membrane permeability, and among the heavy metals, lead is implicated to induce genotoxicity at DNA level in addition to the metal ions that tend to form covalent bonds with DNA sequence (Sharma and Dubey 2005; Enan 2006; Chaoui et al. 1997).

Lead toxicity is the main consequence of plants which rapidly inhibits root growth, probably due to the inhibition of cell division in the root tip (Eun et al. 2000). The high level of lead concentration induces increased oxidative stress by ROS production in plants (Reddy et al. 2005). More than 100 to 200,000 tons of lead for every year is being released from vehicle exhausts in US country. Some amount of lead is taken up by plants, fixation to soil, and flow into water bodies; hence human exposure of lead in the general population is either due to food or drinking water (Goyer 1990).

17.2.3.3 Cadmium Stress (Cd)

Cadmium, a divalent cation, is a most toxic heavy metal although not essential for plant growth and has no described biological function. It is readily taken up by roots, probably in competition with other bivalent ions (Kim et al. 2002). The high concentration of cadmium reduces the concentration of K, Ca, and Mg in the tissue which has been reported in cucumber and tomato plants (Burzynski 1988). The main effect of this metal observed in most plants till date is the inhibition of photosynthesis. This has been attributed to either an indirect action of Cd on plant–water relations, stomatal conductance, and CO_2 availability (Costa et al. 1994; Baryla et al. 2001) or to a more direct effect on chloroplast organization, chlorophyll biosynthesis, electron transport, and activity of Calvin–Benson–Bassham cycle enzymes (Stobart et al. 1985; Chugh and Sawhney 1999). The highest concentration of Cd, along with different plant organs and roots, in most cases, decreases in the following order: roots > leaves > grains or seeds. Cd affects the inhabits of photosynthesis and damages the photosynthetic apparatus, in particular the light-harvesting complex II and photosystems I and II, and cadmium metal stress can exert most of its effect on PSII (Wagner 1993; Krupa et al. 1993; Siedlecka et al. 1997).

17.2.4 Cold Stress

Plants are growing in varied environmental condition with their different life cycle. The climate condition (0–15 °C) is called chilling or cold, and less than (<0 °C) is called freezing. Cold is one of the harmful environmental problem for limiting the plant growth and productivity of geographical distribution of the crops (Zhu et al. 2007). Cold stress is usually a consequence in reduced germination, diminutive seedlings, yellowing of leaves, withering, and reduced tillering. The effects of cold stress on the reproductive period of plant delay heading and result in pollen sterility, which could act as a key factor responsible for the reduction in grain yield (Suzuki et al. 2008).

Temperature threshold for cold damage is depressed even in cold-sensitive crops by prior exposures to suboptimal low temperatures (Anderson et al. 1994). Seeing as the vapor pressure of ice is large amount lower than water at any given temperature, ice formation in the apoplast establishes a vapor pressure gradient between the apoplast and surrounding cells. The unfrozen cytoplasmic water migrates down the gradient from the cell cytosol to the apoplast. It allows contributing to the enlargement of existing ice crystals and causes a mechanical strain on the cell wall and plasma membrane leading to cell rupture (McKersie and Bowley 1997; Olien and Smith 1997).

17.2.5 Heat Stress

Plants are one of the sessile organisms, which do not move for any favorable environments (Lobell and Asner 2003; Lobell and Field 2007). Among the increasing temperature is considered as one of the most unfavorable stresses. The universal air temperature is predicted to increase through 0.2 °C per decade, which will lead to temperatures 1.8–4.0 °C higher than the current level by 2100 (IPPC- 2007).

The temperature is one of the main significance in the regulation of plant phenological development and plant growth (Bahuguna and Jagadish 2015). The high temperature disturbs normal growth and development of the plant; heat stress passes the optimum physiological parameters; therefore, very high temperatures have caused heat stress. The high heat stress varies the level of the plant kingdom, damaging the plant growth morphologically and physiologically and even within genotypes. Heat-induced damages contain changes instability of proteins, enzymes, nucleic acids, biomembranes, and cytoskeletal structures (Asthir 2015).

Heat stresses induce major effect in plant growth and yield; thus it may act directly or indirectly by changing through the environmental conditions (Lobell and Asner 2003; Lobell and Field 2007). Heat stress would induce huge effect in a crop plant during the stage of plant seed germination and seed growth, although the range of temperatures is largely in reduction of crop yields (Johkan et al. 2011; Kumar et al. 2011). Plant system photosynthesis is one of the important mechanisms in plant growth and development which is physiologically heat sensitive (Crafts-Brandner and Salvucci 2002). The C3 and C4 plants' photosynthetic capacity is always influenced in high-temperature condition (Yang et al. 2006).

Heat stress affects chloroplast in carbon metabolism of the stroma and photochemical reactions in thylakoid lamellae and also highly reduces the photosystem II (PSII) activity or terminated the activity (Wang et al. 2009a, b, c; Marchand et al. 2005; Morales et al. 2003); the amount of photosynthetic pigments (Marchand et al. 2005), germination rate of the wheat crop, and further the high heat stress cause cell death and embryo development (Cheng et al. 2009). Toward a high-temperature stress response by the plants reduces their height and biomass of rice cultivars (Mitra and Bhatia 2008). In common, the environmental heat increased by 1 °C might always result 4.1% to 10.0% of decreased cereals grain yield (Wang et al. 2012).

17.3 Stress Tolerant in Crop Plant

Tolerance of plant refers to the capability toward survival and reproduction under environmental stresses (Simms 2000). Plants adjust their physiology, metabolic mechanisms, gene expressions, and developmental activities to cope with the stress effects. Therefore, plant possesses unique and complicated mechanisms to tolerate stresses (Madhava et al. 2006). The general theme of stress is the development of reactive oxygen species at cellular and molecular level, strong oxidants that can do significant damage to membrane systems and DNA.

ROS include superoxide, hydrogen peroxide, and superhydroxide (Scandalios 1993). Antioxidative systems play an important role against stress, together with enzymatic (superoxide dismutases, catalase, peroxidase, phenol oxidase, and ascorbic acid oxidase) and nonenzymatic systems (compounds that are strong reductants such as glutathione, phenols, flavonoids, and polyamines), which are balancing and preventing oxidative damage in stress plant (Foyer et al. 1994). Not only the antioxidative systems produce stress resistant plants, but also some plant phytohormones play an important role in stress.

17.3.1 Role of Phytohormone in Improving Stress-Resistant Crop Production

Plant hormones play essential roles in the capability of plants to acclimatize to varying environments by mediating growth, development, and nutrient allocation. Several classes of small molecules are regulated and coordinated through the action of plant development, which may act close to or remote from their sites of synthesis to mediate genetically programmed developmental changes (Davies 2010). Through the action of these molecules, plants are able to adjust their physiology and biochemistry response to changes in their environment, a critical requirement for their survival as sessile organisms. Hormones have important role in the plant's response to environmental stress, from which the plant attempts to escape by outgrowing the stress (Franklin 2008).

Phytohormone through specific pathways to regulatory sites is responded to the stress at awfully low concentration, and plant hormones affect the biological activities either directly or indirectly. Through biotic or abiotic stress, plants produce increased amounts of phytohormones such as ABA and ethylene. Salicylic acid and jasmonic acid are concerned in some parts of addition, stress responses. These hormones may interact with plant stress tolerance by one another in regulating stress signaling. For example, ethylene has shown to enhance ABA in seeds (Gazzarrini and McCourt 2001). Phytohormone engineering is a method of selection and improves the productivity to combat various environmental stresses through novel and dynamic approaches (Sreenivasulu et al. 2012) (Fig. 17.3).

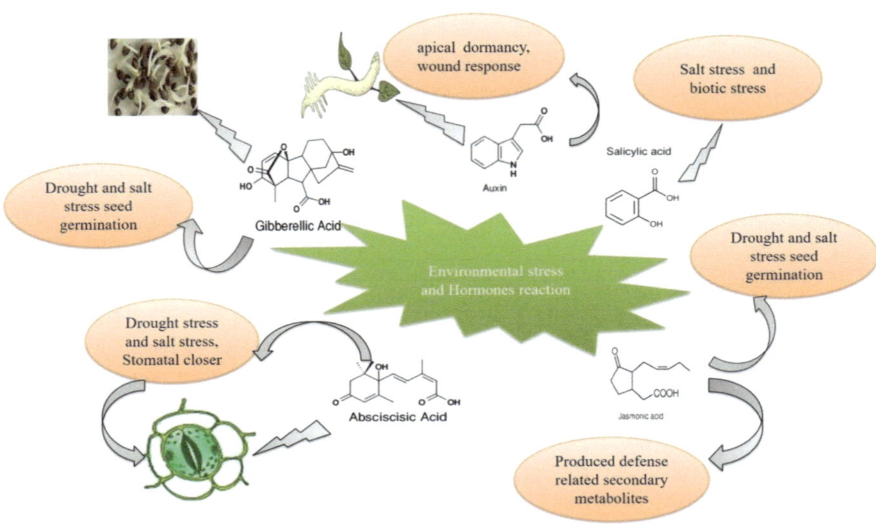

Fig. 17.3 Role of phytohormone on plants under environmental stress

17.3.1.1 Role of ABA in Environmental Stresses

ABA is called stress hormone, and it triggers the activation of a number of plants' physiological and developmental processes, thereby inducing adaptation to the stress conditions (Ton et al. 2009; Finkelstein et al. 2002). It is a significant phytohormone and plays a critical role in response to various stress signals. Role of ABA in abscission of plant leaves and the most studied phytohormone for its response to abiotic stresses. It is an isoprenoid plant hormone produced in the plastidial 2-C methyl-D-erythritol-4-phosphate pathway. ABA plays a vital role in plant physiological processes and developmental stages as well as seed dormancy and development, delay seed germination, stomatal closing, embryo morphogenesis, synthesis of storage proteins and lipids, leaf senescence, and also defense against pathogens (Swamy and Smith 1999; Sreenivasulu et al. 2010).

The abscisic acid has been mediated by abiotic stress such as drought and salinity, and its effect as an increase in root-to-shoot ratio, along with the regulation of stomatal closure, helps plants cope with water stress (Keskin et al. 2010; Taiz and Zeiger 2002). The ABA acts as an endogenous messenger during the plant's water regulation status (Swamy and Smith 1999). It also plays a significant role in regulating water status in the plant through guard cells and growth as well as by induction of genes that encode enzymes and other proteins involved in cellular dehydration tolerance (Luan 2002; Zhu 2002).

ABA-induced stomatal closure requires the commencement of the ABA signal transduction pathway in guard cells; in addition to that, ABA functions as a positive regulator of disease resistance at the preinvasive level. In contrast, defense hormone-triggered resistance was inhibited by ABA in the apoplastic space, and in contrast ABA has a negative effect on disease resistance at the post-invasive level. The understanding of ABA is being a key regulator of abiotic stress responses where been utilized for developing crops with enhanced tolerance under stress conditions. The major aim of studying plant stress responses is to develop crops with improved tolerance to abiotic stresses. Molecular treatment of ABA synthesis or signaling has been done in different crop plants for improved stress tolerance.

ABA-induced expression through an ABA-independent pathway often relies on the presence of cis-acting element called ABRE element (ABA-responsive element) (Thomashow 1999; Shinozaki and Yamaguchi-Shinozaki 2000; Uno et al. 2000). The 9-cis-epoxycarotenoid dioxygenase (NCED) genes and cytochrome P450 CYP707A genes encode key enzymes for ABA biosynthesis and ABA catabolism, respectively. The NCED3 gene is induced by drought stress, and it upregulates endogenous ABA levels in overexpressed transgenic plants, thereby leading to lower transpiration rates (Iuchi et al. 2001; Thompson et al. 2000). The main function of ABA involved the regulation of plant water balance and osmotic stress tolerance; a number of ABA-deficient mutants, namely, aba1, aba2, and aba3, have been reported for *Arabidopsis* (Koornneef et al. 1998). The regulation of ABA signaling in response to the environmental conditions is necessary for the modification of stomatal opening and closure while limiting transpirational water loss and/or pathogen invasion or while optimizing gaseous exchange for photosynthesis (Zheng et al. 2015; Wilson et al. 2009).

17.3.1.2 Cytokinin (CK)

Cytokinin is a plant growth hormone where it plays significant roles in many plant growth and developmental processes (Kang et al. 2012; Nishiyama et al. 2011). Cytokinin is originally discovered in 1950 by Carlos Miller on the basis of their ability to promote plant cell division that is N_6-substituted adenine-based molecules (Miller et al. 1955).

The chemical nature of cytokines is N_6-substituted purine derivatives. The higher plants found predominantly in cytokinin isopentenyladenine (iP), zeatin (Z), and dihydrozeatin (DZ). Plant morphogenesis was hypothesis coined by Skoog and Miller (Skoog and Miller 1957). The free bases and their ribosides (iP, Z, DZ) are considered to be the biologically active compounds. The key role of glycosidic conjugates was involved in cytokinin transport, protection from degradation, and reversible and irreversible inactivation (Letham 1994). The main response of endogenous level of cytokinin in alteration to stress indicates their involvement in abiotic stress (Brien and Benkova 2013), including drought (Kang et al. 2012) and salinity (Nishiyama et al. 2011).

The hypothesis predicted to cytokinin, and with auxin, plays a crucial role in plant morphogenesis, which influences the formation of roots and shoots with their relative growth (Miller et al. 1955). Cytokinin has found to induce delay senescence; therefore, it leads to better adaptation of plants by delaying drought-induced senescence (Bhargava and Sawant 2013; Chernyad'ev 2005).

The level of cytokinin in plants decreases during stress-induced leaf senescence, where the genes, cytokinin synthase, and adenosine phosphate isopentenyltransferase (IPT) involved are downregulated. Consequently, cytokinin oxidase genes are upregulated and recently, it directly involved in leaf senescence through cytokinin signaling (Lim et al. 2007). Drought-induced leaf senescence naturally coincides through decreased endogenous cytokinin concentrations. However, the hypothesis states that low cytokinin content which directly triggers leaf senescence may not be accurate since cytokinin-deficient mutants naturally display delayed chlorophyll deprivation compared to wild types. Therefore, suggesting other factors such as altered source–sink responses or antioxidant profiles could be responsible for the observed senescence during drought stress (Werner et al. 2003).

17.3.1.3 Role of Auxins in Stress

Auxins are one of the plant phytohormones generally distributed in many parts of the plant. This hormone involved in many signaling action growth and development pathways throughout the plants. Auxins have a lot of fundamental role in harmonization of many developments and behavioral processes such as root growth and elongation, apical dormancy, wound response, flowering, and fruit growth and development, in the plants' life cycle (Nemhauser et al. 2000; McSteen et al. 2007; Sorefan et al. 2009).

On artificially prepared different forms of auxin have been applied to regulate the plant growth, exogenously. The IAA is familiar as the key auxin in most of the plants (Woodward and Bartel 2005). The IAA also responds toward salinity during crop plants (Iqbal et al. 2014). Specified macro- and micronutrient elements are very important for plant growth, and it is not amazing that plant roots developed exclusive capabilities to sense and act in response to nutrients accessible in soil. Emerging evidence implicates that auxin is one of the major groups involved in this vital adaptive response, seeing as different plant species may respond in a different way to the lack or excess of a particular nutrient (Niu et al. 2012).

It was consequently anticipated with the intention under drought stress. IAR3 generates bioactive auxin which then stimulates lateral root development and contributes to survival under drought stress condition (Kinoshita et al. 2012). Auxin accumulation is altering, and its redistribution resulted high salt stress in root architecture (Petersson et al. 2009; Wang et al. 2009a, b, c). These reorganizations of auxin arrangement in plant tissues are correlated with reduced growth. In addition, some studies reported a major reduction in IAA concentrations in crop plants such as rice and tomato (Kazan 2013). The IAA reduced the salinity stress in level of 75%, in tomato plant (Dunlap and Binzel 1996).

17.3.1.4 Role of Jasmonate in Stress

Jasmonate is one of the plant hormones which acts as important signaling molecule for various developmental processes and resistance mechanism in plants (Kazan and Manners 2012). The biological relevancies of jasmonate in salt stress tolerance have verified and reported that the exogenous application of JA severely decreases sodium concentration (Kang et al. 2005). In addition, it has been observed that the plants are biological active in high salt stress by increased JA level (Moons et al. 1997b). Jasmonates include an essential role in the regulation of the biosynthesis of numerous secondary metabolites and antioxidants, including terpenoids, alkaloids, and phenylpropanoids. The constitutive commencement of the jasmonate-signaling pathway induced improved production of secondary metabolites in tomato plants. Methyl jasmonate (MeJA) is a metabolite derived from the polyamine and phenylpropanoid pathways; it has a putative role in the plant reproductive process (Chen et al. 2006). Exogenous application of MeJA promotes accumulation of caffeoyl putrescine in tomato leaves. The jasmonate pathway regulates the response to abiotic stress, herbivores, and necrotrophic fungal pathogens and regulates defenses also against insect and, amazingly, defenses against biotrophic pathogens such as the powdery mildews (Ellis and Turner 2001). Jasmonate could respond through the negative interaction between pathogens and wounding (Felton et al. 1999). In citrus plants, transient accumulation of JA was established, which is important for ABA increase that facilitates the continued existence of plants under severe drought conditions (Ollas et al. 2013). MeJA applied at a specific concentration showed a protective effect against Cu and Cd ions and prevented the inhibitory effect of heavy metals on chlorophyll accumulation and photosynthetic activity, indicating that MeJA influences heavy metal toxicity in *Arabidopsis* plants, recorded by Maksymiec and Krupa (2002).

17.3.1.5 Role of Gibberellic Acid in Environmental Stress

The action of several classes of plant hormones which plays an important role in plant development is regulated and coordinated through plant growth (Davies 2010). The gibberellic acid shows positive effects on seed germination, leaf development, elongation of stem, flower development and trichome initiation, and flower and fruit development (Yamaguchi 2008). The GA is a large group of tetracyclic diterpenoid carboxylic acids of which a very small number function as growth hormones in higher plants, the predominant bioactive forms being GA and GA (Sponsel and Hedden 2004). It is well conceived that rapid elongation of the stem is an adaptive response of plants to high ambient temperature, conversely, covered up biosynthesis of gibberellic acid and also compromises high temperature-induced hypocotyl elongation in *Arabidopsis* (Stavang et al. 2009). They are very important for plant life cycle growth-stimulatory functions and promote developmental stage transitions (Colebrook et al. 2014).

Gibberellic acid found to be supportive during enhanced wheat and rice development under saline conditions reported in (Parasher and Varma 1988; Prakash and Prathapasenan 1990). The development of wheat gets decreased with increasing salinity levels except being increased by seed treatment through GA3 (Kumar and Singh 1996). GA3 functions significantly and promotes plant length and fresh and dry weight. These biological functions were markedly hindered by NaCl-induced salt stress in soybean plants (Hamayun et al. 2010). Function of GA decreases stomatal resistance and enhanced the efficiency of plant that utilizes water at lesser salinity level in tomato plants (Maggio et al. 2010). The gibberellic acid signaling was necessary to adapt the plant to unfavorable environmental circumstances and helps to maintain source–sink relationship since salinity causes a decrease in submerged enzyme activities (Iqbal et al. 2011).

17.3.2 Antioxidant Defense Mechanisms in Plant Under Stress

Plants have different types of resistance mechanisms which reduce the levels of environmental stress, along with the tolerance capacity of each plant which mainly depends on the supportive function of all these different mechanisms that include the initiation of both enzymatic and nonenzymatic substances. The enzymatic mechanism includes superoxide dismutase, catalase, peroxidase, ascorbate peroxidase, and glutathione reductase. Nonenzymatic components contain cysteine, reduced glutathione, and ascorbic acid (Gong et al. 2005). During environmental stress tolerance, such as drought, high activities of antioxidant enzymes and high contents of nonenzymatic constituents are important.

There are a variety of antioxidant enzymes or lipid-soluble and water-soluble scavenging molecules removed by reactive oxygen species in plants, and antioxidant enzymes are being the most efficient against oxidative stress (Hasegawa et al. 2000; Farooq et al. 2008). Among enzymatic mechanisms, superoxide dismutase plays an important role and catalyzes the dismutation of two molecules of superoxide into O_2 and H_2O_2 (Lima et al. 2002). The half-life of superoxide radical has less than a second and is dismutated by superoxide dismutase into H_2O_2, a product that is relatively stable and can be detoxified by catalase and peroxidase (Apel and Hirt 2004). Carotenoids and other compounds, such as abietane diterpenes, in less concentration even though their ability to scavenge singlet oxygen and lipid peroxy radicals, under dehydrative forces (Deltoro et al. 1998).

Antioxidants are the first line of defense against free radical damage. They are critical for maintaining optimum health in plant cells. The antioxidant enzymes, peptides, and metabolites involved in the scavenging of active oxygen in plants and their activation are known to increase upon exposure to oxidative stress (Tanaka 1994). The reactive oxygen species (ROS) will be formed in unavoidable consequence of aerobic metabolism. ROS include free radicals such as superoxide anion ($O_2^{\bullet-}$), hydroxyl radical ($^{\bullet}OH$), as well as nonradical molecules like hydrogen peroxide (H_2O_2), singlet oxygen ($^1O_2$), and so forth. Stepwise reduction of molecular oxygen (O_2) by exposure to electron-transfer reactions leads to the production of highly reactive ROS. In plants, the ROS are always formed in mitochondria and plasma membranes or as a by-product of different metabolic pathways localized in different cellular compartments (Foyer and Harbinson 1994; Heyno et al. 2011).

In correspondence, a group of lipophilic antioxidants were involved in scavenging of oxygen free radicals and lipid peroxy radicals (α, β, γ, and δ tocopherols) (Diplock et al. 1989). The in vivo relative antioxidant activity of the tocopherol isomers ($\alpha > \beta > \gamma > \delta$) is due to the methylation pattern and the amount of methyl groups attached to the phenolic ring of the polar head structure (Fukuzawa et al. 1982). So, α-tocopherol with its three-methyl substituent has the highest antioxidant activity of tocopherols (Kamal-Eldin and Appelqvist 1996). Revision of arrangement and providence of ROS using advanced analytical techniques will help in developing a broader view of the role of ROS in plants. The biotechnological approaches to improve understanding of ROS will be helpful in producing plants with an inbuilt capacity of enhanced levels of tolerance.

17.3.3 Molecular Marker for Development of Stress Tolerance Crop

The emergence of new diseases and pests and changing of climate are the major issues that address the requirement for sustainable crop development and resistance to abiotic and biotic stresses (Hasan et al. 2015). The molecular techniques include restriction fragment length polymorphism (RFLP), amplified fragment length polymorphism (AFLP), random amplified polymorphic DNA (RAPD), cleavable amplified polymorphic sequences (CAPS), and single-strand conformation polymorphism (SSCP). Microsatellite or simple sequence repeat polymorphism (SSRP) has been employed for the development of stress-tolerant crop. Genetic maps have been made in many crop plants consuming these markers on a single separating population. Plant's heavy metal stress induced genomic DNA alterations, and the random amplified polymorphic DNA (RAPD) assay helps to detect genomic DNA alterations in plant (Ahmad et al. 2012).

17.3.4 Transcriptomic Analysis of Stress Tolerance in Crops

Environmental stress is a major problem worldwide because of global climate change. Several studies have instituted the environmental stress response in crop plant. Among the transcription factors which regulate the activation or deactivation of other genes involved in upstream signal transduction is the role of transcription factor in abiotic stress responses. The transcriptomic being extensively studied in plant, responses to biotic, abiotic, and various stresses. The complete set of transcripts in a cell is christened transcriptome and their quantity, intended for a specific developmental stage or physiological condition. Transcriptomes provide a clear understanding of efficient elements in genome and reveal the molecular constituents of cells and tissues (Wang et al. 2009a, b, c). Microarrays are the first available method for genome-wide transcript expression profiling and have been used widely in plant biology. Microarray analysis and RNA sequencing are induced transcriptome analysis. Microarrays have been extensively used to generate transcriptional profiles in response to abiotic stresses (Kawasaki et al. 2001; Mun et al. 2017).

17.4 Conclusion

In the increasing human population in the world today, the requirement for agricultural yield will increase rapidly. Several of dramatic changes have been induced by the environmental stress in crop plant due to adapted different geographical nature. Mostly the abiotic stress salinity, drought, heavy metal, heat, cold, and flooding are ultimately affecting the plant growth and limiting the crop production. An urgent need for several developments in agriculture to feed the people's demand is in challenge. Food scarcity and productivity are indispensable thrust areas for researchers to make sustain the production of crops to fulfill the demand of food crops toward forthcoming world population.

References

Abdullah K (2006) Use of water and land for food security and environmental sustainability. Irrig Drain 55:19–222

Adams RM, Hurd BH, Lenhart S, Leary N (1998) Effects of global climate change on agriculture: an interpretative review. Clim Res 11:19–30

Ahmad MA, Guar R, Gupta M (2012) Comparative biochemical and RAPD analysis in two verities of rice (Oryza sativa L.) under arsenic stress by using various biomarkers. J Hazard Mat 217–218:141–148

Anderson MD et al (1994) Differential gene expression in chilling-acclimated maize seedlings and evidence for the involvement of abscisic acid in chilling tolerance. Ann Rev Plant Physiol 105:331–339

Apel K, Hirt H (2004) Reactive oxygen species: metabolism, oxidative stress, and signal transduction. Ann Rev Plant Biol 55:373–399

Ashraf U, Hussain S et al (2017) Alterations in growth, oxidative damage, and metal uptake of five aromatic rice cultivars under lead toxicity. Plant Physiol Biochem 115:461–471

Asthir B (2015) Mechanisms of heat tolerance in crop plants. Biol Plant 59(4):620–628

Bahuguna RN, Jagadish KSV (2015) Temperature regulation of plant phonological development. Environ Exp Bot 111:83–90

Baryla A, Carrier P, Franck F et al (2001) Leaf chlorosis in oilseed rape plants (Brassica napus) grown on cadmium-polluted soil: causes and consequences for photosynthesis and growth. Planta 212:696–709

Bhargava S, Sawant K (2013) Drought stress adaptation metabolic adjustment and regulation of gene expression. Plant Breed 32:21–32

Boyer JS (1982) Plant productivity and environment. Science 218(4571):443–448

Brien JAO, Benkova E (2013) Cytokinin cross-talking during biotic and abiotic stress responses. Front Plant Sci 4:451

Burzynski M (1988) The uptake and accumulation of phosphorus and nitrates and the activity of nitrate reductase in cucumber seedlings treated with Pb and Cd. Acta Soc Bot Pol 57:349–359

Carnesale A, Chameides W, Boesch DF et al (2011) America's climate choices. National Academic Press, Washington, DC, pp 14585–14585. ISBN:978-0-309

Chaoui A, Mazhoudi S, Ghorbal MN, Ferjani EE (1997) Cadmium and zinc induction of lipid peroxidation and effects in antioxidant enzyme activities in bean (Phaseolus vulgaris L.) Plant Sci 127:139–147

Chen H, Jones AD, Howe GA (2006) Constitutive activation of the jasmonate signaling pathway enhances the production of secondary metabolites in tomato. FEBS Lett 580:2540–2546

Cheng L, Zou Y, Ding S et al (2009) Polyamine accumulation in transgenic tomato enhances the tolerance to high temperature stress. J Inte Plant Biol 51:489–499

Chernyad'ev II (2005) Effect of water stress on the photosynthetic apparatus of plants and the protective role of cytokinins: a review. App Biochem Microbiol 41:115–128

Chugh LK, Sawhney SK (1999) Photosynthetic activities of Pisum sativum seedlings grown in presence of cadmium. Plant Physio Biochem 37:297–303

Colebrook EH, Thomas SG, Phillips AL et al (2014) The role of gibberellin signaling in plant responses to abiotic stress. J Exp Biol 217:67–75

Costa G, Michaut JC, Morel JL (1994) Influence of cadmium on water relations and gas exchanges. In: Phosphorus deficient Lupinus albus, plant Physio Bioche 32, pp 105–114

Crafts-Brandner SJ, Salvucci ME (2002) Sensitivity of photosynthesis in a C4 plant, maize, to heat stress. Plant Physio 129:1773–1780

Davies PJ (2010) The plant hormones: their nature, occurrence, and function. In: Plant hormones: biosynthesis, signal transduction, action! (ed. P. J. Davies), 115. Springer, Dordrecht

Davies WJ, Kudoyarova G, Hartung W (2005) Long-distance ABA signaling and its relation to other signaling pathways in the detection of soil drying and the mediation of the plant's response to drought. J Plant Grow Regu 24:285–295

De Ollas C, Hernand B, Arbona V et al (2013) Jasmonic acid transient accumulation is needed for abscisic acid increase in citrus roots under drought stress conditions. Physiol Plant 147:296–306
Deltoro VI, Calatayud A, Gimeno C et al (1998) Changes in chlorophyll a fluorescence, photosynthetic CO2 assimilation and xanthophyll cycle interconversions during dehydration in desiccation-tolerant and intolerant liverworts. Planta 207:224–228
Diplock T, Machlin LJ, Packer L et al (1989) Vitamin E: biochemistry and health implications. Ann N Y Acad Sci 570:372–378
Du X, Zhu YG, Liu WJ et al (2005) Uptake of mercury (Hg) by seedlings of rice (Oryza sativa L.) grown in solution culture and interactions with arsenate uptake. Envi Exp Bot 54:1–7
Dunlap J, Binzel M (1996) NaCl reduces indole-3-acetic acid levels in the roots of tomato plants independent of stress induced abscisic acid. Plant Physio 112:379−384
Eapen D, Barroso ML, Ponce G et al (2005) Hydrotropism, root growth responses to water. Tren Plant Sci 10:44–50
Ellis C, Turner JG (2001) The Arabidopsis mutant cev1 has constitutively active jasmonate and ethylene signal pathways and enhanced resistance to pathogens. Plant Cell 13:1025–1033
Enan MR (2006) Application of random amplified polymorphic DNA (RAPD) to detect the genotoxic effect of heavy metals. Appl Biochem Biotechnol 43:147–154
Eun SO, Youn HS, Lee Y (2000) Lead disturbs microtubule organization in the root meristem of Zea mays. Physiol Plant 110:357–365
Farooq M, Aziz T, Basra SMA et al (2008) Chilling tolerance in hybrid maize induced by seed priming with salicylic acid. J Agro Crop Sci 194:161–168
Felton GW, Kort KL, Bi JL et al (1999) Inverse relationship between systemic resistance of plants to microorganisms and to insect herbivory. Curr Biol 9:317–320
Finkelstein RR, Gampala SS, Rock CD (2002) Abscisic acid signaling in seeds and seedlings. Plant Cell 14:15–45
Fortmeier R, Schubert S (1995) Salt tolerance of maize (Zea mays L.), the role of sodium exclusion. Plant Cell Environ 18:1041–1047
Foyer CH, Harbinson J (1994) Oxygen metabolism and the regulation of photosynthetic electron transport. In: Foyerand CH, Mullineaux P (eds) Causes of photooxidative stresses and amelioration of defense systems in plants. CRC Press, Boca Raton, pp 1–42
Foyer CH, Lelandais M, Kunert KJ (1994) Photooxidative stress in plants. Physiol Plant 92:696–717
Franklin KA (2008) Shade avoidance. New Phytol 179:930–944
Fukuzawa K, Tokumura A, Ouchi SH et al (1982) Antioxidant activities of tocopherols on Fe2+–ascorbate-induced lipid peroxidation in lecithin liposomes. Lipids 17(7):511–513
Gallego S, Benavides M, Tomarow A (2002) Involvement of an antioxidant defense system in the adaptive response to heavy metal ions in Helianthus annus L. Cells Plant Grow Regul 36:267–273
Gazzarrini S, McCourt P (2001) Genetic interaction between ABA, ethylene and sugar signaling pathways. Curr Opin Plant Biol 4:387–391
Gong H, Zhu X, Chen K et al (2005) Silicon alleviates oxidative damage of wheat plants in pots under drought. Plant Sci 169:313–321
Goyer RA (1990) Lead toxicity: from overt to subclinical to subtle health effects. Environ Health Perspect 86:177
Griggs DJ, Noguer M (2002) Climate change 2001: the scientific basis. Contribution of working group I to the third assessment report of the intergovernmental panel on climate change. Weather 57:267–269
Guala SD, Veg FA, Covelo EF (2010) The dynamics of heavy metals in plant–soil interactions. Ecol Model 22:1148–1152
Hamayun M et al (2010) Growth promotion of cucumber by pure cultures of gibberellin-producing Phoma sp. GAH7. World J Microbiol Biotechnol 26(5):889–894
Hasan MM, Rafi MY, Ismail MR et al (2015) Marker-assisted backcrossing: a useful method for rice improvement. Biotech Biotechnol Equip 29:237–254
Hasegawa PM, Bressan RA, Zhu JK et al (2000) Plant cellular and molecular responses to high salinity. Annu Rev Plant Physiol 51:463–499

Heyno E, Mary V, Schopfer P et al (2011) Oxy-gene activation at the plasma membrane: relation between superoxide and hydroxyl radical production by isolated membranes. Planta 234(1):35–45

Inal A, Gunes A (2008) Interspecific root interactions and rhizosphere on salt ions and nutrient uptake between mixed grown peanut/maize and peanut/barley in original saline- sodic-boron toxic soil. Plant Physiol 165:490–503

Iqbal N, Nazar R, Iqbal MRK et al (2011) Role of gibberellins in regulation of source sink relations under optimal and limiting environmental conditions. Curr Sci 100:998–1007

Iqbal N, Umar S, Khan NA et al (2014) A new perspective of phytohormones in. Botany 100:34–42

Iuchi S, Kobayashi M, Taji T et al (2001) Regulation of drought tolerance by gene manipulation of 9-cis-epoxycarotenoid dioxygenase, a key enzyme in abscisic acid biosynthesis in Arabidopsis. Plant J 27:325–333

Johkan M, Oda M, Maruo T et al (2011) Crop production and global warming. In: Casalegno S (ed) Global warming impacts—Case studies on the economy, human health, and on urban and natural environments. InTech, Rijeka, pp 139–152

Kamal-Eldin A, Appelqvist LA (1996) The chemistry and antioxidant properties of tocopherols and tocotrienols. Lipids 31(7):671–701

Kang D, Seo Y, Lee JD et al (2005) Jasmonic acid differentially affects growth, ion uptake and abscisic acid concentration in salt-tolerant and salt sensitive rice cultivars. J Agro Crop Sci 191:273–282

Kang NY, Cho C, Kim NY et al (2012) Cytokinin receptor-dependent and receptor-independent path ways in the dehydration response of Arabidopsis thaliana. Plant Physio 169:1382 1391

Kavamura VN, Esposito E (2010) Biotechnological strategies applied to the decontamination of soils polluted with heavy metals. Biotech Adv 28:61–69

Kawasaki S, Borchert C, Deyholos M et al (2001) Gene expression profiles during the initial phase of salt stress in rice. Plant Cell 13(4):889–906

Kazan K (2013) Auxin and the integration of environmental signals into plant root development. Ann Bot –London 112:1655–1665

Kazan K, Manners JM (2012) JAZ repressors and the orchestration of phytohormone crosstalk. Tren Plant Sci 17:22–31

Keskin BC, Sarikaya AT, Yuksel B et al (2010) Abscisic acid regulated gene expression in bread wheat. Aus. J Crop Sci 4:617–625

Keunen E, Remans T, Bohler S et al (2011) Metalinducedoxidative stress and plant mitochondria. Int J Mol Sci 12:6894–6918

Kim YY, Yang YY, Lee Y (2002) Pb and Cd uptake in rice roots. Physiol Plant 116:368–372

Kinoshita N, Wang H, Kasahara H et al (2012) A-Ala Resistant3, an evolutionarily conserved target of miR167, mediates Arabidopsis root architecture changes during high osmotic stress. Plant Cell 24:3590–3602

Koornneef M, Leon Kloosterziel KM, Schwartz SH et al (1998) The genetic and molecular dissection of abscisic acid biosynthesis and signal transduction in Arabidopsis. Plant Physiol Biochem 36:83–89

Krupa Z, Quist G, Hurner NPA (1993) The effect of cadmium on photosynthesis of Phaseolus vulgaris – A fluorescence analysis. Physiol Plant 88:626–630

Kumar B, Singh B (1996) Effect of plant hormones on growth and yield of wheat irrigated with saline water. Ann Agri Sci 17:209–212

Kumar S, Kaur R, Kaur N et al (2011) Heat-stress induced inhibition in growth and chlorosis in mung bean (Phaseolus aureus Roxb.) is partly mitigated by ascorbic acid application and is related to reduction in oxidative stress. Acta Physiol Plant 33:2091–2101

La Rocca N, Andreoli C, Giacometti GM et al (2009) Responses of the Antarctic microalga Koliellaantartica (Trebouxiophyceae, Chlorophyta) to cadmium contamination. Photosynt 47:471–479

Lang L et al (2017) Quantitative trait locus mapping of salt tolerance and identification of salt-tolerant genes in *Brassica napus* L. Front Plant Sci 8

Lawlor DW, Cornic G (2002) Photosynthetic carbon assimilation and associated metabolism in relation to water deficits in higher plants. Plant Cell Environ 25:275–294

Letham S (1994) Cytokinins: Chemistry, activity and function Mok DWS. Mok MC, (eds). University of Florida Research Center, Boca Raton, pp 57–80

Lim PO, Kim HJ, Nam HG (2007) Leaf senescence. Ann Rev Plant Biol 58:15–36

Lima ALS, DaMatta FM, Pinheiro HA et al (2002) Photochemical responses and oxidative stress in two clones of Coffea canephora under water deficit conditions. Environ Exp Bot 47:239–247

Lobell DB, Asner GP (2003) Climate and management contributions to recent trends in U.S. agricultural yields. Science 299

Lobell DB, Field CB (2007) Global scale climate–crop yield relationships and the impacts of recent warming. Environ Res Lett 2(1):014002

Lobell DB, Burke MB, Tebaldi C et al (2008) Prioritizing climate change adaptation needs for food security in 2030. Science 319:607–610

Lu J, Vecchi GA, Reichler T (2007) Expansion of the Hadley cell under global warming. Geophys Res Lett:34–L06805

Luan S (2002) Signalling drought in guard cells. Plant Cell Environ 25:229–237

Madhava Rao KV, Raghavendra AS, Janardhan Reddy K (2006) Physiology and molecular biology of stress tolerance in plants. Spring:1–14

Maggio A, Barbieri G, Raimondi G, De Pascale S (2010) Contrasting effects of GA treatments on tomato plants exposed to increasing salinity. J Plant Growth Regul 29:63–72

Maksymiec W, Krupa Z (2002) Jasmonic acid and heavy metals in Arabidopsis plants-a similar physiological response to both stressors. J Plant Physiol 159:509–515

Marchand FL, Mertens S, Kockelbergh F et al (2005) Performance of high arctic tundra plants improved during but deteriorated after exposure to a simulated extreme temperature event. Glob Change Biol 11:2078–2089

McKersie BD, Bowley SR (1997) Active oxygen and freezing tolerance in transgenic plants. Plant Cold Hardiness, Molecular Biology, Biochemistry and Physiology, Plenum, New York:203–214

McSteen P, Malcomber S, Skirpan A et al (2007) Barren inflorescence2 encodes a co-ortholog of the PINOID erine/threonine kinase and is required for organogenesis during inflorescence and vegetative development in maize. Plant Physiol 144:1000–1011

Miller CO, Skoog F, Von Saltza MH, Strong FM (1955) Kinetin, a cell division factor from deoxyribonucleic acid1. J Am Chem Soc 77(5):1392–1392

Mitra R, Bhatia CR (2008) Bioenergetic cost of heat tolerance in wheat crop. Curr Sci 94:1049–1053

Moons A, Prinsen E, Bauw G et al (1997) Antagonistic effects of abscisic acid and jasmonates on salt stress-inducible transcripts in rice roots. Plant Cell 9:2243–2259

Moosavi SE, Gharineh MH, Afshari RT et al (2012) Efects of some heavy metals on seed germination characteristics of canola (Brassica napus), wheat (Triticum aestivum) and safflower (Carthamus tinctorious) to evaluate phytoremediation potential of these crops. J Agri Sci 4:1–19

Morales D, Rodriguez P, Dellamico J et al (2003) High-temperature preconditioning and thermal shock imposition affects water relations, gas exchange and root hydraulic conductivity in tomato. Biol Plant 47:203–208

Mun BG, Lee SU, Park EJ et al (2017) Analysis of transcription factors among differentially expressed genes induced by drought stress in Populus davidiana. 3 Biotech 7(3):209

Munns R, James LA (2006) Approaches to increasing the salt tolerance of wheat and other cereals. J Exp Bot 57:1025–1043

Munns R, Tester M (2008) Mechanisms of salinity tolerance. Ann Rev Plant Biolo 59:651–681

Nayyar H, Gupta D (2006) Differential sensitivity of C3 and C4 plants to water deficit stress: association with oxidative stress and antioxidants. Environ Exp Bot 58:106–113

Nemhauser JL, Feldman LJ, Zambryski PC (2000) Auxin and ETTIN in Arabidopsis gynoecium morphogenesis. Develop 127:3877–3888

Nishiyama R, Watanabe Y, Fujita Y et al (2011) Analysis of cytokinin mutants and regulation of cytokinin metabolic genes reveals important regulatory roles of cytokinin's in drought, salt and abscisic acid responses, and abscisic acid biosynthesis. Plant Cell 23:2169–2183

Niu YF, Chai RS, Jin GL et al (2012) Responses of root architecture development to low phosphorus availability: a review. Ann Bot 112:391–408
Olien CR, Smith MN (1997) Ice adhesions in relation to freeze stress. Plant Physiol 60:499–503
Ortiz R, Sayre KD, Govaerts B et al (2008) Climate change: can wheat beat the heat? Agri Ecosys Environ 126:46–58
Pagliano C, Raviolo M, Vecchia FD et al (2006) Evidence for PSII-donor-side damage and photo inhibition induced by cadmium treatment on rice (Oryza sativa L). J Photochem Photobiol B 84:70–78
Parasher A, Varma SK (1988) Effect of pre-sowing seed soaking in gibberellic acid on growth of wheat (Triticumaestivum L.) under saline conditions. Ind. J Biol Sci 26:473–475
Peralta-Videa JR, Lopez ML, Narayan M et al (2009) The biochemistry of environmental heavy metal uptake by plants: implications for the food chain. Int J Biochem Cell Biol 41:1665–1677
Petersson SV, Johansson AI, Kowalczyk M (2009) An auxin gradient and maximum in the Arabidopsis root apex shown by high-resolution cell-specific analysis of IAA distribution and synthesis. Plant Cell 21:1659–1668
Petit JR, Jouzel J, Raynaud D et al (1999) Climate and atmospheric history of the past 420,000 years from the Vostok ice core, Antarctica. Nature 399:429–436
Pilon-Smits E, Pilon M (2000) Breeding mercury-breathing plants for environmental cleanup. Trends Plant Sci 5:235–236
Pons R, Cornejo MJ, Sanz A (2011) Differential salinity-induced variations in the activity of H+– pumps and Na+/H+ antiporters that are involved in cytoplasm ion homeostasis as a function of genotype and tolerance level in rice cell lines. Plant Physiol Biochem 49:1399–1409
Poschenrieder C, Barcelo J (2004) Water relations in heavy metal stressed plants. Springer, Berlin, pp 207–299
Prakash L, Prathapasenan G (1990) NaCl and gibberellic acid-induced changes in the content of auxin, the activity of cellulose and pectin lyase during leaf growth in rice (Oryza sativa L.) Ann Bot 365:251–257
Reddy AM, Kumar SG, Jyonthsnakumari G et al (2005) Lead induced changes in antioxidant metabolism of horsegram (Macrotyloma uniflorum (Lam.) Verdc.) and bengalgram (Cicer arietinum L.) Chemosphere 60:97–104
Sarker AM, Rahman MS, Paul NK (1999) Effect of soil moisture on relative leaf water content, chlorophyll, proline and sugar accumulation in wheat. J Agro Crot Sci 183:225–229
Scandalios JG (1993) Oxygen stress and superoxide dismutases. Plant Physiol 101:7–12
Seckin B, Sekmen AH, Turkan I (2009) An enhancing effect of exogenous mannitol on the antioxidant enzyme activities in roots of wheat under salt stress. J Plant Gro Regul 28:12–20
Shahbaz M, Ashraf M (2013) Improving salinity tolerance in cereals. CRC Crit Rev Plant Sci 32:237–249
Sharma P, Dubey RS (2005) Lead toxicity in plants. Braz J Pp Physiolo 17:35–52
Shinozaki K, Yamaguchi-Shinozaki K (2000) Molecular responses to journal of plant physiology dehydration and low temperature: differences and cross-talk between two stress signaling pathways. Curr Opin Plant Biol 3:17–23
Siedlecka A, Krupa Z, Samuelsson G et al (1997) Primary carbon metabolism in Phaseolus vulgaris under Cd/Fe interaction. Plant Physiol Biochem 35:951–957
Simms EL (2000) Defining tolerance as a norm of reaction. Evol Ecol Res 14:563–570
Skoog F, Miller CO (1957) Symposia of the society for. Exp Biol 11:118–131
Sorefan K, Girin T, Liljegren SJ (2009) A regulated auxin minimum is required for seed dispersal in Arabidopsis. Nature 459:583–586
Sponsel VM, Hedden P (2004) Gibberellin, biosynthesis and inactivation. In: Davies PJ (ed) Plant hormones biosynthesis, signal transduction, action! Springer, Dordrecht, p 6394
Sreenivasulu N, Radchuk V, Alawady A et al (2010) Regulation of abscisic acid contents causes abnormal endosperm development in the barley mutant seg8. Plant J 64:589–603
Sreenivasulu N et al (2012) Contrapuntal role of ABA: does it mediate stress tolerance or plant growth retardation under long-term drought stress. Gene 506(2):265–273

Stavang JA, Gallego-Bartolomé J, Gómez MD et al (2009) Hormonal regulation of temperature-induced growth in rabidopsis. Plant J 60(4):589–601
Stobart AK, Griffiths WT, Ameen-Bukhari I et al (1985) The effect of Cd 2+ on the biosynthesis of chlorophyll in leaves of barley. Physiol Plant 63:293–298
Suzuki K, Nagasuga K, Okada M (2008) The chilling injury induced by high root temperature in the leaves of rice seedlings. Plant Cell Physiol 49:433–442
Swamy PM, Smith B (1999) Role of abscisic acid in plant stress tolerance. Curr Sci 76:1220–1227
Taiz L, Zeiger E (2002) Photosynthesis: physiological and ecological considerations. In: Plant physiology. Sinauer Associates, Sunderland, pp 171–192
Tanaka K (1994) Tolerance to herbicides and air pollutants. In: Foyer CH, Mullineaux PM (eds) Causes of photoxidative stress and amelioration of defense systems in plants. CRC Press, Boca Raton, pp 365–378
Thomashow MF (1999) Plant cold acclimation: freezing tolerance genes and regulatory mechanisms. Ann Rev Plant Physiol Plant Mol Biol 50:571–599
Thompson AJ, Jackson AC, Symonds RC et al (2000) Ectopic expression of a tomato 9-cis-epoxycarotenoid dioxygenase gene causes over-production of abscisic acid. Plant J 23:363–374
Ton J, Flors V, Mauch-Mani B (2009) The multifaceted role of ABA in disease resistance. Tren Plant Sci 14(6):310–317
Turral H, Burke J, Faures JM (2009) Climate change, water and food security. FAO, ISBN: 978-92-5-106795-6
Uno Y, Furihata T, Abe H et al (2000) Arabidopsis basic leucine zipper transcription factors involved in an abscisic acid-dependent signal transduction pathway under drought and high-salinity conditions. Proc Natl Acad Sci U S A 97:11632–11637
Wafaa M, Haggag AHF et al (2015) Agriculture biotechnology for management of multiple biotic and abiotic environmental stress in crops. J Chem Pharm Res 7:882–889
Wagner GJ (1993) Accumulation of cadmium in crop plants and its consequences to human health. Adv Agron 51:173–212
Wang Z, Gerstein M, Snyder M (2009a) RNA-Seq: a revolutionary tool for transcriptomics. Nat Rev Genet 10:57–63
Wang JZ, Cui LJ, Wang Y, Li JL (2009b) Growth, lipid peroxidation and photosynthesis in two tall fescue cultivars differing in heat tolerance. Biol Plant 53:247–242
Wang Y, Li K, Li X (2009c) Auxin redistribution modulates plastic development of root system architecture under salt stress in Arabidopsis thaliana. J Plant Physiol 166:1637–1645
Wang X, Cai J, Liu F et al (2012) High temperature acclimation alleviates the negative effects of post-anthesis heat stress on stem stored carbohydrates remobilization and grain starch accumulation in wheat. J Cereal Sci 55:331–336
Werner T, Motyka V, Laucou V et al (2003) Cytokinin-deficient transgenic Arabidopsis plants show multiple developmental alterations indicating opposite functions of cytokinins in the regulation of shoot and root meristem activity. Plant Cell 15(11):2532–2550
Wilson ID, Ribeiro DM, Bright J et al (2009) Role of nitric oxide in regulating stomatal apertures. Plant Sig Behav 4:467–469
Woodward AW, Bartel B (2005) Auxin: regulation, action, and interaction. Ann Bot 95:707–735
Yamaguchi S (2008) Gibberellin metabolism and its regulation. Ann Rev Plant Physiol 59:225–251
Yang X, Chen X, Ge Q et al (2006) Tolerance of photosynthesis to photoinhibition, high temperature and drought stress in flag leaves of wheat: A comparison between a hybridization line and its parents grown under field conditions. Plant Sci 171:389–397
Zheng Z, Xu X, Crosley RA et al (2015) The protein kinase SnRK2.6 mediates the regulation of sucrose metabolism and plant growth in Arabidopsis. Plant Physiol 153:99–113
Zhu JK (2002) Salt and drought stress signal transduction in plants. Ann Rev Plant Biol 53:247
Zhu J, Dong CH, Zhu JK (2007) Interplay between cold-responsive gene regulation, metabolism and RNA processing during plant cold acclimation. Cur Opi Plant Biol 10(3):290–295

Chapter 18
Strategies of Bioremediation of Heavy Metal Pollutants Toward Sustainable Agriculture

S. Dhanam

Abstract Heavy metals are toxic and hazardous materials to the environment, and production of agricultural products is declined at acute level. The toxic heavy metals are released from various natural and anthropogenic effects like industrialization. Hence, the removing of toxic elements from the environment is key focus in recent scientific scenario. Various methods of removal have been developed since a decade in which phytoremediation is the use of biological interventions of biodiversity for mitigation of the noxious effects caused by environmental pollutants, aid to the cleansing and facilitating to sustain the agriculture. In this review the following techniques and strategies of phytoremediation are discussed such as phytosequestration, phytodegradation, phytovolatilization, phytostabilization, phytoremediation, phytoextraction, rhizofiltration, rhizoremediation, rhizodeposition, and phytohydraulics to remove the heavy metal pollution in contaminated sites including agricultural lands using plants integrating with ecosystem service providers for sustainable agriculture.

18.1 Introduction

Due to global industrialization, war, and nuclear processes, a large amount of toxic compounds have been released into the biosphere. The heavy metals released through the various industries as effluent, nuclear radiation and releases of heavy metals by other processes in the environment may contaminate the soil. The soil contamination was divided into two major classes, i.e., inorganic and organic. Inorganic pollutants comprise heavy metals such as arsenic, cadmium, mercury, and lead, while organic contaminants include petroleum, hydrocarbons, phenolic compounds, fertilizers, herbicides, and pesticides (Hawumba et al. 2010). Bioremediation is a natural process which relies on bacteria, fungi, and plants to alter contaminants as these organisms carry out their normal life functions. Metabolic processes of

S. Dhanam (✉)
Department of Botany, Arignar Anna Government Arts College (Affiliated to Thiruvalluvar University), Villupuram 605 602, Tamil Nadu, India
e-mail: sdhanam2@gmail.com

A. Dhanarajan (ed.), *Sustainable Agriculture towards Food Security*,
https://doi.org/10.1007/978-981-10-6647-4_18

these organisms are capable of using chemical contaminants as an energy source, rendering the contaminants harmless or less toxic products in most cases (Kamaludeen et al. 2003).

18.2 Process of Phytoremediation

18.2.1 Uptake and Transport

Plants and metals interact in the root environment. The contaminant has to be in contact with the roots in order to be uptaken, The same kind of transporters that are used for the macronutrient and micronutrient entrance is also used by plants for the heavy metals uptake. Metals are stored in the subcellular compartments like vacuoles and lignocellulosic material (Marmiroli et al. 2006).

18.3 Phytoextraction

Phytoextraction is based on the mechanism of "hyperaccumulation," while phytostabilization is based on the mechanism of surface complexation, and the latter is involved in the phenomena of metal sorption (Xu et al. 2013). The technique of phytostabilization can be defined as the establishment of a vegetative cover by woody species on contaminated soils to minimize the mobility of heavy metals in polluted soils. In this process, the plant roots and microbial interaction can immobilize organic and inorganic contamination binding them to soil particles in the rhizosphere. Thus, pollutants become less bioavailable, and exposure of livestock, wildlife, and human beings is reduced.

18.4 Phytodegradation

The phytodegradation can occur inside the plant or in the rhizosphere. It is suitable for organic compounds such as solvents, petroleum, aromatic, and volatile compounds in the air. But, it remains rather hard to measure plant effect on PAH (polycyclic aromatic hydrocarbon) degradation, for example, because of all the interactions between soil, microorganisms, and rhizosphere (Newman and Reynolds 2004).

18.5 Phytovolatilization

Phytovolatilization process is the plants ability to absorb and subsequently volatilize the contaminant into the atmosphere. This process is for metal contaminants in groundwater, soils, sediments, and sludges medium. Phytotransformation/phytodegradation process is the breakdown of contaminants taken up by plants through metabolic processes within the plant. This process is for complex organic molecules that are degraded into simpler molecule contaminants in soils, sediments, sludges, and groundwater medium (USPA 2000; Moreno et al. 2008).

18.6 Phytostabilization

Phytostabilization and phytoextraction are the most usual phytoremediation techniques adapted for soils contaminated with heavy metals. Phytostabilization consists of using green plants to reduce the mobility of contaminant agents through revegetation strategies (Garbisu and Alkorta 2001). Some of the advantages associated with this technology are that the disposal of hazardous material/biomass is not required and it is very effective when rapid immobilization is needed to preserve ground and surface waters (Zhang et al. 2009).

18.7 Phytoremediation Groundcovers

These techniques widely used application and have been applied at various remediation projects. They are applied to surface soils, deep soils, and groundwater. These groundcover systems can also be used as certain types of landfill covers that also promote the degradation of the underlying waste. These have been referred to as bioreactor landfills. It is widely applied to soils impacted with recalcitrant compounds.

18.8 Phytoextraction

Phytoextraction has two strategies, viz., (1) natural phytoextraction (natural metal hyperaccumulating plants) and (2) chemically induced phytoextraction in which metal hyperaccumulation in plants is triggered through chemical amendments (Lombi et al. 2001). In natural phytoextraction, hyperaccumulating plants, natural ability to extract high amounts of metals from soil, have efficient mechanism to translocate metals from roots to shoots and can accumulate and tolerate high metal concentrations due to inherent mechanisms to detoxify metals in the tissues. On the other hand, in chemically induced phytoextraction, metal mobilization, root uptake, and translocation to shoots are facilitated by chemical amendments in soil; plants have mostly low tolerance to high metal concentrations thus surviving for a short period following metal hyperaccumulation.

18.9 Rhizofiltration

Rhizofiltration is a process by which inorganic contaminants present in the soil water are absorbed onto plant roots. The application of rhizofiltration is generally employed to mitigate contaminants in the groundwater rather than within the soil matrix. This phenomenon can further assist rhizofiltration applications as it increases metal specificity to binding domains in the plant roots. Plants may contain several phytochelatins to increase binding capacity of several metal ions. A number of metals are known to induce phytochelatin synthesis; however, metal binding by phytochelatins has only been demonstrated for Cu, Zn, Pb, and Cd (Dushenkov and Kapulnik 2000).

18.10 Rhizoremediation

Rhizosphere bioremediation is the enhanced biodegradation of recalcitrant organic pollutants by root-associated bacteria and fungi under the influence of selected plant species. The use of selected vegetation and sound plant management practices increases the total proportion of pollutant degraders in numbers and activity in the rhizosphere, leading to enhanced rhizodegradation. The rhizosphere is the zone of soil around the root in which microbes are influenced by the root system forming a dynamic root-soil interface (Kuiper et al. 2004). It also can establish an improvement of the physical and chemical properties of contaminated soil and an increase in contact between the microbes associated with the roots and the contaminants in soil (Aprill and Sims 1990; Kingsley et al. 1994; Kuiper et al. 2001; Nichols and Lopatin 1997; Schwab et al. 1995).

18.11 Rhizodeposition and Root Exudates

Plants sustain large microbial populations in the rhizosphere by secreting a variety of products (exudates, mucigels, dead cell material, etc.), which is known as rhizodeposition (Stottmeister et al. 2003). The chemical composition of the exudates is very diverse and varies with the plant species, but they normally include sugars, organic acids, and vitamins. It has been shown that this availability of nutrient in the immediate proximity of the roots makes the microbial population much larger in the rhizosphere than in the bulk soil and that these larger populations increase the degradation of organic compounds (Yu et al. 2003; Sun et al. 2004).

18.12 Phytohydraulics

Phytohydraulics is a mechanism to control or minimize of migration of contaminants in groundwater. In this phenomenon plants behave as organic "pumps" to pull in large volumes of the contaminated water (Sridhar et al. 2002). The term phytohydraulics describes the way in which the transpiration stream can transport not only water and dissolved mineral nutrients for benefit of the plant but also can transport contaminants with this water. Once the contaminants are in the upper reaches of the plant, the contaminants may then either phytovolatilize or phytodegrade or a combination (Interstate Technology and Guidance Regulatory Council 2009).

18.13 Phytoremediation Mechanisms of Heavy Metals

Plants have also evolved highly specific mechanisms to translocate and store micronutrients. These same mechanisms are also involved in the uptake, translocation, and storage of toxic elements, whose chemical properties simulate those of essential elements. Thus, micronutrient uptake mechanisms are of great interest to phytoremediation (USDE 1994). A living plant can be seen as a solar-driven pump because it is able to extract and concentrate particular elements from the contaminated environment. After harvesting the plants that are rich in accumulated contaminants, the following processes may be drying, ashing, or composting (Raskin et al. 1997).

18.14 Mechanisms for Metal Remediation

Bioremediation is the microbe-mediated process for clearance or immobilization of the contaminants, including all possible toxins like hydrocarbons, agrochemicals, and other organic toxicants. But for inorganic toxic compounds such as heavy metals, microbes are unable to simplify them into harmless compounds, and they should be used according to their specialization for the type of contaminants. Thus the bioremediation strategy for heavy metals depends on the active metabolizing capabilities of microorganisms. Several microorganisms are known to require varying amounts of heavy metals as essential micronutrients for growth and development (Ahemad 2014). Phytoremediation is a general term including several processes in function of the plant-soil-atmosphere interactions.

18.15 Immobilization in Phytoremediation Toward Cleaning Environment

Immobilization of contaminants can be achieved mainly through adsorption, precipitation, and complexion reactions which result in the redistribution of contaminants from solution phase to solid phase, thereby reducing their bioavailability and transport in the environment (Porter et al. 2004). Immobilization of heavy metals through the addition of lime (Krebs et al. 1999), phosphate (Ebbs et al. 1998) for ameliorating of contaminants have been studied well.

18.16 Mobilization

The value of phytosiderophores in enhancing the availability of heavy metal such as Fe, Cu, and Zn. Plant nutrition and various chelated compounds are available as nutrient sources. The potential value of chelating agents in the remediation of contaminated soils through mobilization of metalloid(s) has been explored (Thayalakumaran et al. 2003; Chen et al. 2004). Especially, remobilization of metalloid(s) with breakdown of organic matter which binds metalloid(s) needs to be investigated, particularly in acidic soils for metals. It cannot be more clearly stated that immobilization remediation is not a "treat and retreat" process and requires a long-term monitoring plan to be successful (Fang et al. 2012).

18.17 Reclamation of Abandoned Mine Sites

Mine wasteland generally comprises the bare stripped area, loose soil piles, waste rock and overburden surfaces, subsided land areas, and other degraded land by mining facilities, among which the waste rocks often pose extreme stressful conditions for restoration. The mining disrupts aesthetics of the landscape and also disrupts soil components such as soil horizons and structure, soil microbe populations, and nutrient cycles; those are crucial for sustaining a healthy ecosystem and hence result in the destruction of existing vegetation and soil profile (Kundu and Ghose 1997).

18.18 Heavy Metals

Heavy metals can be divided into two categories: essential and nonessential. Essential heavy metals such as Mn, Fe, Ni, and Zn are needed by living organisms for their growth, development, and physiological functions. While nonessential heavy metals such as Cd, Pb, Hg, and As are not needed by living organisms for any physiological function (Peng et al. 2009).

Heavy metal phytoremediation, a fast emerging technology, is an eco-friendly, low-tech, cost-effective, green alternative to the problem (Meagher et al. 2000). The specific plant and wild species that are used in this technique accumulate increasing amounts of toxic heavy metals by their roots and transport/translocate them through various plant tissues where they can be metabolized, sequestered, and volatilized (Greenberg et al. 2006; Doty et al. 2000).

18.18.1 Mercury

Mercury biogeochemistry has been affected by humans to such an extent that it is impossible to separate the issue from the issue of mercury pollution. In fact, it is estimated that total global atmospheric mercury has increased by a factor of 2–5 times and the deposition of atmospheric mercury has tripled or quadrupled since the industrial revolution (Keating et al. 1997). The use of plants to physically remove mercury from polluted soils or sediments in an efficient manner is only likely to be feasible using genetically engineered plants. Therefore, a simple model species, *Arabidopsis thaliana*, was engineered with a modified microbial gene, merA, encoding a protein that catalyzes the electrochemical reduction and detoxification of Hg (II) to Hg (0) (Rugh et al. 1996).

18.18.2 Cadmium

Cadmium is a chemical element. It is no known useful role in higher organisms. Cadmium is an extremely toxic metal commonly found in industrial workplaces. Cadmium is an especially mobile element in the soil and is taken up by plants primarily through the roots. The major factors governing cadmium speciation, adsorption, and distribution in soils are pH, soluble organic matter content, hydrous metal oxide content, clay content and type, presence of organic and inorganic ligands, and competition from other metal ions. Cadmium is a heavy metal with high toxicity and has an elimination half-life of 10–30 years (Jan et al. 1999).

18.18.3 Arsenic

Arsenic can combine with other elements to form inorganic and organic arsenicals (National Ground Water Association 2001).The As accumulation and resistance varies between plant species due to genetic differences and diversity in detoxification processes (Meharg and Hartley-Whitaker 2002).

18.18.4 Lead

Lead found within the soil can be classified into six general categories: ionic lead dissolved in soil water, exchangeable, carbonate, oxyhydroxide, organic, or the precipitated fraction. All of these categories combined make up the total soil lead content (Raskin and Ensley 2000). Water soluble and exchangeable lead is the only fractions readily available for uptake by plants. Oxyhydroxides, organic, carbonate, and precipitated forms of lead are the most strongly bound to the soil (Chaney 1998).

18.19 Transgenic Plants in Phytoremediation

The plant species currently being developed for phytoremediation seem capable of effective bioaccumulation of targeted contaminant, but efficiency might be improved through the use of transgenic (genetically engineered) plants. Naturally occurring plant species that can be genetically engineered for improved phytoremediation include *Brassica juncea* for phytoremediation of heavy metals from soil (Dushenkov et al. 1995), *Helianthus annuus* (Dushenkov et al. 1995), and *Chenopodium amaranticolor* (Eapen et al. 2003) for rhizofiltration of uranium.

18.20 Conclusion

Heavy metal pollution is a major hindrance for agricultural production worldwide. Natural metal sites and sites/land contamination with heavy metals by anthropogenic effect are the hazardous to the ecosystem which limits plant growth. Hence, prevention and control of metal pollution in contaminated sites is the indispensable phenomenon to save the soil for sustainable agriculture. In this review, various strategies of phytoremediation techniques have been discussed for thriving new ideas for removing of heavy metals in the contaminated sites. Mechanisms responsible for phytoremediation have also been elucidated for possible implementation on a variety of contaminated and polluted sites which are tool box for removal of metals. Developing countries like India, paying more attention to seek the remedies for land reclamation for agricultural issues and the new strategies for faming policy legislation will make peoples well-being when the protect the environment form pollution.

References

Ahemad M (2014) Remediation of metalliferous soils through the heavy metal resistant plant growth promoting bacteria: paradigms and prospects. Arab J Chem 11:020. https://doi.org/10.1016/j.arabjc

Aprill W, Sims RC (1990) Evaluation of the use of prairie grasses for stimulating polycyclic aromatic hydrocarbon treatment in soil. Chemosphere 20:253–265
Chaney RL (1998) Metal speciation and interactions among elements affect trace element transfer in agricultural and environmental food-chains. Lewis publishers, Chelsea
Chen YH, Li XD, Shen ZG (2004) Leaching and uptake of heavy metals by ten different species of plants during an EDTA-assisted phytoextraction process. Chemosphere 57:187–196
Doty SL, Shang QT, Wilson AM et al (2000) Enhanced metabolism of halogenated hydrocarbons in transgenic plants contain mammalian cytochrome P450 2E1. Pro Nat Acad Sci USA 97:6287–6291
Dushenkov S, Kapulnik Y (2000) Phytofilitration of metals. In: Raskin I, Ensley BD (eds) Phytoremediation of toxic metals – Using plants to clean-up the environment. Wiley, New York, pp 89–106
Dushenkov V, Kumar PBAN, Motto H et al (1995) "Rhizofiltration": the use of plants to remove heavy metals from aqueous streams. Environ Sci Technol 29:1239–1245
Eapen S, Suseelan KN, Tivarekar S, Kotwal SA, Mitra R (2003) Potential for rhizofiltration of uranium using hairy root cultures of *Brassica juncea* and *Chenopodium amaranticolor*. Environ Res 91(2):127–133
Ebbs SD, Brady DJ, Kochian LV (1998) Role of uranium speciation in the uptake and translocation by plants. J Exp Bot 49:1183–1190
Fang Y, Cao X, Zhao L (2012) Effects of phosphorus amendments and plant growth on the mobility of Pb, Cu, and Zn in a multi-metal-contaminated soil. Environ Sci Pollut Res 19:1659–1667
Garbisu C, Alkorta I (2001) Phytoextraction: a cost-effective plant-based technology for the removal of metals from the environment. Bioresour Technol 77:229–236
Greenberg BM, Hunag XD, Gurska Y et al (2006) Successful field tests of a multi process phytoremediation system for decontamination of persistent petroleum and organic contaminants. Proceedings of the twenty ninth Artic and marine oil spill program (AMOP), Canada, pp 389–400
Hawumba JF, Seruwagi P, Hung YT et al (2010) Bioremediation. In: Wang LK, Tay JH, Hung YT (eds) Handbook of environmental engineering. Springer, Dordrecht
Interstate Technology and Guidance Regulatory Council (2009) Phytotechnology technical and regulatory guidance and decision trees. Revised February 2009. Prepared by The Interstate Technology & Regulatory Council Phytotechnologies Team, Tech Reg Update
Jan AS, Roels HA, Emelianov D et al (1999) Environmental exposure to cadmium, forearm bone density, and risk of fractures: prospective population study for the public health and environmental exposure to cadmium. Lancet 353(9159):1140–1144
Kamaludeen SP, Arunkumar KR, Avudainayagam S et al (2003) Bioremediation of chromium contaminated environments. Ind J Exper Biolo 41:972–985
Keating MH, Mahaffey KR, Schoeny R et al (1997) Mercury study report to congress. EPA 452/R-9-003
Kingsley MT, Fredrickson JK, Metting FB et al (1994) Environmental restoration using plant-microbe bioaugmentation. In: Hinchee RE, Leeson A, Semprini L, Ong SK (eds) Bioremediation of chlorinated and polyaromatic hydrocarbon compounds. Lewis Publishers, Boca Raton, pp 287–292
Krebs R, Gupta SK, Furrer G, Schulin R (1999) Gravel sludge as an immobilizing agent in soils contaminated by heavy metals – a field study. Water Air Soil Pollut 115:465–479
Kuiper I, Bloemberg GV, Lugtenberg BJJ (2001) Selection of a plant-bacterium pair as a novel tool for rhizostimulation of polycyclic aromatic hydrocarbon-degrading bacteria. Mol Plant-Microbe Interact 14:1197–1205
Kuiper I, Lagendijk EL, Bloemberg GV et al (2004) Rhizoremediation: a beneficial plant-microbe interaction. Mol Plant-Microbe Interact 17(1):6–15
Kundu NK, Ghose MK (1997) Soil profile characteristic in Rajmahal Coalfield area. Soil Water Conser 25(1):28–32
Lombi E, Zhao FJ, Dunham SJ et al (2001) Phytoremediation of heavy metal-contaminated soils: natural hyperaccumulation versus chemically enhanced phytoextraction. J Environ Qual 30:1919–1926

Marmiroli N, Marmiroli M, Maestri E (2006) Phytoremediation and phytotechnologies: a review for the present and the future. In: Twardowska I et al (eds) Soil and water pollution monitoring, protection and remediation. Springer, Dordrecht, pp 403–416

Meagher RB, Rugh CL, Kandasamy MK, Gragson G, Wang NJ (2000) Engineered phytoremediation of mercury pollution in soil and water using bacterial genes. In: Terry N, Banuelos G (eds) Phytoremediation of contaminated soil and water. Lewis Publishers, Boca Raton, pp 201–221

Meharg AA, Hartley-Whitaker J (2002) Arsenic uptake and metabolism in arsenic resistant and nonresistant plant species. New Phytol 154(1):29–43

Moreno FN, Anderson CWN, Stewart RB et al (2008) Phytofiltration of mercury-contaminated water: volatilisation and plant-accumulation aspects. Environ Exp Bot 62(1):78–85

National Ground Water Association (2001) Copyright 2001. Arsenic

Newman LA, Reynolds CM (2004) Phytodegradation of organic compounds. Curr Opin Biotechnol 15:225–230

Nichols CG, Lopatin AN (1997) Inward rectifier potassium channels. Annual Rev Phys 59(1):171–191

Peng KJ, Luo CL, Chen YH et al (2009) Cadmium and other metal uptake by Lobelia chinensis and Solanum nigrum from contaminated soils. Environ Contam Tox 83:260–264

Porter SK, Scheckel KG, Impellitteri CA, Ryan JA (2004) Toxic metals in the environment: thermodynamic considerations for possible immobilisation strategies for Pb, Cd, As, and Hg. Crit Rev Environ Sci Technol 34:495–604

Raskin I, Ensley BD (2000) Phytoremediation of toxic metals: using plants to clean up the environment. Wiley, New York

Raskin I, Smith RD, Salt DE (1997) Phytoremediation of metals: using plants to remove pollutants from the environment. Plant Biotechnol 8:221–226

Rugh CL, Wilde D, Stack NM et al (1996) Mercuric ion reduction and resistance in transgenic Arabidopsis thaliana plants expressing a modified bacterial merA gene. Proc Natl Acad Sci U S A 93:3182–3187

Schwab AP, Banks MK, Arunachalam M (1995) Biodegradation of polycyclic aromatic hydrocarbons in rhizosphere soil. In: Hinchee RE, Anderson DB, Hoeppel RE (eds) Bioremediation of recalcitrant organics. Battelle Memorial Institute, Columbus, pp 23–29

Sridhar S, Victor FM, Steven CMC (2002) "Phytoremediation": an ecological solution to organic chemical contamination. Ecolog Eng 18:647–658

Stottmeister U, Wiessner A, Kuschk P et al (2003) Effects of plants and microorganisms in constructed wetlands for wastewater treatment. Biotechnol Adv 22:93–117

Sun H, Xu J, Yang S et al (2004) Plant uptake of aldicarb from contaminated soil and its enhanced degradation in the rhizosphere. Chemosph 54(4):569–574

Thayalakumaran I, Vogeler DR, Scotter HJ et al (2003) Leaching of copper from contaminated soil following the application of EDTA. II. Intact core experiments and model testing. Aust J Soil Res 41:335–350

U. S. Department of Energy (1994) Plume focus area, December. Mechanisms of plant uptake, translocation, and storage of toxic elements. Summary report of a workshop on phytoremediation research needs

U. S. Environmental Protection Agency (2000) Introduction to phytoremediation. National Risk Management Research Laboratory, EPA/600/R-99/107

Xu H, Chen Y, Huang H et al (2013) Removal of lead(II) and cadmium (II) from aqueous solutions using spent Agaricus bisporus. Can J Chem Eng 91(3):421–431

Yu Y, Chen YX, Luo YM et al (2003) Rapid degradation of butachlor in wheat rhizosphere soil. Chemosph 50(6):771–774

Zhang H, Zheng LC, Yi XY (2009) Remediation of soil co-contaminated with pyrene and cadmium by growing maize (Zea mays L.) Int J Environ Sci Tech 6:249–258

Chapter 19
Pesticide-Mediated Toxicity in Modern Agricultural Practices

Sivakumar Loganathan and Tamilselvi Murugan

Abstract Pesticides are, at the moment, man's main weapon against insect pests. Pesticides have been used heavily in almost all parts of the world. The average consumption of pesticides in India was only 3.2 g/ha in 1954–55. In 1075–76, this had increased to 4.4 kg/ha, and it keeps on increasing day by day. Proper use of pesticides increases farmers' incomes. Yields of most crops increase by 10–20% when pesticides are used. Most experts agree that removal of pesticides from crop protection would result in an immediate drop in food production. Discontinuation of all pesticide use would reduce the production of crops and livestock by 30% and would increase the prices of farm products by 50–70%. Phenomenal progress has been made in the development of insecticides. But their detrimental effects are numerous, including several acute and chronic illnesses in humans and worsening quality of the environment. Two types of contamination of the ecosystem are recognized: point- and nonpoint-sourcepollution. Pesticides are transported into the aquatic environment and the agricultural field as a nonpoint source most of the time and thereby reach all organisms and interfere in the food chain. So we need to supplement natural modern agricultural practices.

19.1 Introduction

Pesticides are chemicals used to control pests, destroy weeds, control microbes and harmful organisms that spoil agricultural commodities, and control parasites and vectors causing dangerous plant diseases. Pesticide toxicology is the qualitative and quantitative study of the adverse effects of pesticides on organisms.

S. Loganathan (✉)
Division of Toxicology and Molecular Oncology, Department of Environmental Science, School of Life Sciences, Periyar University, Salem, Tamil Nadu, India
e-mail: petsiva@gmail.com

T. Murugan
Department of Zoology, Govt. Arts College (Affiliated to Bharathiar University), Coimbatore 641018, Tamil Nadu, India

A. Dhanarajan (ed.), *Sustainable Agriculture towards Food Security*,
https://doi.org/10.1007/978-981-10-6647-4_19

The adverse effects may include both lethality (i.e., mass mortality of organisms) and sublethal effects such as changes in growth, development, reproduction, survival, physiology, biochemistry, and behavior of the organisms. Pesticide toxicology is also concerned with the concentrations or quantities of pesticides that can be expected to occur in the environment, i.e., water, air, soil, and biota (Headley and Lewis 2015). Therefore, pesticide toxicology is the study of the transport, distribution, transformation, and ultimate fate of pesticides in the environment.

The average utilization of pesticides in India was only 3.2 g/ha in 1954–55. In 1975–76, this had increased to 4.4 kg/ha. Proper use of pesticide increases farmers' incomes and increases yields of most crops by 10–20%. Discontinuation of all pesticide use would reduce the production of crops and livestock by 30% and would increase the prices of farm products by 50–70%. Extensive use of pesticides leaves residues in the environment, and these residues are responsible for disturbance of the ecological balance and health hazards. Damage to the environment by pesticide residue results through two types of interaction: interaction with biotic components and interaction with abiotic components.

The first type of interaction is responsible for causing imbalance in nature and serious health hazards to various forms of animal life (i.e., creation of new pests and establishment of resistant populations). In addition, pesticide residues also cause damage to the natural environment. Persistent insecticides are known to destroy, directly or indirectly, various phytoplankton and zooplankton, affecting the entire interdependent evolutionary ladder because the lower animals are involved in energy transfer to high trophic levels. Once the pesticide enters into the soil, the pesticide affects the microbial population in the soil and affects nutrient availability to plants (Thompson 2017).

Some organochlorine pesticides get deposited in adipose tissues of animals and are concentrated in higher animals through the biomagnification process. The second kind of interaction is a direct physical or chemical reaction with nonbiological components of the environment, such as soil and water. Such a direct reaction may directly or indirectly modify the utility of these components. This type of problem is common with soil pesticides, which change the soil pH, thus affecting microbial growth and micronutrient levels (Sun et al. 2017).

Insecticides were in use before man learned to write, and early stone tablets are said to have referred to red squall as a rat poison. By 1960, tobacco was being used for control of lace bug on pear trees. The modern use of insecticides dates from 1867, when Paris green was first used for Colorado potato beetle control. The discovery of the insecticidal properties of DDT (dichlorodiphenyltrichloroethane) in 1939 revolutionized our concept of insecticides and of insect control. Other families of chemicals are being investigated for possible insecticidal activity. Many pesticides used for agricultural purposes drain into water, causing hazardous effects on aquatic ecosystems. Pesticides travel from agricultural areas to potential catchment areas by two methods: horizontal travel during rain through runoff water and underground travel during rain or at other times (Majid et al. 2016).

On the other hand, two types of contamination of aquatic ecosystems are recognized as point- and nonpoint-source pollution (Murthy 1986). Point sources refer to a single source of contamination, such as effluent from a pesticide-manufacturing or formulation plant. Nonpoint sources refer to contamination of a widespread and diffuse nature, as in the case of agricultural runoff (Ouyang et al. 2016).

19.2 Sources of Pesticides

19.2.1 Nonpoint Sources

19.2.1.1 Agricultural Land Sources

Sediment resulting from soil erosion is regarded as the largest pollutant that affects water quality. More than land use, land form characteristics such as soil texture, soil type (mineral or organic), surface geology, slope, and drainage density—and also soil chemistry—influence the degree of nonpoint-source pollution. Pesticides adsorbed to clay particles are transported considerable distances. When land is exposed to erosive forces such as rain and surface runoff, the hazard of aquatic pollution is much greater (Yang et al. 2016).

Pesticide Mobility

Volatilization Pesticide volatilization can be defined as the movement of pesticide vapors through the air. People such as farm workers and bystanders can be exposed to pesticides by breathing these vapors after an application has occurred. Volatilization is considered different from pesticide movement by spray drift, erosion, or windblown dust/soil particles (Astoviza et al. 2016). Exposure to pesticide vapors due to postapplication volatilization generally occurs from three main sources. Temperature increases the rate of volatilization. The rate of evaporation of pesticides from the soils is related to the vapor pressure, water content of the soil, adsorption of the chemicals by the soil, and soil characteristics. Wind and water carry pesticide residues in the gaseous phase, in solution, or as a soil–pesticide complexes.

Adsorption Soil moisture and organic matter content determine the extent of adsorption. The amount of insecticide leaching is inversely related to the organic matter content.

19.2.1.2 Urban Land Sources

Urban areas that have construction sites with exposed soil contribute the highest amounts of nonpoint pollution loads.

Sewage

Urban sewage is another nonpoint source of pesticide residues in the aquatic environment.

Used Pesticide Containers

Careless handling of spraying equipment and used containers transfers pesticide residues to the aquatic environment, as in the case of washing and cleaning of spraying equipment in various water sources.

19.2.2 Point Sources

19.2.2.1 Effluent from Pesticide-Manufacturing Plants

In the aquatic environment the effluent from pesticide-manufacturing plants contains between a few and 100 lb/day of pesticides.

19.2.2.2 Joint Action of Pesticide Mixtures

In nature, organisms are exposed not to a single toxicant but often to numerous chemicals present either in considerable amounts or in traces. For discussion we take the interaction between any two or a group of toxicants. It was realized that interactions between chemicals may either increase or decrease the overall effect, i.e., the resulting action is more or less than the simple combined effect, or there may not be any interaction at all between the two. Further, the situation is complicated when one of the compounds is not toxic by itself but, in its presence, the other compound may act as an enhancer.

The joint action is defined as additive, more than additive, or antagonistic. If half of the concentration of toxicant A needed to produce a given response and half of the concentration of toxicant B needed to produce the same response together produce that response, the action of toxicants A and B is additive. If this combination causes more than that response, the resultant action is more than additive, and if it causes less than that response, then it is less than additive, antagonistic, or without interaction.

Sprague (1970) explained that supposing that one toxic unit of toxicant A or B is needed to produce a particular response, then if the same response is produced by combinations of toxicants A and B at concentrations that meet within the square, the toxicants are helping each other and the resultant action is called "joint action." Akobundu et al. (1975) used the term *synergism* when both components are active and the combined effects are more than the sum of the two individual effects.

19.2.2.3 Enhancement

Enhancement occurs if one of the components is inactive but somehow increases the effect of the other. More recently, models have been proposed that take into account the underlying components of joint action, including the effect of one substance on another with respect to the site of action, elimination, metabolism, competition for receptor sites, speed of action, and interaction.

19.3 Observations in Fish

Various reports on the combined toxicity of pesticides to fish may be examined from the above discussion.

19.3.1 Pre-exposure

It is not clear whether pre-exposure (also called preconditioning) confers any protection for the organism. Exposure of spot to 0.01 and OL.ug/L toxaphene for 5 months made the fish more susceptible to the action of the same compound, as evidenced by the increased acute toxicity of toxaphene (lowered LC50 values) to pre-exposed fish in comparison with those that were not previously exposed. Previous exposure of goldfish to alkyl benzene sulfonate (ABS) made them more susceptible to the toxic effects of dieldrin and DDT. Higher mortality was recorded in the group that was pre-exposed to ABS.

19.3.2 Observations on the Joint Toxicity of Pesticides in Fish

The possible interaction between methyl parathion and the cotton defoliant DEF (S,S,S-tributyl phosphorotrithioate) was tested. Gambusias were exposed to 0.5 mg/L DEF or 5 mg/L methyl parathion, or both. There was 8% mortality of the fish exposed to ethyl parathion, 89% mortality of the fish exposed to both the toxicant and no mortality of the fish exposed to both the toxicant, and no mortality of fish exposed to DEF. It was concluded that there was potentiation of the toxicity of DEF by methyl parathion.

19.4 Pesticide-Induced Morpho-anatomical Changes

19.4.1 Morphological Changes

Many morphological changes have been reported following the exposure of fish either to high concentrations for a short time or to sublethal concentrations for longer periods. Some of these changes may apparently change the body coloration or may be an erosion of something, distortion of vision, etc. Exposure of *Oreochromis mossambicus* to any pesticides for 15 days resulted in darkening of the skin, formation of a brown short on the forehead, swelling of the eyes, and erosion of the fin margins.

Many pesticides, irrespective of the group to which they belong, have been reported to induce vertebral damage and skeletal deformities. Mehrle et al. (1982) suggested that a decrease in vertebral mechanical properties is an early indicator of contaminant stress. They also suggested that containment-induced competition for vitamin C between collagen metabolism in bone and microsomal mixed function oxidase would cause vertebral damage. The competition for vitamin C would decrease the water bodies. Commonly, fish are subjected to long-term stress arising from exposure to sublethal concentrations.

In the long run, these sublethal concentrations may prove more deleterious than lethal concentrations because subtle and small effects on aquatic organisms, especially fish, may alter their behavior, vitamin C, and collagen content of the bone with a concomitant increase in the ratio of bone minerals to collagen, resulting in increased fragility of the bone.

19.5 Pesticide-Induced Behavioral Changes

19.5.1 General Behavioral Changes

The behavior or activities of an organism represent the final integrated result of a diversity of biochemical and physiological processes. Thus, single behavioral parameters are generally more comprehensive than a physiological or biochemical parameter. In addition, behavior patterns are known to be highly sensitive to changes in the steady state of an organism. Hence, an alteration in the behavior of the organism due to stress is a diagnostic tool for identifying the ecological effects of the release of a toxicant into the environment.

Sherer (1975) considered that behavioral tests may sometimes ascertain lower thresholds than physiological techniques because the response results from an intact, integrated, functional system. Bull and McInerney (1974) studied behavioral changes in salmon exposed to several concentrations of fenitrothion. Various behavioral changes occurred with zhrobl exposure. Marked nipping, etc., was noticed. On the other hand, comfort behaviors such as thrusts, coughs, etc., increased with increasing concentrations of the toxicant. Some exposed individuals were unable to maintain position and were swept downstream. Some freshwater fish exposed to phosphate esters became hypersensitive to disturbance, fed leers, and showed impaired swimming ability (Pyle and Ford 2017).

19.5.2 Avoidance Responses

Fish can sometimes sense the presence of xenobiotic chemicals in water and tend to avoid them. In an experiment to test the avoidance reactions of *O. mossambicus* to pesticides, the fish avoided DDT, endrin, and malathion. Most of the fish preferred on with mediate lethal concentration of 0.1 mg/L which is a sort of a sensory trap for the high.

19.5.3 Swimming and Hypersensitivity

Swimming activity of fish in sublethal concentrations of pesticides may be greater than that of fish in pesticide-free water. When the high was exposed to lethal concentrations of pesticides, the fish exhibited erratic movements and sometimes lost balance. The number of opercula beats increased with increasing concentrations of pesticides (Banaee et al. 2014).

19.5.4 Schooling Behaviors

Weis and Weis (1982) reported that pesticides had a marked effect on the schooling behavior of fish. Increased uptake of the toxicant resulted in disruption of schooling behavior. This disrupted schooling resulted in increased predation of the fish. Apart from disruption of schooling behavior, pesticides affect the spawning migration of fish.

19.5.5 Coughing

Coughing in fish has been described as an interruption of the normal ventilating cycle, with more rapid expansion and contraction of the filament and opercular cavities, and serves the purposes of clearing the gills of accumulated debris and coughing up concentrations of pesticides. The frequency of cough was proportional to the sublethal toxicant concentration and predicted chronic effects at levels near the maximum acceptable toxicant concentration. Thus, the cough response is a useful and sensitive tool for evaluation of the quantity of toxicant in the riotous medium.

19.6 Pesticide-Induced Biochemical Changes

Many pesticides have been reported to produce a number of biochemical changers in fish at both lethal and sublethal levels. Changes in ionic concentrations, organic constituents, enough the activity, and osmoregulation in fish have been attributed to pesticides (Melo et al. 2015).

19.6.1 Effects of Pesticides on Osmotic and Ionic Regulation

Osmoregulation is the process by which the total electrolyte content and water volume in an organism are held relatively constant. It is well known that any chemical that affects osmoregulation in marine or freshwater fish will alter the percentage of water in tissue. Changes in osmoregulation in fish exposed to a toxicant are generally determined by measuring the blood sodium chloride or total osmolality.

Researchers have studied the effects of pesticides on ion balances in freshwater and marine fish. In seawater-adapted fish, exposure to a wide variety of different pesticides induces osmoregulation dysfunction, which is reflected in a rise in plasma NaCl and osmolality, while the opposite occurs in freshwater-adapted fish. Evidently a large majority of pollutants affect osmoregulation or ionic regulation (Vernberg and Vernberg 2013).

19.7 Effects of Pesticides on Feeding Energetics

Intake of food is an important factor governing various physiological functions such as growth and reproduction. Growth represents a net outcome of a series of processes such as digestion, assimilation, metabolism, and excretion. In this context, proposed a hypothetical model for average partitioning of dietary energy in carnivorous fish. In this hypothetical partitioning of dietary energy, in a fish with an intake of 100 calories, 20 calories are lost as feces and 7 calories as metro groups waste. The cost of digestion and assimilation of the food is 14 calories for splenetic dynamic fountain, leaving net energy of 59 calories to be used by the organism. This net energy is partitioned between maintenance activity, metabolism, and growth. Energy metabolism and growth compete for the net energy. Thus, if metabolism is elevated, growth will be limited unless the intake of food is increased. Certain environmental pollutants reduce the feeding rate, growth, and food conversion efficiency of fish (Narra 2016; Zheng et al. 2016 and Hoseini et al. 2016). Effects of pesticides on feeding energetics have been studied in certain commercially freshwater fish such as *O. mossambicus*, *Channa punctata*, *Mystus vittatus*, *Cyprinus carpio*, *Lepidocephalichthys thermalis*, etc.

Many pesticides, irrespective of the group to which they belong, have been reported to reduce the feeding rate, absorption rate, growth rate, metabolic rate, absorption efficiency, and food conversion efficiency. Intake of food may be affected by the effect of pesticides on the appetite of the fish. Reduction of the growth rate and conversion efficiency may be due to utilization of body organic reserve substances for the purpose of energy requirements during stress conditions (Giaquinto et al. 2017).

19.8 Effects of Pesticides on Liver Function

The liver is of key importance when considering the effects of polluting chemicals on fish. It is the primary organ for biotransformation of organic xenobiotics and probably also for excretion of household chemicals. Since many of these organic chemicals tend to accumulate to high concentrations in the liver, liver cells are exposed to higher levels of harmful chemicals than may be present in the environment or in other organs of the fish. The liver serves a number of functions related to other physiological activities such as endocrine control of reproduction and interconversion of foodstuffs.

Therefore, the potential effects of xenobiotics on the liver are numerous but often seen only by industry. Chemical laboratories have long used measurement of blood plasma as a diagnostic tool for assessing liver damage in humans. Alkaline phosphatase, glutamic oxaloacetic transaminase, and glutamic pyruvic transaminase are normally found in low concentrations in blood, but their concentrations may be increased by leakage from damaged liver cells, increasing their catalytic activity. Liver cells are particularly rich in transaminase because this organic is the major one for interconversion of foodstuffs.

When a hepatotoxic pesticide is administered to either freshwater or marine fish, increased plasma glutamic oxaloacetic transaminase and glutamic pyruvic transaminase from samies is glutamic usually observed. (Sabra and Mehana 2015).

19.8.1 The Liver Has a Wide Range of Functions

Some pesticides affect liver function, which is usually reflected in a change in blood chemistry. For example, DDT and enduring caused a 50% elevation in serum cholesterol when mullet were exposed to insecticides for 96th. This elevation in cholesterol could be explained by either enhanced production by the liver or inhibition of cholesterol excretion. Regarding the first possibility, harmful cortisol inhibits cholesterol synthesis and this hormone becomes elevated in fish under stress.

It is well known that the liver is responsible for elimination, in bile, of breakdown products, primarily bilirubin. If this process is impaired, the condition of jaundice may develop and be reflected in a rise in the bilirubin concentration in plasma. Exposure of fish to organic pesticides produced a marked elevation in plasma concentrations. Concomitant inhibition of the enzyme uridine diphosphate (UDP) glucuronosyltransferase (UGT) in liver tissue was also noted. Since this enzyme is required for the normal configation reactions by which bilirubin is prepared for exertion, its inhibition may be the primary cause of jaundice.

Ascorbic acid is better known as vitamin C. It plays a number of known biochemical roles. It is essential for normal synthesis of collagen. Other less well-studied functions include involvement in the metabolism of adrenal steroids by the liver, oxidation of tyrosine and, in conjunction with superoxide dismutase and catalase, helping to prevent build-up of free radicals in cells. Mayer et al. (1977) exposed catfish to various phenols in the diet for 150 days. They noted dose-dependent depression of backbone collagen and fin deformities.

However, when ascorbate was added to the diet at several different doses, the extent of collagen depression and the incidence of deformities in the fish exposed to toxaphene were decreased roughly in proportion to the ascorbate dose. Toxaphene was shown to cause a depletion of vitamin C in the spine but not in the liver. Vitamin C supplementation also reduced the whole-body residues of toxaphene (Singh and Singh 2017). This suggested that this molecule is used by the liver in detoxifying toxaphene but at the expense of collagen. The depletion of tissue ascorbic acid during chronic exposure to pollutants may depend on the ability of the species of fish to synthesize this vitamin. For example, a week of exposure to pesticide (monocrotophos) caused a 45% loss of liver ascorbic acid in *Oreochromis mossambicus*.

It was also observed that liver ascorbic acid was not affected much by natural stressors such as temperature and salinity. Thus, measurements of this trace nutrient in the liver may have potential for specifically indicating chemical pollution. From the stoned point of the fish, depletion of ascorbic acid can result in pathological conditions including delayed wound healing, anemia, skin lesions, fin necrosis, etc. (Thomas et al. 1990; Hilton et al. 1977).

19.9 Development of Resistance

When an insecticide is used to control a pest, not all members of the population against which it is used are killed by the toxic material. Some survive. To control the surviving members, a higher concentration of insecticide would be required, and a stage may be reached when the insecticide is totally ineffective against the pest.

According to Georghion and Taylor (1977), the number of species of insects and acarines in which resistant strains have been reported has increased from in 1908 to 364 in 1975. Agriculturists are compelled to use higher doses and more frequent applications to kill the same numbers of pests. This leads to disruption of the ecosystem. Some strains of insects and acarines have developed resistance to arsenic, DDT, and other chlorinated hydrocarbons, followed by organophosphates (OPs), carbamates and, more recently, pyrethroids and all compounds available commercially for pest control (Naqqash et al. 2016).

The development of resistance has been termed the most serious problem for modern pest control. According to the World Health Organization (1976), resistance is probably the biggest single obstacle in the struggle against vector-borne disease and is mainly responsible for preventing malaria eradication in many countries. Resistance has been reported not only to the most recent insecticides but also to insect growth regulators, chemosterilants, and even biological agents.

Reports in the literature show that an insect's development of resistance to an insecticide does not occur until after the insect has been exposed to the chemical for several generations. It took 25 generations of intensive selection for insects to develop resistance to IC compounds. DDT resistance develops after an initial latent period of several generations before it increases steeply, whereas cyclodiene resistance develops without delay (Elhag 2016). The resistance to carbamate takes the same time as the resistance but develops a little faster when it builds on a basis of OP resistance.

The rate of development of resistance in an unexposed population is initially very low. But as the frequency of major genes for resistance in the surviving population is gradually increased, the insect becomes better organized genetically to exist in the contaminated environment. The more intense the selection pressure, the more rapid the development of resistance, provided that the number of survivors is large enough to maintain genetic variability. The persistence of the resistance genes prevents introduction of insecticides against populations that have apparently reverted to full susceptibility as a result of release of pressure from insecticides.

Several ecological factors may influence the development of resistance to insecticides in a population. It may depend on the relative isolation of populations from each other or the degree of exchange of genetic material, and variations of ecological tissue factors such as the season, size, growth, and generations per year. The reproductive potential of the population may influence the development of resistance. Behavioral factors also influence the development of resistance. Several species of insects are known to have inherited ability to detect the pressure of specific insecticides and escape before taking up lethal quantities.

19.9.1 Types of Resistance

There are three types of resistance:

1. Cross-resistance
2. Multiple resistance
3. Multiplicate resistance

The term *cross-resistance* refers to resistance of a strain of insects to compounds other than the selecting agent, due to the same biochemical mechanism.

Multiple resistance occurs when two or more distinct mechanisms are present together, each protecting against different possessing.

When two or more mechanisms coexist in the same organism and protect it against the same position, the animal is considered to possess *multiplicate resistance*. Such resistance usually results from simultaneous or consecutive use of several insecticides. In countries where many different insecticides have been used organist houses, resistance may be both multiple and multiplicate. Insects with a high level of resistance to chlorinated hydrocarbons may exhibit a high level of cross-resistance to carbamates. Where resistance to an OP is already present, exposure to an alternative OP may result in development of resistance to the new compound.

19.9.2 Mechanisms of Resistance

Considerable information is available regarding the biochemistry, physiology, and genetics of resistance of some species of arthropods. Sawicki (1979) classifies the resistance mechanism into three types:

1. Delayed entry of the toxicant due to factors such as decreased permeability of the cuticle
2. Decreased sensitivity of the site of action
3. Increased detoxification of the insecticide or decreased activation by metabolic processes

19.9.3 Mechanisms of Resistance to Organochlorine Compounds

It is a well-known fact that chlorinated insecticides are highly toxic because of their effects on the nervous system and they may also interfere with various metabolic processes in animals. One mechanism for DDT resistance was found to be metabolic conversion to DDE (dichlorodiphenyldichloroethylene) of DDT dehychlorinase. A gene responsible for this process is situated on chromosome II in the housefly. Some resistant strains of insects have increased detoxification of DDT while others exhibit absorption of the compound. Lindone is metabolized in insect tissues, but the mechanism is not clear.

Dehydrochlorination is involved in these processes. In the house fully agency responsible for slower penetration of DDT is located on chromosome III, which also carries the gene responsible for resistance to knockdown. The gene responsible for conversion of both DDT and DDE to the more easily excreted polar metabolic sites is situated on the fifth chromosome in the housefly.

The housefly also possesses other types of resistance against DDT. Knockdown resistance in the housefly is believed to be result of decreased nerve sensitivity, which causes resistance to DDT and pyrethroids. Changes in the ability to metabolize glucose may be a factor in some cases of insect resistance to chlorinated insecticides (Zhu et al. 2016).

The mechanism of cyclodiene resistance is not well studied. Some researchers agree that resistance to insecticides does not involve reduced penetration, increased metabolism, or enhanced excretion of insecticides when compound with the same processing die—susceptible insects (Schaefer and Sun 1967) believe that they did in resistance in house—these may be due to insensitivity at the neuromuscular receptor site.

19.9.4 Organophosphate Compounds

The biochemical mechanism for the toxicity of OP compounds was first established in mammals. Now it is known that metabolism of these compounds in mammals and insects is the same. The mammalian liver and the insect's body fat resemble one another biochemically in several aspects. Though insects possess a variety of mechanisms resisting the toxic effects of insecticides, an important mechanism is enhanced detoxification. OP poisoning in insects is also known to occur through inhibition of acetylcholine esterase. In a few cases the property of resistance is due to the possession of acetylcholinesterase, which is insensitive to OP and carbamate insecticides. Slower penetration and increased metabolism of the chemical are responsible for resistance to OP compounds in larvae of the mosquito *Culex* spp. and not the insensitivity of acetylcholinesterase to OP (Wu et al. 2016).

19.9.4.1 Three Factors Control or Modify Resistance to Organophosphate Compounds

These factors are:

1. Decreased penetration through the cuticle
2. Detoxification mechanisms
3. Decrease in the sensitivity of cholinesterase

Among these factors, increased detoxification is believed to be the main cause of insect resistance to OPs. Genetic studies show that resistance to OP compounds is high by complicated though it may be caused by a single gene. In *Drosophila* spp., the main OP resistance gene is called "a" because of its association with reduced aliesterase activity located on chromosome IV. Three different detoxifying mechanisms in the parathion resistant house this are controlled by different genes on chromosome II. Delayed penetration is many resistant strains of house these is controlled by the gene "Pen" on chromosome III.

19.9.5 Carbamates

These are active anticholinesterases, which cause characteristic span atoms of cholinergic stimulation similar to those caused by OP insecticides. In house by the OP oxidative mechanisms may be responsible for resistance. The occurrence of high levels of oxidases in the microsomes of carbonate resistant house this supports this theory. Resistance to carbamate insecticides in the housefly has sometimes been attributed to a single gene and sometimes to polygenic systems. Plapp (1970) concluded that in the housefly this resistance to carbamate insecticides is due to the interaction of genes located on chromosomes II, III, and IV.

References

Akobundu IO, Duke WB, Sweet RD et al (1975) Basis for synergism of atrazine and alachlor combinations of Japanese millet. Weed Sci 23:43–48

Astoviza MJ, Cappelletti N, Bilos C, Migoya MC, Colombo JC (2016) Massive airborne endosulfan inputs related to intensive agriculture in Argentina's pampa. Chemosphere 144:1459–1466

Banaee M, Sureda A, Zohiery F, Hagi BN, Garanzini DS (2014) Alterations in biochemical parameters of the freshwater fish, *Alburnus mossulensis*, exposed to sub-lethal concentrations of fenpropathrin. Int J Aquat Biol 2(2):58–68

Bull CJ, McInerney JE (1974) Behaviour of juvenile coho salmon (Oncorhynchus kisutch) exposed to Smithion (Fenitrothion), an organophosphate insecticide. J Fish Res Board Can 31:1867–1872

Elhag EA (2016) Evolution of insecticide resistance in the greenhouse whitefly. Doctoral dissertation, University of Khartoum

Georghion GP, Taylor CE (1977) Oprational influence in evaluation of insecticide resistance. J Econ Entomol 70:653–658

Giaquinto PC, de Sá MB, Sugihara VS, Gonçalves BB, Delício HC, Barki A (2017) Effects of glyphosate-based herbicide sub-lethal concentrations on fish feeding behavior. Bull Environ Contam Toxicol 98(4):460–464

Headley JC, Lewis JN (2015) The pesticide problem: an economic approach to public policy. Routledge, London

Hilton JW, Cho CY, Slinger SJ (1977) Factors affecting the stability of supplemental ascorbic acid in practical trout diets. J Fish Board Can 34(5):683–687

Hoseini SM, Rajabiesterabadi H, Kordrostami S (2016) Chronic exposure of *Rutilus rutilus* *c*RC*aspicus* fingerlings to ambient copper effects on food intake, growth performance, biochemistry and stress resistance. Toxicol Ind Health 32(2):375–383

Majid MA, Hossain ABMZ, Cappuyns V, Wyseure GCL (2016) Transport characteristics of heavy metals, metalloids and pesticides through major agricultural soils of Bangladesh as determined by TDR. Soil Res 54:970–984

Melo KM, Oliveira R, Grisolia CK, Domingues I, Pieczarka JC, de Souza Filho J, Nagamachi CY (2015) Short-term exposure to low doses of rotenone induces developmental, biochemical, behavioral, and histological changes in fish. Environ Sci Pollut Res 22(18):13926–13938

Murthy MN (1986) Interfuel substitution and derived demands for inputs in the manufacturing sector in India. J Quant Econ 2:119–135

Naqqash MN, Gökçe A, Bakhsh A, Salim M (2016) Insecticide resistance and its molecular basis in urban insect pests. Parasitol Res 115(4):1363–1373

Narra MR (2016) Single and cartel effect of pesticides on biochemical and haematological status of *Clarias bBatrachus*: a long-term monitoring. Chemosphere 144:966–974

Ouyang W, Jiao W, Li X, Giubilato E, Critto A (2016) Long-term agricultural non-point source pollution loading dynamics and correlation with outlet sediment geochemistry. J Hydrol 540:379–385

Plapp FW (1970) On the molecular biology of insecticide resistance. In: Biochemical toxicology of insecticides. Academ Press, New York, pp 179–192

Pyle G, Ford AT (2017) Behaviour revised: contaminant effects on aquatic animal behaviour. Aquat Toxicol 182(201701):226–228

Sabra FS, Mehana ESED (2015) Pesticides toxicity in fish with particular reference to insecticides. Asian J Agric Food Sci 3(01). ISSN: 2321–1571

Schaefer CH, Sun YP (1967) A study of dieldrin in the house fly central nervous system in relation to dieldrin resistance. J Econ Entomol 60(6):1580–1583

Sherer GK (1975) Tissue interaction in chick liver development: a re-evaluation. I. Epithelial morphogenesis: the role of vascularity in mesenchymal specificity. Devi Biol 46:281–295

Singh DK, Singh NS (2017) Endosulfan a cyclodiene organochlorine pesticide: possible pathways of its biodegradation. In: Microbe-induced degradation of pesticides. Springer International Publishing, Cham, pp 105–130

Sprague JB (1970) Measurement of pollutant toxicity to fish II. Utilizing and applying bioassay results. Water Res 4:3–32

Sun X, Zhu L, Wang J, Wang J, Su B, Du Z, Guo P (2017) Effects of endosulfan on the populations of cultivable microorganisms and the diversity of bacterial community structure in Brunisolic soil. Water Air Soil Pollut 228(4):169

Thomas PETER (1990) Molecular and biochemical responses of fish to stressors and their potential use in environmental monitoring. Am Fish Soc Symp 8:9–28

Thompson PB (2017) The spirit of the soil: agriculture and environmental ethics. Routledge, London

Vernberg FJ, Vernberg WB (eds) (2013) Pollution and physiology of marine organisms. Elsevier, New York

World Health Organization (1976) Epidemiology of onchocerciasis. Report of a WHO Expert Committee. 597 Geneva, Switzerland

Weis P, Weis JS (1982) Toxicity of methyl mercury, mercury chloride and lead in killi fish Fundulus hetoctitus, from Southampton, NY. Environ Res 28:364–374

Wu ZM, Chu HL, Wang G, Zhu XJ, Guo XX, Zhang YM, Xing D, Yan T, Zhao M-H, Dong Y-D, Zhao T-Y, Li CX (2016) Multiple-insecticide resistance and classic gene mutations to Japanese encephalitis vector *Culex tritaeniorhynchus* from China. J Am Mosq Control Assoc 32(2):144–151

Yang B, Huang K, Sun D, Zhang Y (2016) Mapping the scientific research on non-point source pollution: a bibliometric analysis. Environ Sci Pollut Res 24:1–15

Zheng S, Chen B, Qiu X, Chen M, Ma Z, Yu X (2016) Distribution and risk assessment of 82 pesticides in Jiulong River and Estuary in South China. Chemosphere 144:1177–1192

Zhu F, Lavine L, O'Neal S, Lavine M, Foss C, Walsh D (2016) Insecticide resistance and management strategies in urban ecosystems. Insects 7(1):2

Chapter 20
Solid State Fermentation Utilizing Agro-Industrial Waste for Microbial Pigment Production

Chidambaram Kulandaisamy Venil, Nur Zulaikha Binti Yusof, and Wan Azlina Ahmad

Abstract Microbial pigments due to their better biodegradability and higher compatibility with the environment offer promising avenues for various industrial applications like food, cloth, painting, cosmetics, pharmaceuticals, plastics, etc. Nevertheless, the current bacterial pigment productions are not effective to meet their industrial needs. Current research on microbial pigments signify that genetic engineering for strain improvement, optimization of bioprocess modeling and utilizing cheap agro-industrial residues as substrates are key developmental strategies to maximize pigment production. Solid-state fermentation (SSF) has been reassessed as an alternative to submerged fermentation and could be a possible strategy for the cost-effective production of microbial pigments. The investment for SSF is usually lower than that of submerged fermentation, since it uses waste agricultural residues. This chapter summarizes the effective way to produce microbial pigments in agro-industrial waste (SSF) for its wide-ranging industrial applications and commercial viability.

20.1 Introduction

Natural pigments and synthetic dyes are extensively used in various fields of everyday life such as food production, textile industries, paper production, and agricultural practices (Cserhati 2006). According to green technology, less toxic products and more natural starting material are favorable for today's production lines. It is well known that some synthesized dye manufacturing is prohibited due to the carcinogenicity of the precursor or product and also the effect of disposal of their industrial wastes in the ecosystem. It is therefore, essential to explore various natural

C.K. Venil (✉) • N.Z.B. Yusof • W.A. Ahmad
Biotechnology Laboratory, Department of Chemistry, Faculty of Science,
Universiti Teknologi Malaysia, Skudai, Johor Bahru 81310, Malaysia
e-mail: ckvenil@gmail.com

A. Dhanarajan (ed.), *Sustainable Agriculture towards Food Security*,
https://doi.org/10.1007/978-981-10-6647-4_20

Table 20.1 Comparison of price for the production of prodigiosin from *Serratia marcescens* UTM1 using commercial and agricultural-based growth medium

Prodigiosin production	Medium/raw materials	Quantity used (kg)	Cost (USD)	Total cost (USD)
Synthetic medium	Nutrient broth	0.4	58.61	58.61
Agricultural-based substrate	Agriculture-based substrate	0.2	0.10	5.91
	Nutrient broth	0.044	5.81	

sources of colorants and their applications. Now the industry is able to produce some microbial pigments for applications in food, cosmetics, or textiles (Dufosse 2009). Numerous previous studies reported that microorganisms of the genus *Monascus* produce red pigments, which are used as coloring agents in food and textiles (Santis et al. 2005; Nagia and El-Mohamedy 2007; Babitha 2009; Velmurugan et al. 2010).

Biotechnology industries are in continual quest for the discovery of novel bacterial pigments and the enhancement of productivity of value products to retain their global competitiveness (Venil et al. 2013). Development of microbial strains that can utilize cheap and renewable substrates will make the price of pigments competitive with synthetic pigments. Therefore, discovering cheap substrates for pigment production is believed to reduce the production cost (Table 20.1). Although the price of bacterial pigment will be relatively higher compared to the synthetic dyes, the production cost can be reduced via:

- Use of agricultural wastes such as pineapple wastes, sugarcane bagasse, and molasses as growth medium for cultivation of bacteria
- Use of locally isolated wild-type bacterial strains – eliminates the cost for any genetic alterations, etc.
- Use of simple extraction techniques

The bacterial pigments will offer good opportunities due to their enhanced environmental acceptability and superior performance characteristics; classical or conventional grades are expected to continue to dominate the organic market.

20.2 Solid-State Fermentation

Solid-state fermentation (SSF) is a low-cost fermentation process, particularly suitable to the agro-industrial residues as the substrates for the bioprocesses. It deals with the controlled growth of the microorganisms, mainly on the surface of water-insoluble substrates, in the presence of varying amounts of available water (Sindhu et al. 2015).

20.2.1 Applications of SSF

Recently on SSF have been highlighted to investigated for application of SSF for the development of bioprocess such as bioremediation and biodegradation of hazardous wastes, detoxification of biological of agro-industrial wastes, nutritional enrichment from biotransformation of crops and crop-residues and production of value-added products such as biologically active secondary metabolites, including antibiotics, alkaloids, plant growth factors, enzymes, organic acids, bio-pesticides, including myco-pesticides and bio-herbicides, bio-surfactants, biofuel, aroma compounds and other various economic applications of SSF offer the potential of significantly improving and raising living standards with only a low technology input requirement (Mitchell 1992). The research work on since a few years, new applications of SSF in the environmental control have been developed including the bioremediation and biodegradation of hazardous compounds and the detoxification of agro-industrial residues (Cannel and Moo-Young 1980).

Comparative studies between submerged liquid fermentation (SLF) and SSF have proved higher yields, and other advantages for products made by SSF have been studied extensively by some researchers (Cannel and Moo-Yound 1980; Raimbault 1998; Hesseltine 1972; Aidoo et al. 1981; Steinkraus 1984; Kumar and Lonsane 1989).

20.3 Production of Microbial Pigments Utilizing Agro-Industrial Waste from Different Industry

Pigment production on an industrial scale is not economical since the cost of technology used is still high, and researches have led to find out low-cost substrate as alternative to reduce production cost, and application of agro-industrial residues in bioprocesses on the one hand provides alternative substrates and on other helps in solving pollution problems, which their disposal may otherwise cause.

Agro-industrial wastes such as rice bran, wheat bran, coconut oil cake, sesame oil cake, palm kernel cake, groundnut oil cake, cassava powder, spent brewing grain, and jackfruit seed powder have been used as substrates as source of carbon, starch, vitamins, and minerals for the production of pigments (Nadzri 2012).

There may be many other factors affecting pigmentation by the bacteria, and a thorough understanding of the effect of these factors and regulation of biosynthetic pathways for pigment production will help to develop a controlled bioprocess for the enhanced production of the desired pigment, thus opening the new avenues for further research in this field (Bhat et al. 2013).

Microbes can synthesize carotenoids when cultivated in commercial medium, containing various refined carbon sources, such as glucose (Mata-Gomez et al. 2014), xylose (Polulyakh et al. 1991), cellobiose, sucrose, glycerol, and sorbitol; nevertheless this type of medium represents high costs (Wang et al. 2001). As a

Table 20.2 Production of microbial pigments utilizing agro-industrial waste

Microbes	Media	Pigment	References
Monascus ruber	Broken rice-based medium, packed bed of long grain rice-based medium, jackfruit seed powder-based medium	Red and yellow pigments	Vidyalakshmi et al. (2009)
Monascus purpureus	Corn meal, coconut residue, peanut meal, soybean meal-based medium	Red pigment	Nimnoi and Lumyong (2011)

result, studies on carotenogenesis have led to find out low-cost substrate as alternative to reduce production costs.

The natural carbon substrates that have been used for growing for microbial pigment production are studied by various investigators; some of the examples are the following: from grape juice (Buzzini and Martin 1999), grape must (Vazquez and Martin 1998), peat extract and peat hydrolyzate (Tinoi et al. 2006), hydrolyzed mung bean waste flour (Fontana et al. 1997), sugar cane juice (An et al. 2001), sugar cane, and sugar beet molasses (Bhosale and Gadre 2001) (Table 20.2).

Solid-state fermentation has always found elevated applications in the production of antibiotics, surfactants, and other value-added products like enzymes, secondary metabolites, biopesticides, aroma compounds, etc. at industrial and commercial level. Wheat wastes are indicated as an important by-product of wheat processed industrially and present as an appropriate solid medium for microorganisms in SSF systems (Nigam and Pandey 2009). Different kinds of microorganisms are capable of synthesizing astaxanthin pigment such as a green algae *Haematococcus pluvialis*, *Brevibacterium* spp. and *Mycobacterium lacticola* bacteria, *Rhodotorula* sp., *Sporidiobolus* sp., *Xanthophyllomyces dendrorhous*, and *Yarrowia lipolytica* yeasts (Rodríguez et al. 2010).

20.3.1 Production of Microbial Pigment from Agricultural Wastes by Fermentation

A number of different kinds of pigments are produced by microorganisms using fermentation process, which may be submerged fermentation or solid-state fermentation. In submerged fermentation process, yellow pigments were obtained from *Rhodotorula rubra* by whey medium containing coconut water (Kaur et al. 2008), carotenoids from *Sporidiobolus salmonicolor* (Valduga et al. 2009), *Rhodotorula glutinis* by tomato waste-based medium (Silveira et al. 2008), *Rhodosporidium paludigenum* by loquat kernel extract-based medium (Taskin and Erdal 2001), and red pigment from *Penicillium purpurogenum* (Mendez et al. 2011) and *Serratia marcescens* (Gulani et al. 2012). The red and yellow pigments were taken from *Monascus purpureus* (Wongsorn et al. 2011). Prodiginines' pigments were from *Streptomyces* sp. (Xueping et al. 2012) and phycocyanin from *Pseudomonas aeruginosa* by cottonseed meal (El-Foulya et al. 2014). From solid-state fermentation, red and yellow pigmentproduced from *Monascus ruber* and *Monascus purpureus* using corncob powder-based medium (de Araujo et al. 2010). The red pigments were obtained from *Monascus ruber* by using various media such as packed bed of long grain rice-based medium (Said et al. 2010) and jackfruit seed powder-based medium (Kamalam et al. 2012).

20.4 Conclusion

Number of industries belonging to the agriculture generate a huge waste biomass in the environment, is of great interest to alter the rich nutrient compounds dependingon source of agro-industrial waste. By using microbial metabolism, waste can be fermented, and development of value-added products is in much attention in recent scenario of research to produce microbial pigments. Hence, in this context, utilization of agro-waste not only eliminates environmental pollution, and also fermentation-derived microbial pigment is more commercial with ecofriendly way of disposing the waste and production of microbial pigments.

Acknowledgments The authors are thankful to the Universiti Teknologi Malaysia for the "Visiting Researcher" Fellowship (Q.J090000.21A4.00D20) and to Dr. C. K. Venil. The authors would like to thank Universiti Teknologi Malaysia for the Research University grants (Q.J.130000.2526.07H03, Q.J.130000.2526.10J38, R.J.130000.7826.4F454) and Ministry of Agriculture, Malaysia, for the Techno fund grant (TF0310F080) for the financial support of research activities.

References

Aidoo KE, Hendry R et al (1981) Estimation of fungal growth in solid state fermentation. Eur J Appl Microbiol Biotechnol 12:6–9

An G, Jang B, Cho M (2001) Cultivation of the carotenoid-hyper- producing mutant 2A2n of the red yeast *Xanthophyllomyces dendrorhous* (*Phaffia rhodozyma*) with molasses. J Biosci Bioeng 92:121–125

Babitha S (2009) Microbial pigments. In: Nigam PS, Pandey A (eds) Biotechnology for agro-industrial residues utilisation, vol 5. Springer, Heidelberg, pp 147–162

Bhat SV, Khan SS, Amin T (2013) Isolation and characterization of pigment producing bacteria from various foods for their possible use as biocolours. Int J Rec Sci Res 4:1605–1609

Bhosale P, Gadre RV (2001) β -carotene production in sugarcane molasses by a *Rhodotorula glutinis* mutant. J Ind Microbiol Biotechnol 26:327–332

Buzzini P, Martin A (1999) Production of carotenoids by strains of *Rhodotorula glutinis* cultured in raw materials of agro-industrial origin. Bioresour Technol 71:41–44

Cannel E, Moo-Young M (1980) Solid state fermentation systems. Process Biochem 15:2

Cserhati T (2006) Liquid chromatography of natural pigments and synthetic dyes. J. Chromatogr Libr, 71. 1st edn, Elsevier Science de Araujo HWC, Fukushima K, Takaki GMC (2010) Prodigiosin production by Serratia marcescens UCP 1549 using renewable-resources as a low cost substrate. Molecules 15:6931–6940

Dufosse L (2009) Pigments, microbial. Encycl Microbiol 4:457–471

El-Foulya MZ, Sharafb AM, Shahina AAM, El-Bialya HA, Omara AM (2014) Biosynthesis of pyocyanin pigment by Pseudomonas aeruginosa. J Rad Res Appl Sci. https://doi.org/10.1016/j.jrras.2014.10.007. (in press)

Fontana JD, Chocial MB, Baron M (1997) Astaxanthinogenesis in the yeast *Phaffia rhodozyma*. Optimization of low-cost culture media and yeast cell wall lysis. Appl Biochem Biotechnol 63–65:305–314

Gulani C, Bhattacharya S, Das A (2012) Assessment of process parameters influencing the enhanced production of prodigiosin from Serratia marcescens and evaluation of its antimicrobial, antioxidant and dyeing potentials. Malays J Microbiol 8:116–122

Hesseltine CW (1972) Solid state fermentation. Biotechnol Bioeng 14:517–532

Kamalam JN, Anburaj J, Kuberan T, Sundaravadivelan C, Kumar P, Starlin T, Selvi AT, Devi MV (2012) DNA amplification and characterization of pigment producing gene from Monascus rubber. Eur J Exp Biol 2:427–435

Kaur B, Chakraborty D, Kaur H (2008) Production and stability analysis of yellowish pink pigments from Rhodotorula rubra MTCC 1446. Internet J Microbiol 7:1. https://doi.org/10.5580/245b

Kumar PRK, Lonsane BK (1989) Solid state fermentation: physical and nutritional factors influencing the gibberellic acid productions. Appl Microbiol Biotechnol 34:145–148

Mata-Gomez LC, Montanez JC, Zavala AM et al (2014) Biotechnological production of carotenoids by yeasts: an overview. Microb Cell Fact 13:12

Mendez A, Perez C, Montanez CJ, Mrtinez G, Aguilar NC (2011) Red pigment production by Penicillium purpurogenum GH2 is influenced pH and temperature. Biomed Biotechnol 12:961–968

Mitchell DA (1992) In solid substrate cultivation. Elsevier Sci. Publ. Ltd, London/New York, pp 17–20

Nadzri NSB (2012) The optimization of red pigment by Monascus purpureus FTC 5356 in solid state fermentation. B. Tech thesis, Universiti Malaysia Pahang

Nagia FA, El-Mohamedy RSR (2007) Dyeing of wool with natural anthraquinone dyes from *Fusarium oxysporum*. Dyes Pigments 75:550–555

Nigam PS, Pandey A (2009) Solid-state fermentation technology for bioconversion of biomass and agricultural residues. In: Nigam PS, Pandey A (eds) Biotechnology for agro-industrial residues utilization. Springer, Dordrecht, pp 197–221

Nimnoi P, Lumyong S (2011) Improving solid-state fermentation of Monascus purpureus on agricultural products for pigment production. Food Bioprocess Technol 4:1384–1390

Polulyakh OV, Podoprigova OI, Eliseev SA (1991) Biosynthesis of torulene and torularhodin in the yeast *Phaffia rhodozyma*. Prikl Biokhim Microbiol 27:541–545

Raimbault M (1998) General and microbiological aspects of solid substrate fermentation. Electron J Biotechnol 1(3):1–15

Rodríguez LA, Toro ME, Vazquez F (2010) Bioethanol production from grape and sugar beet pomaces by solid-state fermentation. Int J Hydrogen Energy 35:5914–5917

Said FM, Chisti Y, Brooks J (2010) The effects of forced aeration and initial moisture level on red pigment and biomass production by Monascus ruber in packed bed solid state fermentation. Int J Environ Sci Dev 1:1–4

Santis DD, Moresi M, Gallo AM et al (2005) Assessment of the dyeing properties of pigments from *Monascus purpureus*. J Chem Technol Biotechnol 80:1072–1079

Silveira ST, Daroit DJ, Brandelli A (2008) Pigment production by Monascus purpureus in grape waste using factorial design. Food Sci Technol 41:170–174

Sindhu R, Pandey A, Binod P (2015) Solid state fermentation for the production of poly(hydroxyalkanoates). Chem Biochem Eng Q 29:173–181

Steinkraus KH (1984) Solid state (solid substrate) food/beverages fermentations involving fungi. Acta Biotechnol 4:83–88

Taskin M, Erdal S (2001) Production of carotenoids by Rhodotorula glutinis MT-5 in submerged fermentation using the extract from waste loquat kernels as substrate. J Sci Food Agric 91:1440–1445

Tinoi J, Rakariyatham N, Deming RL (2006) Utilization of mustard waste isolated for improved production of astaxanthin by *Xanthophyllomyces dendrorhous*. J Ind Microbiol Biotechnol 33:309–314

Valduga E, Valerio A, Tatsch PO, Treichel Furigo HA, Luccio MD (2009) Assessment of cell disruption and carotenoids extraction from Sporidiobolus salmonicolor (CBS 2636). Food Bioprocess Technol 2:234–238

Vazquez M, Martin AM (1998) Optimization of *Phaffia rhodozyma* continuous culture through response surface methodology. Biotechnol Bioeng 57:314–320

Velmurugan P, Kamala-Kannan S, Balachandar V et al (2010) Natural pigment extracted from five filamentous fungi for industrial applications and dyeing of leather. Carbohyd Polym 79:262–268

Venil CK, Zakaria ZA, Ahmad WA (2013) Bacterial pigments and their applications. Process Biochem 48(7):1065–1079

Vidyalakshmi R, Paranthaman R, Murugesh S, Singaravadivel K (2009) Microbial bioconversion of rice broken to food grade pigments. Global J Biotechnol Biochem 4:84–87

Wang SL, Zhang X, Zhang H et al (2001) Effects of some additives on the growth and carotenoids content of *Rhodotorula*. Food Sci Technol 2:20–21

Wongsorn H, Wongjewboot I, Kongruang S (2011) Solvent stability of ultrasonic mutants of Monascus purpureus pigments. Int J Biosci Biochem Bioinfo 1:206–211

Xueping L, Zhang G, Zhu T, Li D, Gu Q (2012) Strain and culture medium optimization for production enhancement of prodiginines from marine-derived Streptomyces sp. GQQ-10. J Ocean Univ China 11:361–365

MIX
Papier aus verantwortungsvollen Quellen
Paper from responsible sources
FSC® C105338

If you have any concerns about our products,
you can contact us on
ProductSafety@springernature.com

In case Publisher is established outside the EU,
the EU authorized representative is:
Springer Nature Customer Service Center GmbH
Europaplatz 3, 69115 Heidelberg, Germany

Printed by Libri Plureos GmbH
in Hamburg, Germany